National Center for Construction Education and Research

Masonry Level One

PEARSON
Prentice Hall

Upper Saddle River, New Jersey
Columbus, Ohio

contren® Learning Series

National Center for Construction Education and Research

President: Don Whyte
Director of Product Development: Daniele Dixon
Masonry Project Manager: Daniele Dixon
Production Manager: Debie Ness
Quality Assurance Coordinator: Jessica Martin
Editor: Tara Cohen
Desktop Publisher: Laura Parker

The NCCER would like to acknowledge the contract service provider for this curriculum: Topaz Publications, Liverpool, New York.

This information is general in nature and intended for training purposes only. Actual performance of activities described in this manual requires compliance with all applicable operating, service, maintenance, and safety procedures under the direction of qualified personnel. References in this manual to patented or proprietary devices do not constitute a recommendation of their use.

Copyright © 2004 by the National Center for Construction Education and Research (NCCER), Gainesville, FL 32614-1104, and published by Pearson Education, Inc., Upper Saddle River, NJ 07458. All rights reserved. Printed in the United States of America. This publication is protected by Copyright and permission should be obtained from the NCCER prior to any prohibited reproduction, storage in a retrieval system, or transmission in any form or by any means, electronic, mechanical, photocopying, recording, or likewise. For information regarding permission(s), write to: NCCER Product Development, P.O. Box 141104, Gainesville, FL 32614-1104.

10 9 8 7 6 5 4 3 2
ISBN 0-13-109160-3

Preface

This volume was developed by the National Center for Construction Education and Research (NCCER) in response to the training needs of the construction, maintenance, and pipeline industries. It is one of many in NCCER's *Contren® Learning Series*. The program, covering training for close to 40 construction and maintenance areas, and including skills assessments, safety training, and management education, was developed over a period of years by industry and education specialists.

NCCER also maintains a National Registry that provides transcripts, certificates, and wallet cards to individuals who have successfully completed modules of NCCER's *Contren® Learning Series*, when the training program is delivered by an NCCER Accredited training Sponsor.

The NCCER is a not-for-profit 501(c)(3) education foundation established in 1995 by the world's largest and most progressive construction companies and national construction associations. It was founded to address the severe workforce shortage facing the industry and to develop a standardized training process and curricula. Today, NCCER is supported by hundreds of leading construction and maintenance companies, manufacturers, and national associations, including the following partnering organizations:

PARTNERING ASSOCIATIONS

- American Fire Sprinkler Association
- American Petroleum Institute
- American Society for Training & Development
- American Welding Society
- Associated Builders & Contractors, Inc.
- Association for Career and Technical Education
- Associated General Contractors of America
- Carolinas AGC, Inc.
- Carolinas Electrical Contractors Association
- Citizens Democracy Corps
- Construction Industry Institute
- Construction Users Roundtable
- Design-Build Institute of America
- Electronic Systems Industry Consortium
- Merit Contractors Association of Canada
- Metal Building Manufacturers Association
- National Association of Minority Contractors
- National Association of State Supervisors for Trade and Industrial Education
- National Association of Women in Construction
- National Insulation Association
- National Ready Mixed Concrete Association
- National Systems Contractors Association
- National Utility Contractors Association
- National Technical Honor Society
- North American Crane Bureau
- North American Technician Excellence
- Painting & Decorating Contractors of America
- Plumbing-Heating-Cooling Contractors National Association
- Portland Cement Association
- SkillsUSA
- Steel Erectors Association of America
- Texas Gulf Coast Chapter ABC
- U.S. Army Corps of Engineers
- University of Florida
- Women Construction Owners & Executives, USA
- Youth Training and Development Consortium

Some features of NCCER's *Contren® Learning Series* are:

- An industry-proven record of success
- Curricula developed by the industry for the industry
- National standardization providing portability of learned job skills and educational credits
- Credentials for individuals through NCCER's National Registry
- Compliance with Apprenticeship, Training, Employer, and Labor Services (ATELS) requirements for related classroom training (CFR 29:29)
- Well-illustrated, up-to-date, and practical information

Acknowledgments

This curriculum was revised as a result of the farsightedness and leadership of the following sponsors:

Carolinas AGC
CEFGA
Hans on America
McGee Brothers Company

Nester Brothers, Inc.
Pyramid Masonry
Southern Brick Institute
Warren County High School

This curriculum would not exist were it not for the dedication and unselfish energy of those volunteers who served on the Authoring Team. A sincere thanks is extended to:

Tommy Caldwell
Kenneth Cook
Bryan Light
Sam McGee

Lee Morris
Sam Nagel
Michael Perry
Arnold Schueck

Contents

28101-04 Introduction to Masonry 1.i

28102-04 Masonry Tools and Equipment 2.i

28103-04 Measurements, Drawings, and Specifications 3.i

28104-04 Mortar 4.i

28105-04 Masonry Units and Installation Techniques 5.i

Masonry Level One . . . . .Index

Module 28101-04

Introduction to Masonry

COURSE MAP

This course map shows all of the modules in the first level of the *Masonry* curriculum. The suggested training order begins at the bottom and proceeds up. Skill levels increase as you advance on the course map. The local Training Program Sponsor may adjust the training order.

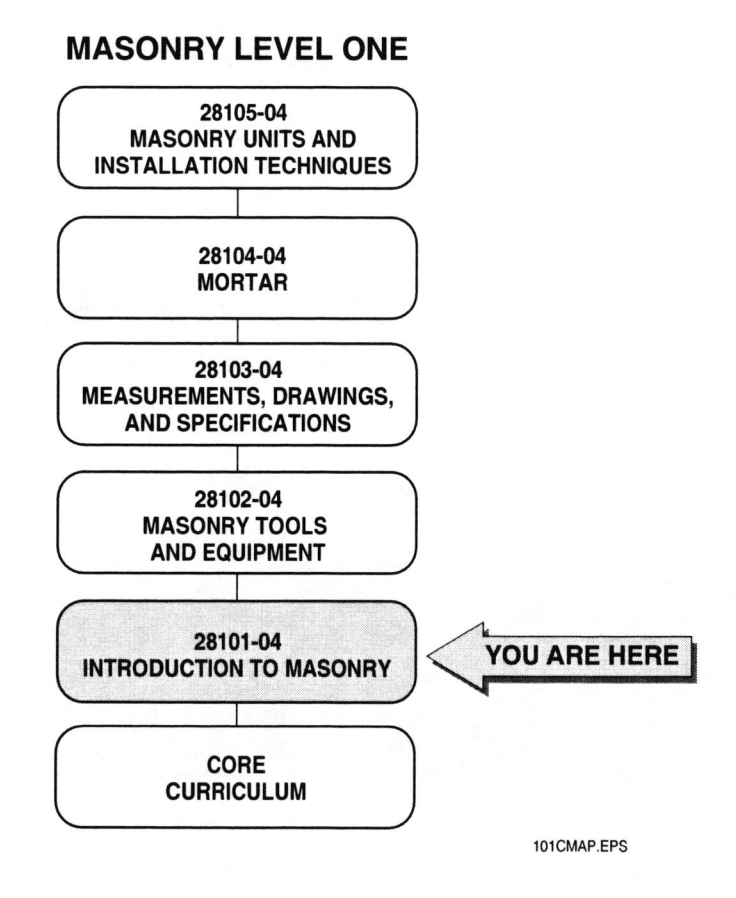

MASONRY LEVEL ONE

- 28105-04 MASONRY UNITS AND INSTALLATION TECHNIQUES
- 28104-04 MORTAR
- 28103-04 MEASUREMENTS, DRAWINGS, AND SPECIFICATIONS
- 28102-04 MASONRY TOOLS AND EQUIPMENT
- 28101-04 INTRODUCTION TO MASONRY ← YOU ARE HERE
- CORE CURRICULUM

101CMAP.EPS

Copyright © 2004 National Center for Construction Education and Research, Gainesville, FL 32614-1104. All rights reserved. No part of this work may be reproduced in any form or by any means, including photocopying, without written permission of the publisher.

MODULE 28101-04 CONTENTS

1.0.0 INTRODUCTION 1.1
2.0.0 THE HISTORY OF MASONRY 1.2
3.0.0 MASONRY TODAY 1.4
    3.1.0 Clay Products 1.4
    3.1.1 Solid Masonry Units/Brick 1.5
    3.1.2 Hollow Masonry Units/Tiles 1.6
    3.1.3 Architectural Terra-Cotta 1.7
    3.1.4 Brick Classifications 1.7
    3.2.0 Brick Masonry Terms 1.7
4.0.0 CONCRETE PRODUCTS 1.8
    4.1.0 Block 1.11
    4.2.0 Concrete Brick 1.11
    4.3.0 Other Concrete Units 1.12
5.0.0 STONE 1.13
6.0.0 MORTARS AND GROUTS 1.13
7.0.0 MODERN CONSTRUCTION TECHNIQUES 1.14
    7.1.0 Wall Structures 1.14
    7.2.0 Modern Techniques 1.17
8.0.0 MASONRY AS A CAREER 1.18
    8.1.0 Career Stages 1.18
    8.2.0 Apprentice 1.18
    8.3.0 Journeyman 1.19
    8.4.0 Supervisors, Superintendents, and Contractors .. 1.20
    8.5.0 The Role of NCCER 1.20
9.0.0 KNOWLEDGE, SKILLS, AND ABILITY 1.20
    9.1.0 Knowledge 1.20
    9.1.1 Job-Site Knowledge 1.21
    9.1.2 Learning More 1.21
    9.2.0 Attitude and Work 1.21
    9.2.1 Dependability 1.21
    9.2.2 Responsibility 1.21
    9.2.3 Adaptability 1.21
    9.2.4 Pride 1.21
    9.3.0 Quality 1.22
10.0.0 BASIC BRICKLAYING 1.22
    10.1.0 Preparing Mortar 1.22
    10.2.0 Spreading Mortar 1.23
    10.3.0 Picking Up Mortar 1.23
    10.3.1 Holding the Trowel 1.23

MODULE 28101-04 CONTENTS (CONTINUED)

10.3.2	*Picking Up Mortar from a Board*	1.24
10.3.3	*Picking Up Mortar from a Pan*	1.24
10.4.0	Spreading, Cutting, and Furrowing	1.25
10.4.1	*Spreading*	1.25
10.4.2	*Cutting or Edging*	1.25
10.4.3	*Furrowing*	1.26
10.5.0	Buttering Joints	1.26
10.6.0	General Rules	1.27
11.0.0	**SAFETY PRACTICES**	**1.28**
11.1.0	The Cost of Job Accidents	1.28
11.2.0	Wearing Safety Gear and Clothing	1.29
11.3.0	Hazards on the Job	1.30
11.4.0	Falling Objects	1.30
11.5.0	Mortar and Concrete Safety	1.31
11.6.0	Flammable Liquid Safety	1.32
11.7.0	Material Handling	1.33
11.7.1	*Materials Stockpiling and Storage*	1.33
11.7.2	*Working Stacks*	1.33
11.8.0	Gasoline-Powered Tools	1.34
11.9.0	Powder-Actuated Tools	1.34
11.10.0	Pressure Tools	1.35
11.11.0	Weather Hazards	1.36
11.11.1	*Cold Weather*	1.36
11.11.2	*Hot Weather*	1.37
12.0.0	**FALL PROTECTION**	**1.37**
12.1.0	Guardrails	1.38
12.2.0	Personal Fall-Arrest Systems	1.39
12.2.1	*Body Harnesses*	1.40
12.2.2	*Lanyards*	1.40
12.2.3	*Deceleration Devices*	1.41
12.2.4	*Lifelines*	1.41
12.2.5	*Anchoring Devices and Equipment Connectors*	1.42
12.2.6	*Selecting an Anchor Point and Tying Off*	1.42
12.2.7	*Using Personal Fall-Arrest Equipment*	1.43
12.3.0	Safety Net Systems	1.43
12.4.0	Rescue After a Fall	1.44
13.0.0	**FORKLIFT SAFETY**	**1.44**
13.1.0	Before You Operate a Forklift	1.45
13.1.1	*Training and Certification*	1.45
13.1.2	*Pre-Shift Inspection*	1.45
13.1.3	*General Safety Precautions*	1.45

MODULE 28101-04 CONTENTS (CONTINUED)

13.2.0	Traveling	1.47
13.2.1	*Stay Inside*	1.47
13.2.2	*Pedestrians*	1.47
13.2.3	*Passengers*	1.47
13.2.4	*Blind Corners and Intersections*	1.47
13.2.5	*Keeping the Forks Low*	1.48
13.2.6	*Horseplay*	1.48
13.2.7	*Travel Surface*	1.48
13.3.0	Handling Loads	1.48
13.3.1	*Picking Up Loads*	1.48
13.3.2	*Traveling with Loads*	1.48
13.3.3	*Traveling with Long Loads*	1.48
13.3.4	*Placing Loads*	1.49
13.3.5	*Placing Elevated Loads*	1.49
13.3.6	*Tipping*	1.49
13.3.7	*Using a Forklift to Rig Loads*	1.50
13.3.8	*Dropping Loads*	1.50
13.3.9	*Obstructing the View*	1.50
13.4.0	Working on Ramps and Docks	1.51
13.4.1	*Ramps*	1.51
13.4.2	*Docks*	1.51
13.5.0	Fire and Explosion Hazards	1.51
13.5.1	*Flammable and Combustible Liquids*	1.52
13.5.2	*Flammable Gases*	1.52
13.5.3	*Fire Fighting*	1.52
13.6.0	Pedestrian Safety	1.52

SUMMARY ... 1.53
REVIEW QUESTIONS ... 1.54
PROFILE IN SUCCESS ... 1.55
GLOSSARY ... 1.57
REFERENCES ... 1.58
ACKNOWLEDGMENTS ... 1.59

Figures

Figure 1	Notre Dame Cathedral, Paris	1.2
Figure 2	Herringbone pattern	1.2
Figure 3	Radial arch	1.3
Figure 4	Roman arch	1.4
Figure 5	Standard brick	1.5
Figure 6	Common bond patterns	1.5
Figure 7	Special brick shapes	1.6
Figure 8	Masonry units and mortar joints	1.8
Figure 9	Cavity walls	1.8
Figure 10	Common concrete block	1.10
Figure 11	Parts of a block	1.11
Figure 12	Concrete brick	1.12
Figure 13	Common pre-faced concrete units	1.12
Figure 14	Manhole and vault unit	1.13
Figure 15	Stone facing used as decorative trim	1.13
Figure 16	A block wall faced with stone	1.13
Figure 17	Types of masonry construction	1.15
Figure 18	Reinforced walls	1.16
Figure 19	Mason's trowels	1.17
Figure 20	Example of apprenticeship training recognition	1.19
Figure 21	Masonry mortar	1.22
Figure 22	Mixing mortar	1.23
Figure 23	Holding the trowel	1.23
Figure 24	Picking up mortar from a board	1.24
Figure 25	Picking up mortar from a pan	1.24
Figure 26	Spreading mortar	1.25
Figure 27	Cutting an edge	1.26
Figure 28	A furrow	1.26
Figure 29	A buttered joint	1.27
Figure 30	Placing the brick	1.27
Figure 31	Checking the level	1.27
Figure 32	Hidden costs of accidents	1.28
Figure 33	Dressed for masonry work	1.29
Figure 34	Crane hazard on the job	1.31
Figure 35	Emergency hand signals	1.31
Figure 36	Palletized brick	1.33
Figure 37	Stacking brick	1.34
Figure 38	Powder-actuated fastening tool	1.34
Figure 39	Proper and improper safety harness use	1.38

Figure 40	Guardrails	1.39
Figure 41	Full-body harnesses with sliding back D-rings	1.40
Figure 42	Harness with front chest D-ring	1.40
Figure 43	Lanyard with a shock absorber	1.41
Figure 44	Rope grab and retractable lifeline	1.41
Figure 45	Vertical lifeline	1.42
Figure 46	Horizontal lifeline	1.42
Figure 47	Eye bolt	1.42
Figure 48	Double locking snaphook	1.42
Figure 49	Forklift	1.44
Figure 50	Forklift operator's daily checklist	1.46
Figure 51	Forklift working in a storage area	1.47
Figure 52	Forklift with forks in low position	1.48
Figure 53	Center of gravity	1.49
Figure 54	Combined center of gravity	1.50

Tables

Table 1	Mortar Composition	1.14
Table 2	Powder Charge Color-Coding System	1.34

MODULE 28101-04

Introduction to Masonry

Objectives

When you have completed this module, you will be able to do the following:

1. Discuss the history of masonry.
2. Describe modern masonry materials and methods.
3. Explain career ladders and advancement possibilities in masonry work.
4. Describe the skills, attitudes, and abilities needed to work as a mason.
5. State the safety precautions that must be practiced at a work site, including the following:
   - Safety practices
   - Fall-protection procedures
   - Forklift-safety operations
6. Perform the following basic bricklaying procedures:
   - Mixing of mortar
   - Laying a mortar bed
   - Laying bricks
7. Put on eye protection, respiratory protection, and a safety harness.
8. Use the correct procedures for fueling and starting a gasoline-powered tool.

Recommended Prerequisites

Core Curriculum

Required Trainee Materials

1. Pencil and paper
2. Appropriate personal protective equipment

1.0.0 ♦ INTRODUCTION

You are beginning the study of masonry, one of the world's oldest and most respected crafts. Masonry construction has been around for thousands of years. The remains of stone buildings date back 15,000 years, and the earliest manufactured bricks unearthed by archaeologists are more than 10,000 years old. These bricks were made of hand-shaped, dried mud. Among the most well-known works of **masons** are the pyramids of ancient Egypt and Notre Dame Cathedral in Paris (*Figure 1*).

Masons build structures of **masonry units**. Masonry units are blocks of brick, concrete, **ashlar**, glass, tile, **adobe**, and other materials. In the most common forms of masonry, a mason assembles walls and other structures of clay brick or **concrete masonry units (CMUs)** using **mortar** to bond the units together.

In this module, you will learn about the basic materials, tools, and techniques used by masons. With the guidance of your instructor, you will learn to mix mortar and lay brick. At first glance, building with masonry units may appear simple. It's not. The first challenge is to lay the units perfectly straight and level. The next challenge is to do it quickly. The production level will vary depending on whether the work is commercial or residential, and the type of work being done.

Laying 600 to 800 bricks a day is a common requirement. In some companies, however, a skilled bricklayer may be expected to lay 1,400 bricks a day. In a contest held by U.S. Brick in Dallas, Texas, in 1996, a bricklayer from McGee Brothers Masonry of Charlotte, North Carolina, laid more than 1,400 bricks in an hour, setting a new world record.

Figure 1 ◆ Notre Dame Cathedral, Paris.

2.0.0 ◆ THE HISTORY OF MASONRY

Brick is the oldest manufactured building material, invented thousands of years ago. The Hanging Gardens of Babylon hung down from brick towers. These hand-formed mud bricks were reinforced with straw and dried in the sun. They were stacked with wet mud between them. Sometimes they were covered with another coat of mud, which was decorated. This was a common and effective building technique for centuries. Some 8,000-year-old bricks have been recovered from the biblical city of Jericho. These sun-dried bricks have a row of thumbprints along their tops. Later, bricks had the king's name and the date stamped on the top; this practice is very useful to today's archaeologists working to date their excavations. Today's bricks are stamped with the manufacturer's name in the same place on the top of the brick. Handmade clay brick is still in use in some parts of the Near East and Africa.

Early bricks led to early brick architecture. Someone had the idea of laying brick in different patterns instead of simply stacking them. *Figure 2*

DID YOU KNOW?
Ancient Wonders

The famous Lion Gate in Istanbul, Turkey is a fine example of ancient brick artistry. It is made of fired and glazed bricks sculpted into high relief before firing. The kiln-firing techniques refined in ancient India, Babylonia, and Rome are still used today.

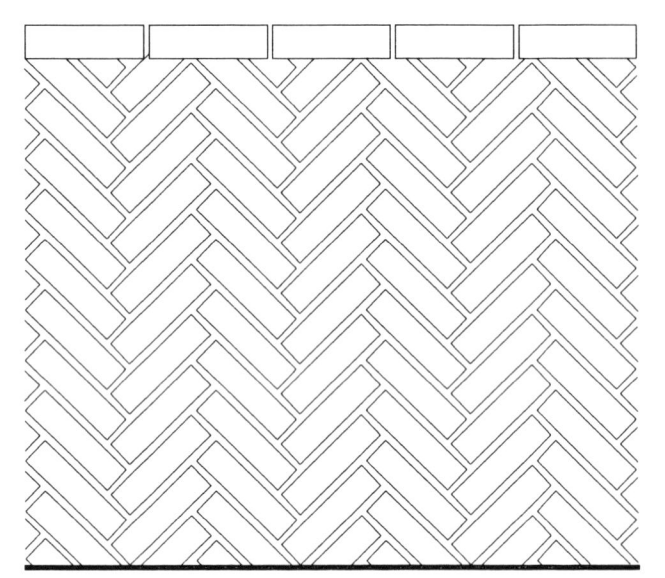

Figure 2 ◆ Herringbone pattern.

shows a herringbone pattern, seen in ancient walls still standing today. These walls used mortar to hold the bricks together. The first mortar was wet mud. Along with firing and glazing brick, the Babylonians developed two new types of mortar. These were based on mixing lime or pitch (asphalt) with the mud.

A later boost for brick architecture was the development of the dome and arch (*Figure 3*). With domes and arches, early masons could build larger and higher structures, with more open space inside.

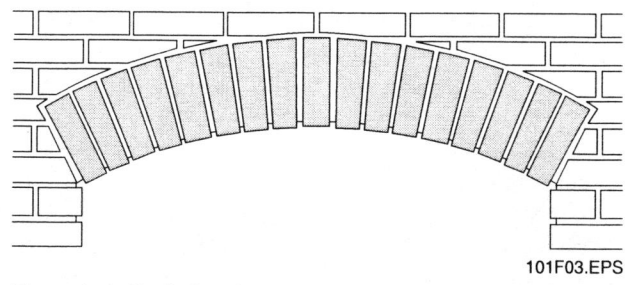

Figure 3 ◆ Radial arch.

The Romans refined arches and domes and built large-scale brickyards. They covered the Roman Empire with roads that brought Roman bricks, mortar, and Roman designs for arches (*Figure 4*) and domes, along with Roman civilization, to the known world. The Romans refined the Babylonian lime mortar by developing a form of cement that was a waterproof mortar. This mortar was useful for both brick and stone construction. It was also applied as a finish coat to the exterior of the surface as an early form of stucco.

The Romans standardized the sizes of their brick. Roman brick is a recognized standard size for bricks today. The Romans produced highly ornate brick architecture using specialized brick shapes and varied brick colors and glazes.

Brick in Construction

Brick is found in all types of construction, from tract houses to stately mansions. Brick is also used in the construction of banks, schools, churches, and office buildings. Brick not only provides an attractive appearance, it creates a sense of permanence.

INTRODUCTION TO MASONRY

Figure 4 ◆ Roman arch.

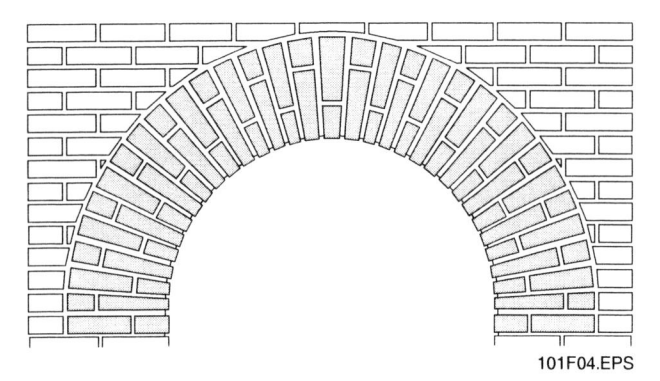

When the Normans conquered England in 1066, they built many castles in England. They imported brick as well as the masons to lay the brick. This construction boom boosted the trade economy of Europe. It also boosted the status of masons. The Norman brick is still a recognized and widely used size of brick.

As the demand for more elaborate construction grew, the need for skilled workers became greater. By the middle 1300s, masons had organized into early unions, known as guilds, across most of Europe.

Guilds controlled the practice of the craft by monitoring the skill level of the craftworker. Early masons' guilds recognized three levels of work:

- The rough mason was the equivalent of the apprentice of today. They worked where the guild would allow, under supervision of masters or journeymen, and served an apprenticeship of three to five years. They were examined by the guild before being allowed to work independently as masons.
- The journeyman, or mason, was a skilled worker allowed to work on finer jobs without supervision. Journeymen were also free to move to wherever the work was. The local guild set the time, as long as ten years, for this stage. At the end, the journeyman could take an examination before the guild and be awarded the status of master. Not all masons reached master status.
- The master mason ran the business, designed structures, employed journeymen, and trained apprentice masons. The master masons also held office, served on boards, and ran the affairs of the masons' guild.

The local guilds continued to control the practice of the masons' craft for centuries. They monitored training, judged disputes, and shared knowledge among members. They collected dues, provided some support to widows and the ill, and celebrated special masons' holidays.

Nonmembers were not allowed to know the secrets of the guilds. Guilds could choose to recognize or not recognize masons certified by other local guilds. Secret signs and passwords were developed so that masons with high skills could recognize each other in different places.

In the early 1700s, Masonic Temples and Orders of Freemasonry were founded in Europe. These organizations were political and spiritual, but based on many of the ideas of the masons' guilds. The pyramid seal on the back of the American dollar bill is an isolated legacy of the masons' guilds.

3.0.0 ◆ MASONRY TODAY

Masons are still recognized as premier craftworkers at any construction site. Their work takes advantage of twentieth-century technology. The two main types of masonry units manufactured today are made of clay or concrete. Clay products are commonly known as brick and tile; concrete products are commonly known as block or concrete masonry units (CMUs).

3.1.0 Clay Products

Brick has been developed and improved upon for centuries. The modern age in brick manufacture started with the first brickmaking machine. It was powered by a steam engine and patented in 1800. The process has not changed much. The clay is mined, pulverized, and screened. It is mixed with water, formed, and cut into shape. Some plants extrude the clay, punch holes into it, then cut it into shape. Any coating or glazing is applied before the units are air dried. After drying, the brick is fired in a kiln. Because of small variations in materials and firing temperatures, not all bricks are exactly alike. Even bricks made and fired in the same batch have variations in color and shading.

The brick is slowly cooled to prevent cracking. It is then bundled into **cubes** and shipped. A cube traditionally holds 500 standard bricks, or 90 blocks, although manufacturers today make cubes of varying sizes.

Today, there are over 100 commonly manufactured **structural** clay products. The **American Society for Testing and Materials (ASTM) International** has published standards for masonry design and construction. The standards cover performance specifications for manufactured masonry units. ASTM has also specified standard sizes for various kinds of brick. *Figure 5* shows the standard sizes for today's most commonly used brick. The first six types are the most widely used. The sizes shown are actual dimensions. Bricks are also identified by nominal sizes, which include the thickness of the mortar **joint**. The nominal size of the modular brick, for example, is 4" × 8" × 2⅜".

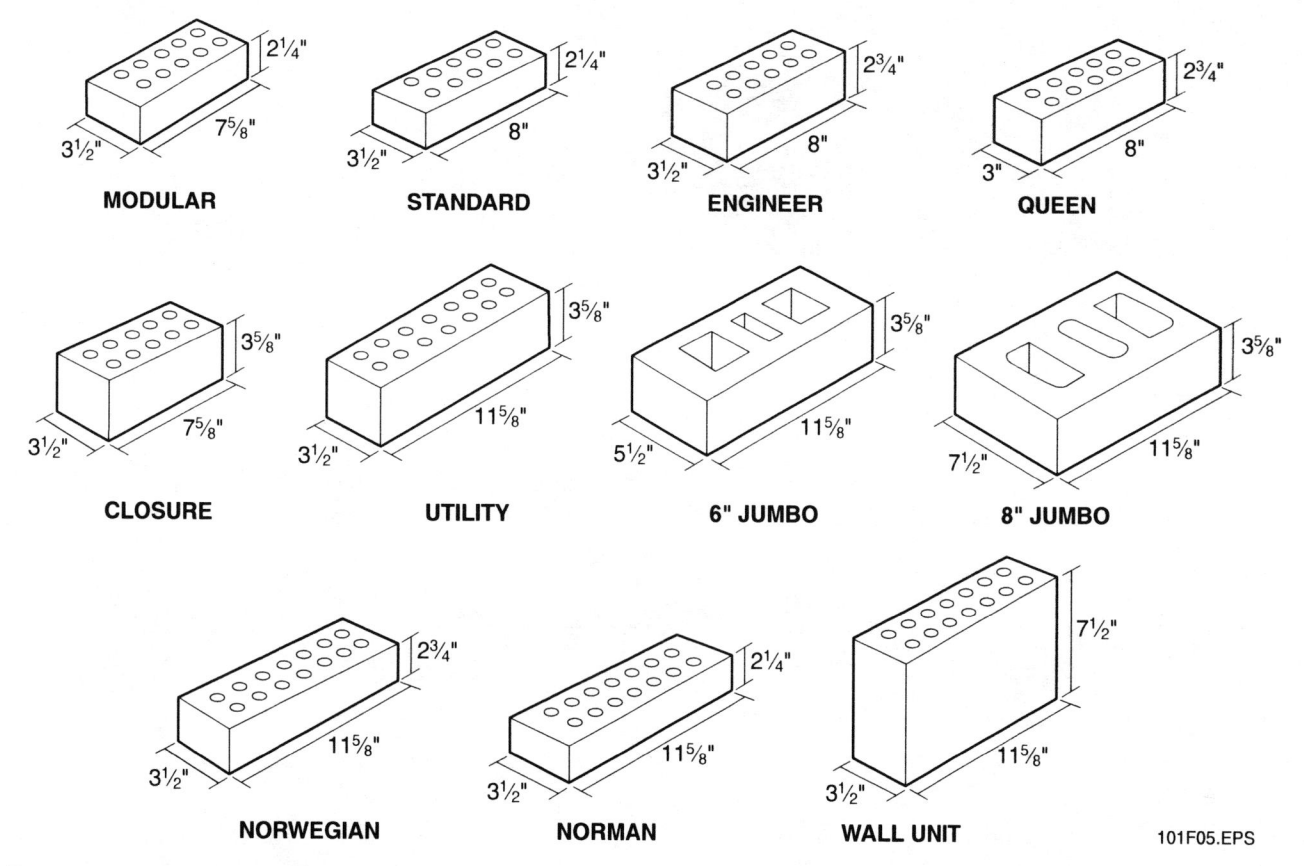

Figure 5 ❖ Standard bricks.

Structural clay products include the following:

- Solid masonry units, or brick
- Hollow masonry units, or tile
- Architectural terra-cotta units

The next sections provide more information about these products.

3.1.1 Solid Masonry Units/Brick

Brick is classified as solid if 25 percent or less of its surface is open (void). Brick is further divided into the following classifications: building, **facing**, hollow, paving, ceramic glazed, thin veneer, sewer, and manhole. ASTM standards exist for all of these types of brick. Fire brick has its own standard, but is not considered a major type classification.

Brick comes in modular and nonmodular sizes, in colors determined by the minerals in the clay or by additives. There is also a variety of face textures and a rainbow of glazes. The variety is dazzling. Brick can be laid in structural bonds to create patterns in the face of a wall or walkway. *Figure 6* shows several examples of commonly used bond patterns. Some bond patterns are traditional in some parts of the country. The herringbone pattern shown earlier is still popular for walkways.

Brick is also made in special shapes to form arches, sills, copings, columns, and stair treads. Custom shapes can be made to order for architectural or artistic use. *Figure 7* shows some commonly manufactured special shapes of brick.

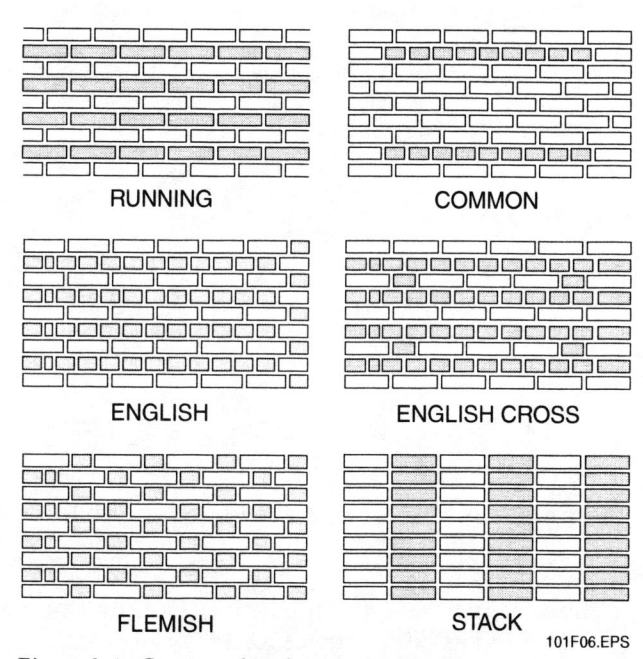

Figure 6 ❖ Common bond patterns.

INTRODUCTION TO MASONRY 1.5

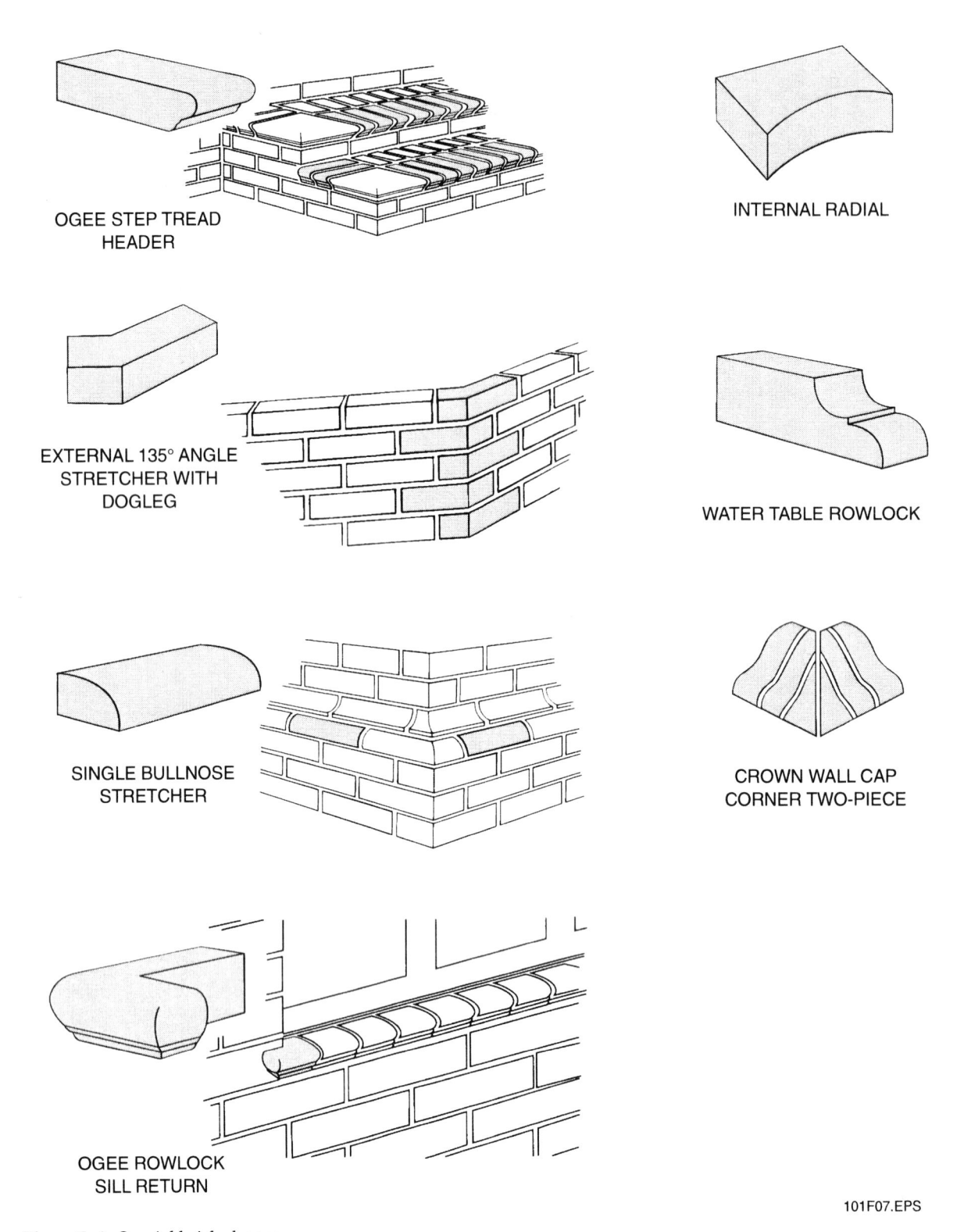

Figure 7 ◆ Special brick shapes.

3.1.2 Hollow Masonry Units/Tiles

Hollow masonry units are machine-made clay tiles extruded through a die and cut to the desired size. Less than 75 percent of the surface area of a hollow masonry units is solid. Hollow units are classified as either structural clay tile or structural clay facing tile.

Structural clay tile comes in many shapes, sizes, and colors. It is divided into loadbearing and nonbearing types. Structural tile can be used for loadbearing on its side or on its end. In some applications, structural tile is used as a backing **wythe** behind brick. Nonbearing tile is designed for use as fireproofing, furring, or ventilating partitions.

Structural clay facing tile comes in modular sizes as either glazed or unglazed tile. It is designed for interior uses where precise tolerances are required. A special application of clay facing tile is as an acoustic barrier. The acoustic tiles have a holed face surface to absorb sound. Clay facing tile can also be patterned by shaping the surface face.

3.1.3 Architectural Terra-Cotta

Architectural terra-cotta is a made-to-order product with an unlimited color range. High-temperature fired ceramic glazes are available in an unlimited color range and unlimited arrangements of parts, shapes, and sizes.

Architectural terra-cotta is classified into anchored ceramic veneer, adhesion ceramic veneer, and ornamental or sculptured terra-cotta. Anchored ceramic veneer is thicker than 1", held in place by **grout** and wire anchors. Adhesion ceramic veneer is 1" or thinner, held in place by mortar. Ornamental terra-cotta is frequently used for **cornices** and column **capitals** on large buildings.

3.1.4 Brick Classifications

As previously stated, the three general types of structural brick-masonry units are solid, hollow, and architectural terra-cotta. All three can serve a structural function, a decorative function, or a combination of both. The three types differ in their formation and composition, and are specific in their use. Bricks commonly used in construction include the following:

- *Building bricks* – Also called common, hard, or kiln-run bricks, these bricks are made from ordinary clays or shales and fired in kilns. They have no special scoring, markings, surface texture, or color. Building bricks are generally used as the backing **courses** in either solid or cavity brick walls because the harder and more durable kinds are preferred.
- *Face bricks* – These are better quality and have better durability and appearance than building bricks because they are used in exposed wall faces. The most common face brick colors are various shades of brown, red, gray, yellow, and white.
- *Clinker bricks* – These bricks are oven-burnt in the kiln. They are usually rough, hard, durable, and sometimes irregular in shape.
- *Pressed bricks* – These bricks are made by the dry-press process rather than by kiln-firing. They have regular smooth faces, sharp edges, and perfectly square corners. Ordinarily, they are used as face bricks.
- *Glazed bricks* – These have one surface coated with a white or other color of ceramic glazing. The glazing forms when mineral ingredients fuse together in a glass-like coating during burning. Glazed brick is particularly suited to walls or partitions in hospitals, dairies, laboratories, and other structures requiring sanitary conditions and easy cleaning.
- *Fire bricks* – These are made from a special type of fire clay to withstand the high temperatures of fireplaces, boilers, and similar constructions without cracking or decomposing. Fire brick is generally larger than other structural brick, and often is hand-molded.
- *Cored bricks* – These bricks have three, five, or ten holes extending through their beds to reduce weight. Three holes are most common. Walls built entirely from cored bricks are not much different in strength than walls built entirely from solid bricks. Both have about the same resistance to moisture penetration. Whether cored or solid, use the more easily available brick that meets building requirements.

Source: U.S. Army FM5-428

3.2.0 Brick Masonry Terms

You need to know the specific terms that describe the position of masonry units and mortar joints in a wall (*Figure 8*). These terms include the following:

- *Course* – One of several continuous, horizontal layers (or rows) of masonry units bonded together
- *Wythe* – A vertical wall section that is the width of one masonry unit
- *Stretcher* – A masonry unit laid flat on its bed along the length of a wall with its face parallel to the face of the wall
- *Header* – A masonry unit laid flat on its bed across the width of a wall with its face perpendicular to the face of the wall; generally used to bond two wythes
- *Rowlock* – A header laid on its face or edge across the width of a wall
- *Bull stretcher* – A rowlock brick laid with its bed parallel to the face of the wall
- *Bull header* – A rowlock brick laid with its bed perpendicular to the face of the wall
- *Soldier* – A brick laid on in a vertical position with its face perpendicular to the courses in the wall

Source: U.S Army FM5-428

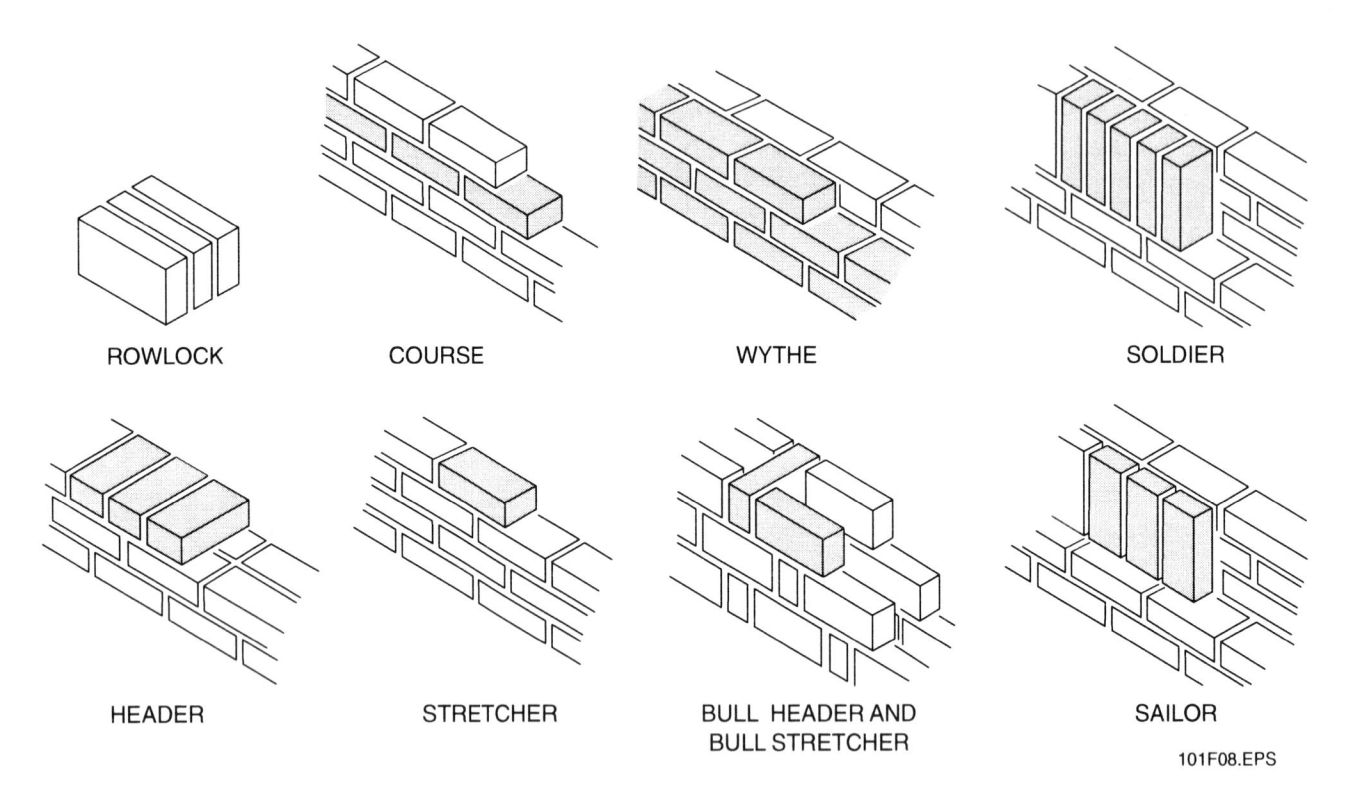

Figure 8 ◆ Masonry units and mortar joints.

4.0.0 ◆ CONCRETE PRODUCTS

CMUs have not been around as long as brick. The first CMUs were developed in 1850, when Joseph Gibbs was trying to develop a better way to build masonry cavity walls.

Masonry is not waterproof, only water resistant. Thick walls slow down moisture so that it does not reach the inside of the wall, but thick walls are expensive to build. Cavity walls were invented to handle this problem. Over the years, many types of cavity walls have been designed to slow down or prevent moisture from reaching the inside surfaces. Today, several different designs are used based on the availability of local materials and environmental requirements. Most of them use either brick or a combination of brick and block.

Figure 9 shows examples of both the old and newer style cavity walls. Cavity walls are made of two courses of masonry units, with a 2" to 4" gap between them. Water can get through the outside wall and run down inside the cavity without wetting the inside wall.

In his search for a faster way to build a cavity wall, Gibbs developed a block with air cells in it. This idea was refined and patented by several other people. In 1882, someone took advantage of new materials developments to make a hollow block of portland cement.

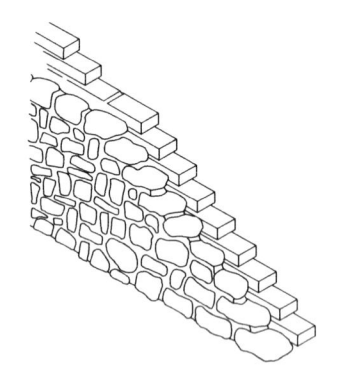

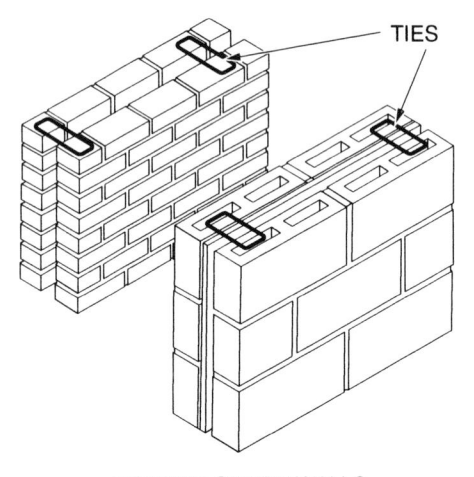

Figure 9 ◆ Cavity walls.

In 1900, Harmon Palmer patented a machine that made hollow concrete block. The blocks were 30" long, 10" high, and 8" wide. Even though these blocks were heavy and very hard to lift, cavity walls became cheaper and faster to build. Over time, other machines were developed to produce smaller blocks. These blocks were easier to handle, but were still very heavy for their size.

In 1917, Francis Straub patented a block made with the cinders left after burning coal. He used the cinders to replace the sand and small **aggregates** in the concrete mix. This new cinderblock was lighter, cheaper, and easier to handle. Straub's block made it possible to build a one-course wall with a built-in cavity very quickly and inexpensively. Faster machinery was developed to keep up with the demand for this new masonry material.

The demand for block increased with the rise of engineered masonry in the U.S. The production of concrete block surpassed that of clay brick in the 1950s. Since the 1970s, there have been more walls built of concrete block in the U.S. than those of clay brick and all other masonry materials together.

Blocks are made of water added to portland cement, aggregates (sand and gravel), and **admixtures**. The cinders have been replaced today by other lightweight aggregates. Admixtures affect the color and other properties of the cement, such as freeze resistance, weight, and speed of setting.

The block is machine-molded into shape. It is compacted in the molds and cured, typically using live steam. After curing, the blocks are dried and aged. The moisture content is checked. It must be a specified minimum amount before the blocks can be shipped for use. *Figure 10* shows commonly used sizes and shapes of concrete block.

Not all blocks are CMUs. Concrete units fall into classifications based on intended use, size, and appearance. ASTM standards exist for the following types of masonry units:

- Loadbearing and nonbearing concrete block
- Concrete brick
- Calcium silicate face brick
- Pre-faced or pre-finished facing units
- Manholes and catch basin units

BUILDING BLOCKS

Block Construction

Concrete block is often used in commercial construction. In some parts of the country, it is also used in the walls of residential construction in place of wood framing. The block can be painted on the outside or faced with brick, stucco, or other finish material.

INTRODUCTION TO MASONRY

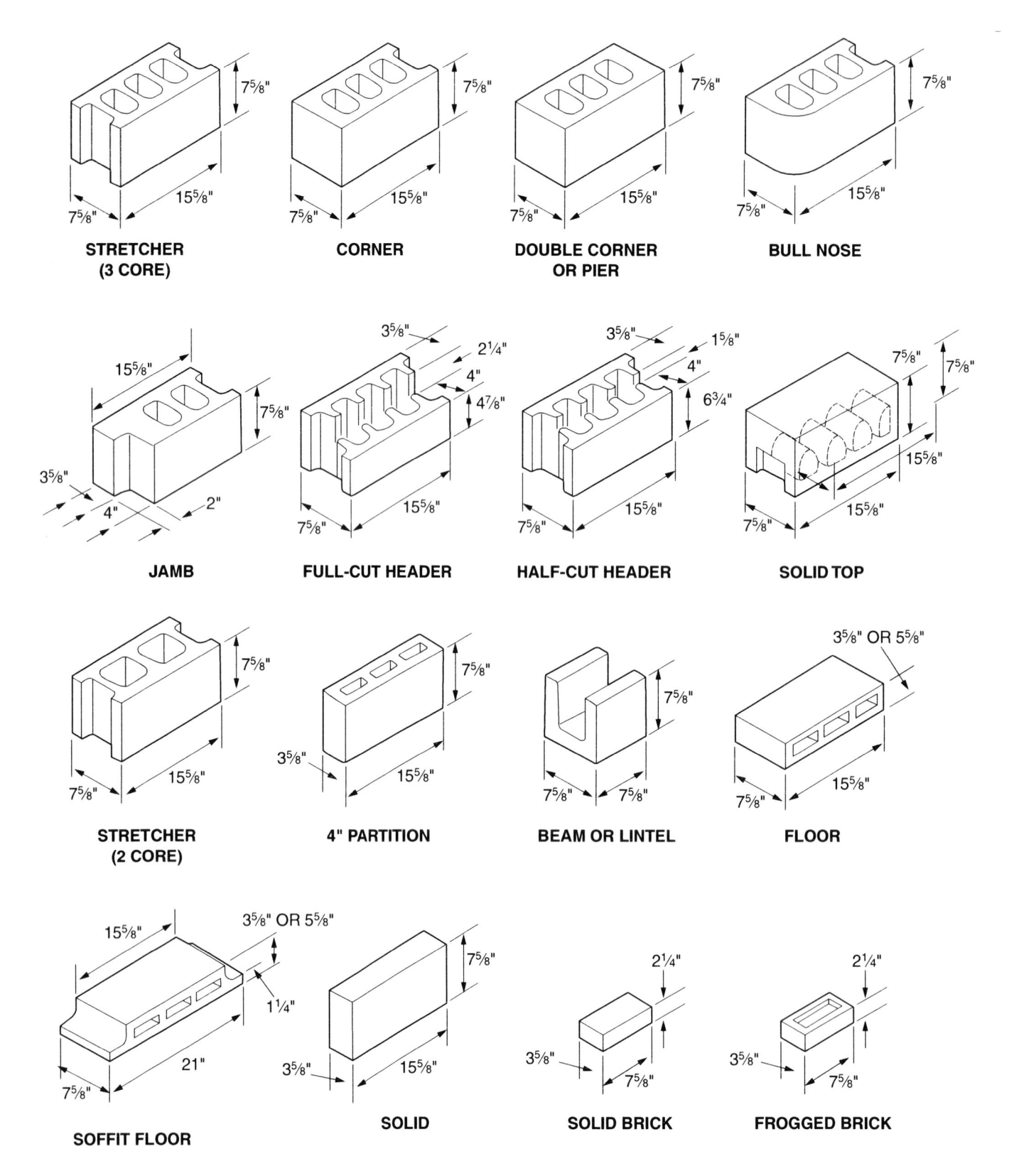

NOTE: Dimensions are actual unit sizes. A 7⅝" × 7⅝" × 15⅝" unit is an 8" × 8" × 16" nominal-size block.

Figure 10 ◆ Common concrete block.

4.1.0 Block

Concrete block is a large unit, typically 8" × 8" × 16", with a hollow core. Blocks come in modular sizes, in colors determined by the cement ingredients, the aggregates, or any admixtures. A variety of surface and mixing treatments can give block varied and attractive surfaces. Newer finishing techniques can give block the appearance of brick, rough stone, or cut stone. Like clay masonry units, block can be laid in structural pattern bonds. *Figure 11* shows the names of the parts of block.

Block takes up more space than other building units, so fewer are needed. Block bed joints usually need mortar only on the shells and webs, so there is less mortaring as well.

Concrete block comes in three weights: normal, lightweight, and aerated. Lightweight block is made with lightweight aggregates. The loadbearing and appearance qualities of the first two weights are similar; the major difference is that lightweight block is easier and faster to lay. Normal-weight block can be made of concrete with regular, high, and extra-high strengths. The last two are made with different aggregates and curing times. They are used to limit wall thickness in buildings over ten floors high. Aerated block is made with an admixture that generates gas bubbles inside the concrete for a lighter block.

Concrete blocks are classified as hollow or solid. Like clay products, less than 75 percent of the surface area of a hollow unit is solid. Common hollow units have two or three cores. The hollow cores make it easy to reinforce concrete block walls. Grout alone, or steel reinforcing rods combined with grout, can be used to fill the hollow cores. Reinforcement increases loadbearing strength, rigidity, and wind resistance. Less than 25 percent of the surface area of a solid block is hollow. Normal and lightweight solid units are intended for special needs, such as structures with unusually high loads, drainage catch basins, manholes, and firewalls. Aerated block is made in an oversize solid unit used for buildings.

Loadbearing block is used as backing for veneer walls, bearing walls, and all structural uses. Both regular and specially shaped blocks are used for paving, retaining walls, and slope protection. **Nonstructural** block is used for screening, partition walls, and as a veneer wall for wood, steel, or other backing. Both kinds of blocks come in a variety of shapes and modular sizes.

4.2.0 Concrete Brick

The length and height dimensions of regular concrete brick are the same as those of standard clay brick. The thickness is an additional $\frac{1}{8}$". A popular type is slump brick, shown in *Figure 12*. Slump brick is made from very wet concrete. When the mold is removed, the brick bulges because it is not dry enough to completely hold its shape. Slump brick looks like ashlar and adds a decorative element to a wall.

Concrete brick is produced in a wide range of textures and finishes. It is available in specialized shapes for copings, sills, and stairs, just as clay brick is. Concrete brick is more popular in some areas of the country because it is less expensive.

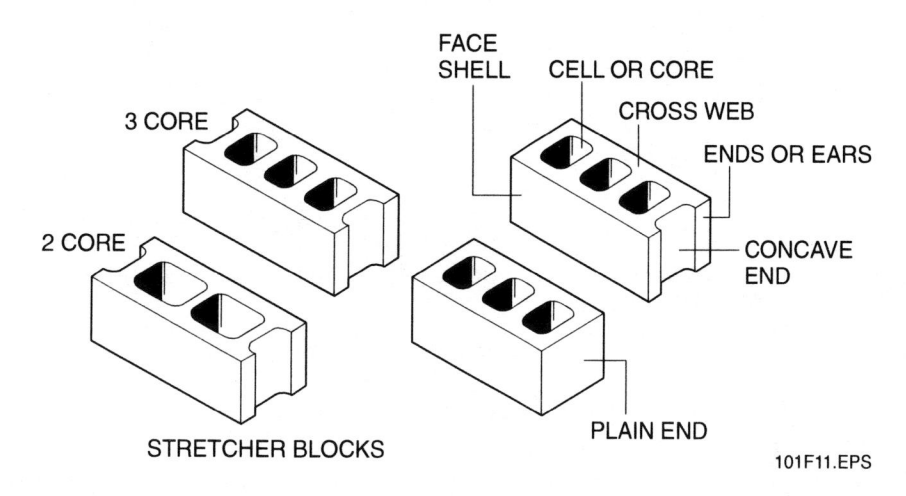

Figure 11 ◆ Parts of a block.

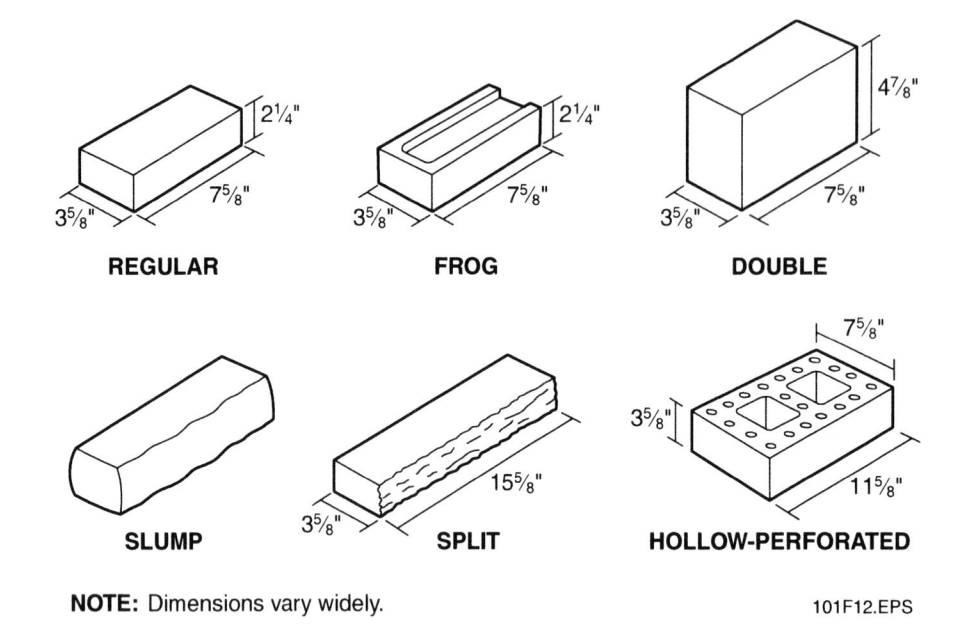

Figure 12 ◆ Concrete brick.

4.3.0 Other Concrete Units

Concrete pre-faced or pre-coated units are coated with colors, patterns, and textures on one or two face shells. The facings are made of resins, portland cement, ceramic glazes, porcelainized glazes, or mineral glazes. The slick facing is easily cleaned. These units are popular for use in gyms, hospital and school hallways, swimming pools, and food processing plants. They come in a variety of sizes and special-purpose shapes, such as coving and bullnose corners. *Figure 13* shows commonly used concrete pre-faced units.

Concrete manhole and catch basin units are specially made with high-strength aggregates. They must be able to resist the internal pressure generated by the liquid in the completed compartment. *Figure 14* shows the shaped units manufactured for the top of a catchment vault. These blocks are engineered to fit the vault shape and are cast to specification. They are made with interlocking ends for increased strength.

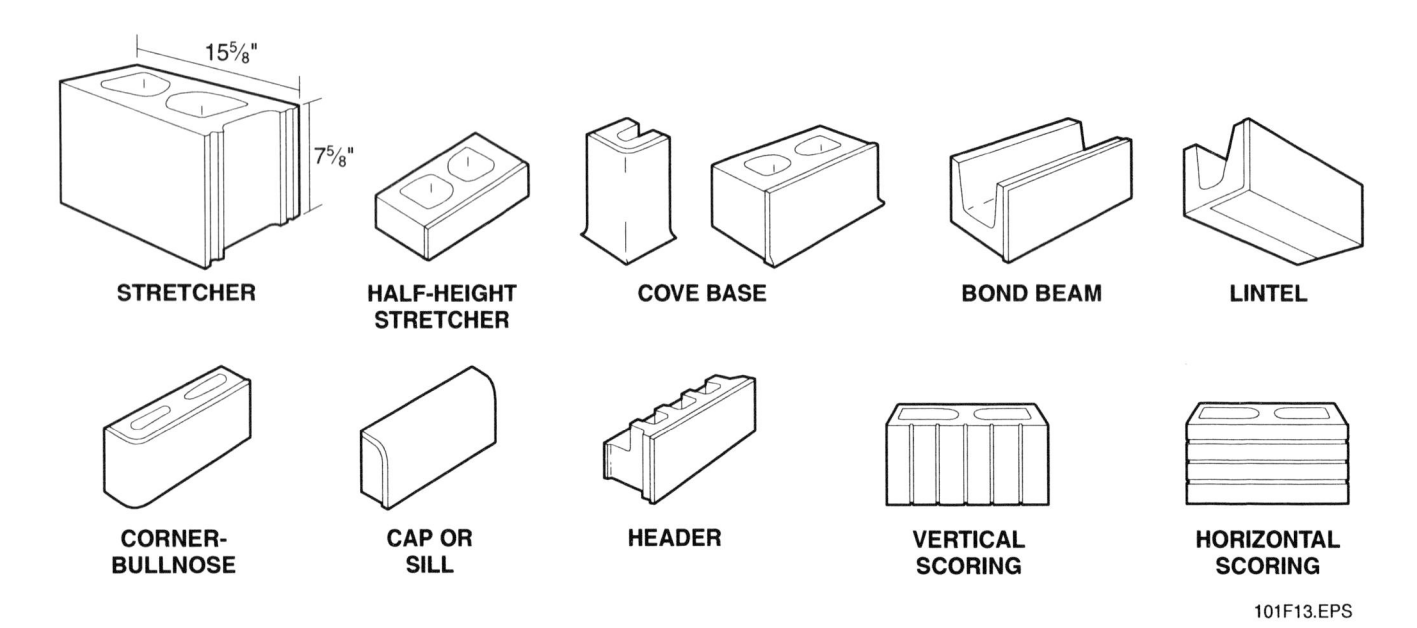

Figure 13 ◆ Common pre-faced concrete units.

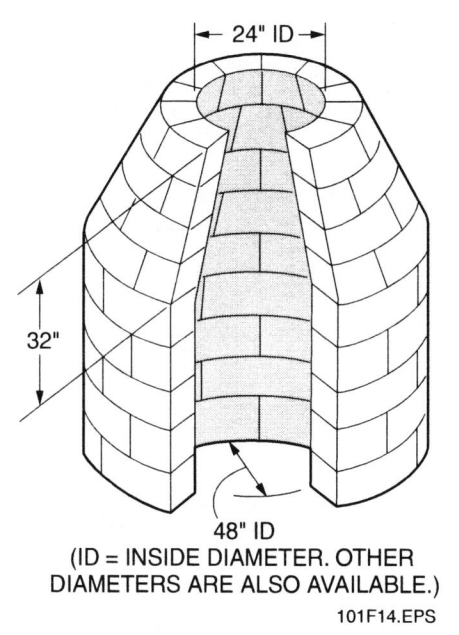

Figure 14 ◆ Manhole and vault unit.

5.0.0 ◆ STONE

Stone was once used in the construction of all types of buildings, especially churches, schools, and government buildings. Today, it is more commonly used as a decorative material such as the stone trim shown on the home in *Figure 15*.

Rubble and ashlar are used for dry stone walls, mortared stone walls, retaining walls, facing walls, slope protection, paving, fireplaces, patios, and walkways.

Rubble stone is irregular in size and shape. Stones collected in a field are rubble. Rubble from quarries is left where shaped blocks have been removed. It is also irregular with sharp edges. Rubble can be roughly squared with a brick hammer to make it fit more easily.

Ashlar stone is cut at the quarry. It has smooth bedding surfaces that stack easily. Ashlar is usually granite, limestone, marble, sandstone, or slate. Other stone may be common in different parts of the country.

Flagstone is used for paving or floors. It is 2" or less thick and cut into flat slabs. Flagstone is usually quarried slate, although other stone may be popular in different areas of the country.

Stone is often used as a veneer over brick or block. The wall shown in *Figure 16* is an example. The brownstone buildings in New York and the grey stone buildings of Paris are veneer over brick. Many of the government buildings and monuments in Washington, D.C., are of stone veneer construction.

Stone, including flagstone, can be laid in a variety of decorative patterns. Concrete masonry units are made in shapes and colorings to mimic every kind of ashlar. These units are called cast stone and are more regular in shape and finish than natural stone. Cast stone has replaced natural stone in many commercial projects because it is more economical. ASTM specifications cover cast stone and natural stone.

6.0.0 ◆ MORTARS AND GROUTS

The first mortar was wet mud, and it is still in use in some parts of the world. Mortar is no longer made of mud, but sometimes it is still called mud.

Figure 15 ◆ Stone facing used as decorative trim.

Figure 16 ◆ A block wall faced with stone.

INTRODUCTION TO MASONRY

By the Roman period, sand was a common additive. Burnt limestone, or quicklime, was added as an ingredient around the first century B.C.E. Experimenting with waterproofing mortar, the Romans added volcanic ash and clay. The resulting cement made a strong, waterproof mortar. This made it possible to build aqueducts, water tanks, water channels, and baths that are still in use today. Unfortunately, some of the Roman formula was lost over time.

In 1824, portland cement was patented by Joseph Aspdin, a mason. He was trying to recreate the waterproof mortar of the Romans. By 1880, portland cement had become the major ingredient in mortar. The new, waterproof portland cement mortar began to replace the old lime and sand mixture.

Portland cement is made of ground earth and rocks burned in a kiln to make clinker. The clinker is ground to become the cement powder. Mixed with water, lime, and rocks, the cement becomes concrete. Mortar is somewhat different from concrete in consistency and use. The components and performance specifications are different also.

Modern mortar is mixed from portland cement or other cementitious material (something that has the properties of cement), along with lime, water, sand, and admixtures. The proportions of these elements determine the characteristics of the mortar.

The two main types of mortar are as follows:

- Cement-lime mortars are made of portland cement, hydrated lime, sand, and water. These ingredients are mixed at the job site by the mason.
- Masonry cement mortars are premixed with additives. The mason only adds sand and water. The additives affect flexibility, drying time, and other properties.

Mortar is mixed to meet four sets of performance specifications, as listed in *Table 1*.

- *Type M* – With high compressive strength, Type M mortar is typically used in contact with earth for foundations, sewers, and walks. This varies with geographic location.
- *Type S* – With medium strength, high bonding, and flex, Type S mortar is used for reinforced masonry and veneer walls.
- *Type N* – With heavy weather resistance, Type N mortar is used in chimneys, **parapets**, and exterior walls.
- *Type O* – With low strength, Type O mortar is used in nonbearing applications. It is not recommended for professional use.

Another type of mortar, Type K, has no cement materials but only lime, sand, and water. This type of mortar is used for the preservation or restoration of historic buildings.

Grout is a mixture of cement and water, with or without fine aggregate. Wet enough to be pumped or poured, it is used in reinforcement to bond masonry and steel together. It gives added strength to a structure when it is used to fill the cores of block walls.

Table 1 Mortar Composition

Mortar Type	Minimum Compressive Strength, PSI at 28 days	Minimum Water Retention,%	Maximum Air Content,%**
M	2,500	75	12***
S	1,800	75	12***
N	750	75	14***
O	350	75	14***

* Adapted from *ASTM C270*.
** Cement-lime mortar only (except where noted).
*** When structural reinforcement is incorporated in cement-lime or masonry cement mortar, the maximum air content shall be 12% or 18%, respectively.

Note: The total aggregate shall be not less than 2¼ and not more than 3½ times the sum of the volumes of the cement and lime used.

7.0.0 ♦ MODERN CONSTRUCTION TECHNIQUES

This section introduces modern structures and modern construction techniques. There are several types of structures you must learn to build. These structures and the techniques used to build them are basic to the craft.

7.1.0 Wall Structures

Masonry structures today take many forms in residential, commercial, and industrial construction. Modern engineering has added loadbearing strength so masonry can carry great weight without bulk. In addition to load bearing, masonry offers these advantages:

- Durability
- Ease of maintenance
- Design flexibility
- Attractive appearance
- Weather and moisture resistance
- Competitive cost

Modern engineering and ASTM standards have been applied directly to everyday masonry work. There are six common classifications of structural wall built with masonry. *Figure 17* shows some of these walls. Masonry walls can fit into more than one classification.

Solid walls, as shown in *Figure 17A*, are built of solid masonry units with full mortar joints. Solid units have voids of less than 25 percent of their surface. These walls can have one or two loadbearing wythes tied together with mortar.

Hollow walls, as shown in *Figure 17B*, are solid walls built of masonry units with more than 25 percent of their surface hollow. These can also have one or two loadbearing wythes tied together with mortar.

Cavity walls, as shown in *Figure 17C*, have two wythes with a 2" to 4½" space between them. Sometimes insulation is put in the cavity. The wythes are tied together with metal ties. Both wythes are loadbearing.

Veneer walls, as shown in *Figure 17D*, are not loadbearing. A masonry veneer is usually built 1" to 2" away from a loadbearing stud wall or block wall. Veneer walls are used in high-rise and residential construction.

Composite walls have different materials in the facing (outer) and backing (inner) wythes. The wythes are set with a 1" air space between them and are tied together by metal ties. Unlike a veneer wall, both wythes of a composite wall are loadbearing.

Reinforced walls (*Figure 18*) have steel reinforcing embedded in the cores of block units or between two wythes. The steel is surrounded with grout to hold it in place. This very strong wall is used in high-rise construction and in areas subject to earthquake and high winds. Sometimes, grout is used alone for reinforcement. The grout is pumped into the cores of the blocks or into the cavity between the wythes.

Contemporary masonry systems are designed not as barriers to water, but as drainage walls. Penetrated moisture is collected on flashing and expelled through **weepholes**. Design, workmanship, and materials are all important to the performance of masonry drainage walls.

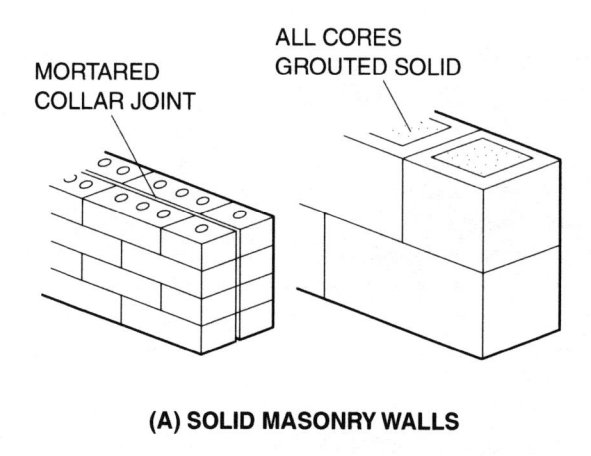

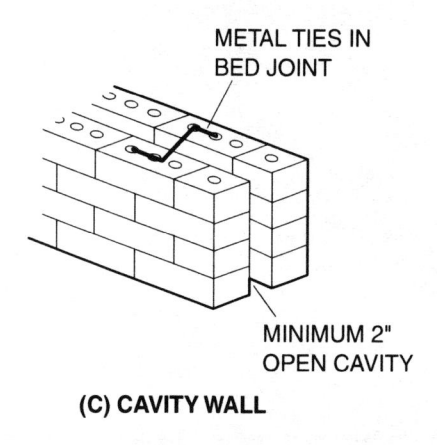

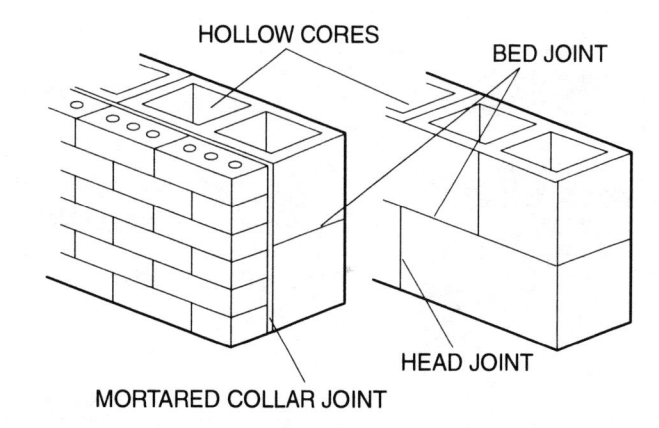

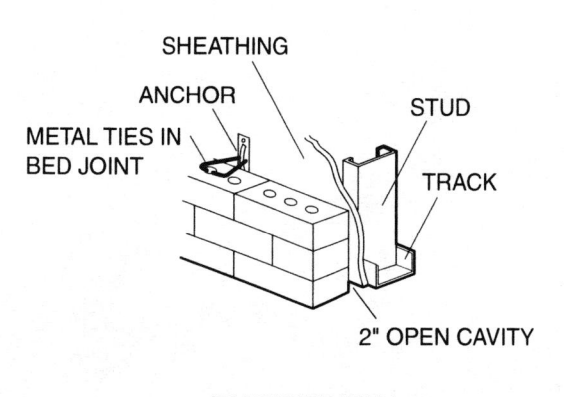

Figure 17 ◆ Types of masonry construction.

INTRODUCTION TO MASONRY

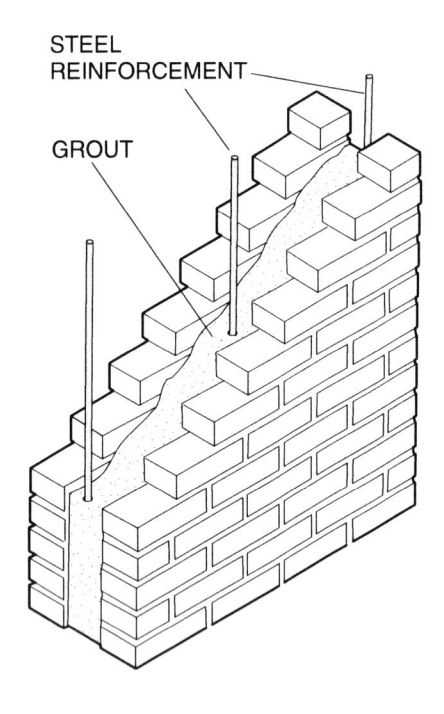

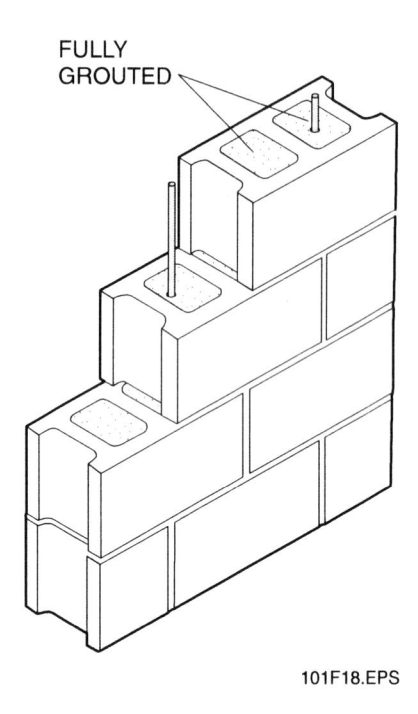

Figure 18 ♦ Reinforced walls.

Curtain Walls

Skyscrapers and other tall buildings were once built with individual bricks. This method has not been used in many years, however. Today, the exterior of a tall building is made by attaching curtain wall sections to a steel or concrete structure. Manufactured curtain wall panels faced with brick are now used when a brick appearance is desired. The curtain wall panel shown here combines 2"-thick concrete brick with a heavy-gauge steel frame and insulated stainless steel anchors.

7.2.0 Modern Techniques

Masons use a number of specialized hand and power tools. As you will learn in the module on tools and equipment, there are many kinds of special trowels (*Figure 19*) and at least six kinds each of hammers, chisels, and steel joint finishing tools. There are seven kinds of measuring and leveling tools. Power tools include several kinds each of saws, grinders, splitters, and drivers. Mortar can be mixed by hand, using special equipment, or in a power mixer. Cranes, hoists, and lifts bring the masonry units to the masons working on one of four types of steel scaffolding.

While masonry tools have changed over the centuries, on thing has not: the relation between the mason and the masonry unit. The mason uses this twentieth-century wealth of tools and equipment to perform the following tasks:

- Calculate the number and type of units needed to build a structure
- Estimate the amount of mortar needed
- Assemble the units near the work station
- Lay out the wall or other architectural structure
- Cut units to fit, as needed
- Mix the appropriate type and amount of mortar
- Place a bed of mortar on the **footing**
- **Butter** the **head joints** and place masonry units on the bed mortar
- Check that each unit is level and true
- Lay courses in the chosen bond pattern or create a new pattern
- Install ties as required for loadbearing
- Install flashing and leave weepholes as required for moisture control
- Clean excess mortar off the units as the work continues
- Finish the joints with jointing tools
- Give the structure a final cleaning
- Complete the work to specification, on time

Masonry work is still very much a craft. The relation between the mason and the masonry unit is personal. The straightness and levelness of each masonry unit in a structure—brick, block, or stone—depend on the hands and the eyes of the mason. These things have not changed in 10,000 years.

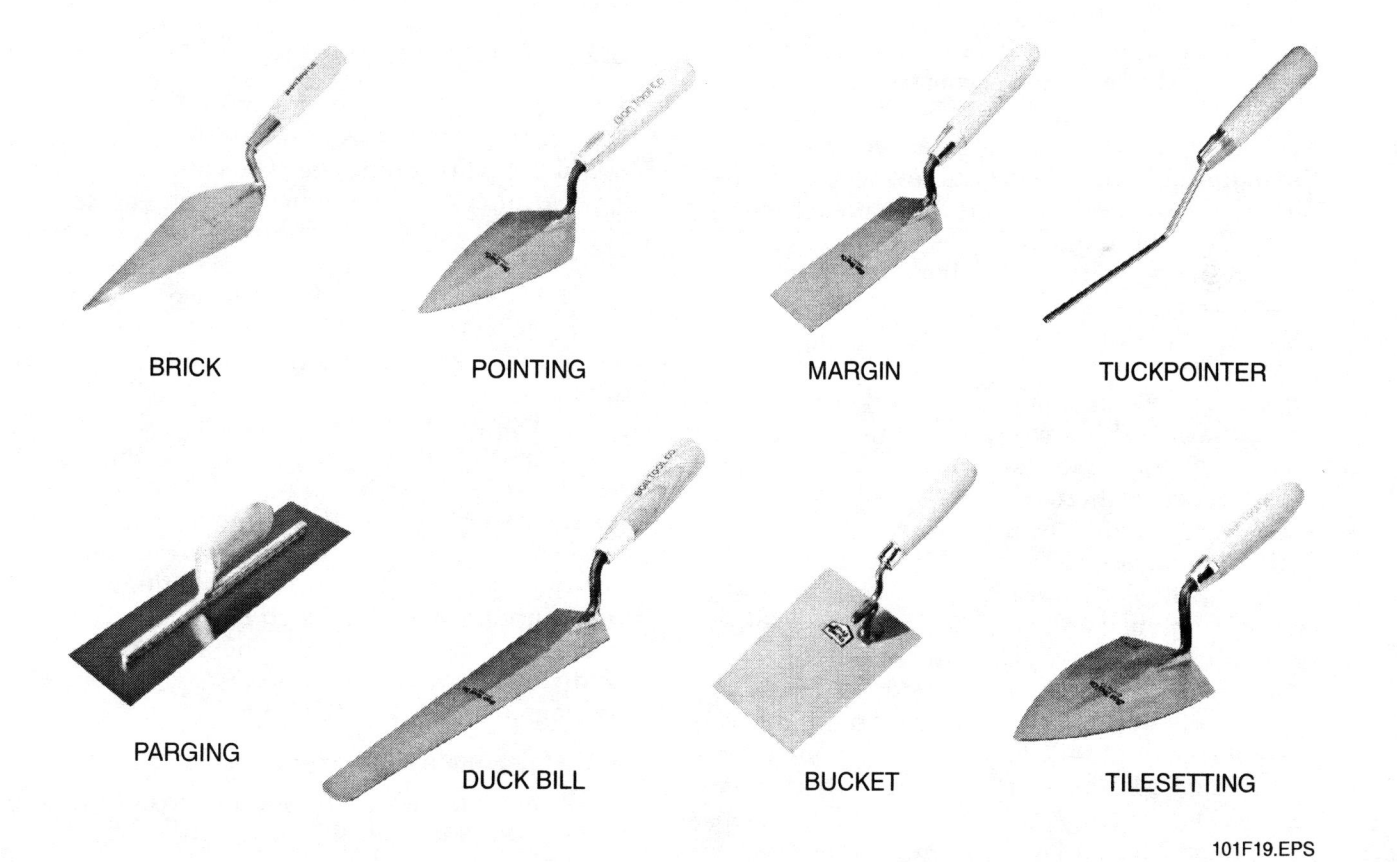

BRICK   POINTING   MARGIN   TUCKPOINTER

PARGING   DUCK BILL   BUCKET   TILESETTING

Figure 19 ◆ Mason's trowels.

INTRODUCTION TO MASONRY

The tradition of masonry calls for a bit of art, too. The mason gets trained by work to see the subtle shadings and gradations of color. He or she learns to create a pattern and to select the right unit to complete the pattern or the shading. The mason grows skilled in building something that is both enduring and attractive.

8.0.0 ♦ MASONRY AS A CAREER

Masonry offers a rewarding career for people who want to work with their hands. As masons, they will be skilled workers who understand the principles and practices of masonry construction. They will earn good pay and be rewarded for initiative. They will have opportunity for advancement.

Masons will continue to play an important part in building homes, schools, offices, and commercial structures. They can add artistic elements to their work and create beauty. They can be proud of their skills and the fact that they produce something people need.

Masons work on different projects, so each job is different and never boring. If they like to travel, masons can find good jobs all over the country. They can be independent and creative while working outdoors. Masons will be in demand as long as buildings are being constructed.

Masons can find work on large construction projects for commercial buildings, as well as projects for building homes, patios, sidewalks, or walls. They can also specialize in repair work, cleaning, and **tuckpointing** old buildings. They can specialize in restoring historic brick buildings, which is a recognized craft specialty in some parts of the country. Historically, they have been well paid.

Masonry is more than physical labor. It is a skilled occupation that calls for good hand-eye coordination, balance, and strength. It also requires good mental skills. This means ongoing study, concentration, and continued learning in an environment free from substance abuse.

Because masonry is a highly skilled craft, it takes time to learn. Your learning starts with this course, combined with, or followed by, an apprentice's job. Masons usually work as part of a team, so you will also learn to be a good team player. Masons work outdoors and do a lot of lifting and bending, so you will learn to keep yourself in good shape. Masons work on high scaffolding, so you will learn safety rules and practices. Masons bring their skills and tools wherever they go.

8.1.0 Career Stages

Masons were among the first workers to band together. During the Middle Ages, they formed influential groups that still shape trade practices. Today, as in the past, masons' organizations recognize several stages of skill:

- Helper
- Apprentice
- Journeyman
- Foreman
- Superintendent
- Contractor

The helper is a laborer, not a mason. The helper carries masonry materials, tools, and mortar and gets things for the mason. The helper mixes mortar, cleans tools, and learns by watching the mason at work. Sometimes, helpers decide they want to become masons. If they do, they may enter an apprenticeship program. Apprentices are at the beginning level of the masonry career path. Their training will lead them to full participation in the mason's trade and the opportunity for higher job levels.

8.2.0 Apprentice

An apprentice is a person who has signed an apprenticeship agreement with a local joint apprenticeship committee. The committee works with local contractors who have agreed to take apprentices. The U.S. Department of Labor regulates the apprenticeship process. In most states, the state department of labor is also involved, as state labor regulations provide guidelines on legal age requirements, pay, hours, and other aspects of apprenticeship.

The length of the apprenticeship will vary. The U.S. Department of Labor program is three years and 4,500 hours with a minimum of 432 hours of classroom instruction. Programs such as the *Contren® Learning Series* are used for classroom instruction or as part of apprenticeship training.

The apprentice is assigned to work with a contractor and to take classes. The apprentice must study and work under supervision. The apprentices agree to do the following:

- Perform the work assigned by the contractor.
- Abide by the rules and regulations of the contractor and the committee.
- Complete the hours of instruction.

- Keep records of work experience, training, and instruction.
- Learn and use safe working habits.
- Work with the assigned contractors for the entire apprenticeship period, unless reassigned by the committee.
- Conduct themselves in an ethical manner, realizing that time, money, and effort are being spent to afford them this opportunity to become a skilled worker.
- Remain free from drug and alcohol abuse.

A typical three-year apprenticeship is divided into six periods of six months each. The first six months is a trial period. The committee reviews the apprentice's performance and may end the agreement.

The apprentice attends classes and works under the supervision of a journeyman mason. As part of the supervised work, the apprentice learns to lay masonry units and perform other craftwork. The apprentice's pay increases for each six-month period as skill and performance increase. At the end of the period, the apprentice receives a certificate of completion. This type of certificate (*Figure 20*) is known and accepted everywhere in the United States. The apprentice is now a journeyman mason.

8.3.0 Journeyman

Unlike an apprentice, a journeyman is a free agent who can work for any contractor. A journeyman can work without close supervision and is skilled in most tasks. The successful journeyman knows that the end of the apprenticeship is not the end of learning.

Journeymen are people with an excellent trade. They earn good wages in a trade that is always in demand. They have the satisfaction of creating and the opportunity to grow as masonry artists. They also have the opportunity to grow as layout persons, trainers, and supervisors.

Figure 20 ◆ Example of apprenticeship training recognition.

An experienced and skilled journeyman can work as a layout person. For a pay premium, the layout person lays out the work and lays the leads. Less experienced masons and apprentices work between the leads set by the layout person. Experienced and skilled journeymen also train apprentices and supervise their work. With further experience, journeymen can supervise crews.

Journeymen can continue to learn by studying and handling more complex tasks. They can continue to develop their skills as they work. Further education in masonry innovations and techniques is available as is training in leadership and supervision.

8.4.0 Supervisors, Superintendents, and Contractors

Supervisors are responsible for managing and supervising a group of workers. This job requires a high degree of knowledge about masonry and leadership skills. Supervisors are typically responsible for training workers in safety measures and keeping work areas safe. They also train workers in new techniques and easier ways of working. They solve daily problems, keep on top of materials and supplies, and make sure workers meet job schedules. They check work to ensure it is done to standards. Supervisors may be called crew leaders or forepersons depending on the company that hires them.

Superintendents have several supervisors reporting to them. Usually, the superintendent is the lead person on a large job. For a smaller company, the superintendent may be in charge of all the work in the field for the contractor. The superintendent oversees the work of the supervisors and makes major decisions about the job under construction. The superintendent must have strong masonry, leadership, and business skills.

A masonry contractor owns the company. He or she bids on jobs, organizes the work and the workers, inspects the work, confers with the clients, and runs the business. The contractor needs to be able to plan ahead to keep up with change.

Contractors, along with journeymen, supervisors, and superintendents, need to keep up with the latest materials and methods. Like apprentices, they need to keep on studying their trade.

8.5.0 The Role of NCCER

This course is part of a curriculum produced by the National Center for Construction Education and Research (NCCER). Like every course in NCCER's curriculum, it was developed by the construction industry for the construction industry. NCCER develops and maintains a training process which is nationally recognized, standardized, and competency-based. A competency-based program requires you to show that you can perform specific job-related tasks safely to receive credit. This approach is unlike other apprenticeship programs that are based on a required number of hours in the classroom and on the job.

The construction industry knows that the future construction workforce will largely be recruited and trained in the nation's secondary and postsecondary schools.

Schools know that to prepare their students for a successful construction career they must use the curriculum that is developed and recognized by the industry. Nationwide, thousands of schools have adopted NCCER's standardized curricula.

The primary goal of NCCER is to standardize construction craft training throughout the country so that both you and your employer will benefit from the training, no matter where you or your job are located. As a trainee in a NCCER-accredited program, you will be listed in the National Registry. You will receive a certificate for each level of training you complete, which can then travel with you from job to job as you progress through your training. In addition, many technical schools and colleges use NCCER's programs.

9.0.0 ♦ KNOWLEDGE, SKILLS, AND ABILITY

Becoming a good mason takes more than the ability to lay a masonry unit and level it. A competent mason is one who can be trusted to perform the required work and meet the project specifications. This mason must have the necessary knowledge, skills, and ability, as well as good attitudes about the work itself, about safety, and about quality.

9.1.0 Knowledge

Masons need to know how to handle all aspects of masonry work. They need to know how to do all of the following:

- Read and interpret drawings and specifications
- Calculate and estimate quantities, lengths, weights, and volumes
- Select the proper materials for the job
- Lay masonry units into structural elements
- Work productively alone or as part of a team
- Assemble and disassemble scaffolding
- Keep tools and equipment in good repair and safe condition
- Follow safety precautions to protect themselves and other workers on the job

9.1.1 Job-Site Knowledge

Masons need to be skilled in applying their knowledge to the challenges they face each day on the job. The best way to do the work at a particular job site will depend on the layout of the work, what is happening around the masonry site, and the conditions surrounding the project.

Most masonry work is done outside in temperature and weather variations. You must be able to work under these conditions and not be distracted by them. You must know how to react to changing conditions around you.

Much of this knowledge can be learned as you work, if you will pay attention. Notice what others do and ask questions. Ask your supervisor questions, too. Learn to respond to conditions at the job site.

9.1.2 Learning More

Masons need to keep on learning after they finish their apprenticeships. They need to keep updating their skills all the time. The environment, tools, and expectations about masonry have evolved and will continue to change. Craftworkers and contractors alike will need to change the way they think about their work and how they do it.

National, regional, and local organizations offer continuing education for masons. Technical seminars, training sessions, publications, and classes are often free or low cost. They can bring you the latest information about tools, materials, and methods. To succeed, you must be alert to change and willing to learn new ways.

9.2.0 Attitude and Work

Attitude can build an invisible bridge, or build an invisible wall, between us and others. No one wants to hang around a grouch or count on someone who is not dependable. No one minds helping someone who can do something in return or working with a friendly, cooperative partner. On top of knowledge, skills, and ability, you need the right attitude. Your attitude comes from how you think and feel about your work and yourself.

9.2.1 Dependability

You must be dependable. Masonry work, like all construction, is a closely timed operation. Once started, it cannot stop without waste of material and money. Employers need workers who report to work on time. An undependable, absent worker will slow or stop masonry work and cost the project time and money. An undependable worker will not be able to depend on having a job for very long.

9.2.2 Responsibility

You must be responsible for doing the assigned work in a proper and safe manner, be responsible enough to work without supervision, and work until the task is complete.

Being responsible for your own work includes admitting your mistakes. It also includes learning from your mistakes. Nobody is expected to be perfect. Everyone is expected to learn and to grow more skilled.

Employers are always in need of workers who are ambitious and want to become leaders. Being responsible for what others do may be your career goal. The path to that goal starts with being responsible for what you do.

9.2.3 Adaptability

On any construction project, a large amount of work must be done in a short time. Planning and teamwork are needed in order to work efficiently and safely. Supervisors sometimes form teams of two or more workers to do specific tasks. You may work in a team to erect a scaffold, then work alone for most of the day, then team with someone else to do a cleanup.

On a job site, you may find yourself teaming with different people at different times. Being a team player becomes important. Team players accept instruction and direction. They communicate clearly, keep an eye out for potential problems, and share information. They meet problems squarely with constructive ideas, not criticism.

All team players treat each other with respect. Everyone must be willing to work together. Everyone must be willing to bring their best attitude to the team. Team members need to be able to depend on each other. Team priorities must be more important than individual priorities.

9.2.4 Pride

Pride in what you do comes from doing high-quality work in a timely manner and from knowing you are doing your best. Being proud of what you do can overflow into other areas. Proud workers take pride in their personal appearance. Their work clothes are clean, safe, neat, and suitable. Proud masons take pride in their tools. They have a complete set of well-maintained tools and other special equipment they need to do their jobs. They keep their tools safe and orderly and know how to use the right tool for the work at hand.

Proud masons work so that they can continue to be proud of what they do and how they do it. Being proud of what you do is an important part of being proud of who you are.

BUILDING BLOCKS

Ethical Principles for Members of the Construction Trades

Honesty: Be honest and truthful in all dealings. Conduct business according to the highest professional standards. Faithfully fulfill all contracts and commitments. Do not deliberately mislead or deceive others.

Integrity: Demonstrate personal integrity and the courage of your convictions by doing what is right even when there is great pressure to do otherwise. Do not sacrifice your principles for expediency, be hypocritical, or act in an unscrupulous manner.

Loyalty: Be worthy of trust. Demonstrate fidelity and loyalty to companies, employers, fellow craftspeople, and trade institutions and organizations.

Fairness: Be fair and just in all dealings. Do not take undue advantage of another's mistakes or difficulties. Fair people display a commitment to justice, equal treatment of individuals, tolerance for and acceptance of diversity, and open-mindedness.

Respect for others: Be courteous and treat all people with equal respect and dignity regardless of sex, race, or national origin.

Law abiding: Abide by laws, rules, and regulations relating to all personal and business activities.

Commitment to excellence: Pursue excellence in performing your duties, be well-informed and prepared, and constantly endeavor to increase your proficiency by gaining new skills and knowledge.

Leadership: By your own conduct, seek to be a positive role model for others.

9.3.0 Quality

The latest ideas about quality in work are not new. Those who work in masonry construction and finishing have been concerned about quality for thousands of years. The walls unearthed at Jericho were laid true and still stand true.

The quality of masonry depends on many factors. When building a wall, you may have little control over its design or the choice of masonry units. But you do have control over the quality of the completed job. A wall out of level or with poorly finished joints is your responsibility.

The quality of the finished masonry structure depends directly on your knowledge, skill, and ability. Good work will be easily seen by all. Poor work will be seen even more easily. Given the durability of masonry, the quality of the work will be a monument to your skill for a very long time. The skilled, proud mason always strives for the highest quality that can be achieved.

10.0.0 ♦ BASIC BRICKLAYING

In this section, you will learn the basic elements of bricklaying. When you have completed the section, you should be able to set up a job, mix mortar, and lay bricks as directed by your instructor.

10.1.0 Preparing Mortar

Mortar (*Figure 21*) is a mixture of portland cement, lime, sand, and water. The first three ingredients are determined by the type of mortar being mixed. Water is added until the mix is at the proper consistency. The ability to mix mortar properly, and to produce the same consistency time after time, can only be developed through practice.

You will probably mix your first few batches by hand in a wheelbarrow, pan, or mortar box. Assuming that you will be using a wheelbarrow, proceed as follows:

Step 1 Place half the sand in the wheelbarrow and make an even spread over the bottom.

Step 2 Add the required amount of cement and lime (or masonry cement) to the wheelbarrow.

Step 3 Blend the dry ingredients with a hoe (*Figure 22*), then pull the mix to one end of the wheelbarrow.

Step 4 Add half the water to the empty end of the wheelbarrow. Begin mixing the water and dry material with short push-pull strokes of the hoe.

Step 5 Add water to obtain the required consistency. When the mix is right, the mortar will stick to a trowel when the trowel is turned upside down.

Figure 21 ◆ Masonry mortar.

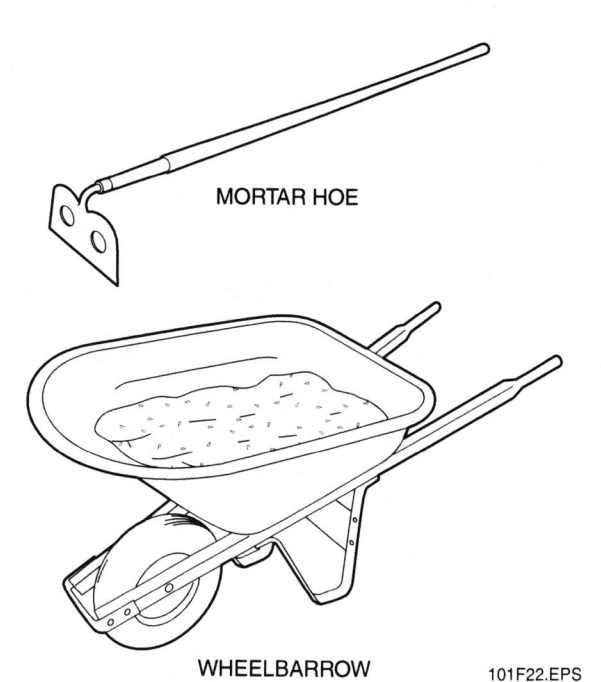

Figure 22 ◆ Mixing mortar.

10.2.0 Spreading Mortar

After the mortar is mixed, pick it up on your trowel and spread it. Filling and emptying the trowel is an important skill. Applying the mortar, or spreading it, is the next step.

The following sections describe holding the trowel, picking up the mortar, and laying it down. At this point, you will learn something to be experienced rather than memorized. The techniques in the next sections should be practiced until you feel comfortable using them.

10.3.0 Picking Up Mortar

There are several ways of using the trowel to pick up mortar. This section will introduce you to a general method for picking up mortar from a board and a general method for picking up mortar from a pan. There are many different ways to do these tasks. The instruction here begins with some tips on holding the trowel. All of these instructions are only approximations of the work itself. You can only learn this through watching a skilled mason and practicing until you feel comfortable with these movements.

10.3.1 Holding the Trowel

Pick up your trowel by the handle. Put your thumb along the top of the handle with the tip on the handle, not the shank, as shown in *Figure 23*.

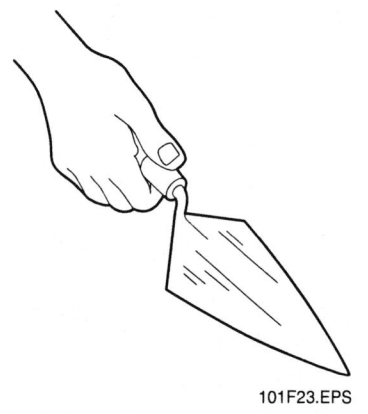

Figure 23 ◆ Holding the trowel.

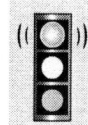

> **WARNING!**
> Keep your thumb off the shank to keep it out of the mortar. Mortar is caustic and can cause chemical burns.

Keep your second, third, and fourth fingers wrapped around the handle of the trowel. Keep the muscles of your wrist, arm, and shoulder relaxed so you can move the trowel freely.

Most of your work with the trowel will require holding the blade flat, parallel to the ground, or rotating the blade so it is perpendicular to the ground.

Rotating the blade gives you a cutting edge. The best edge for cutting is the edge on the side closest to your thumb. It is best this way because you can see what you are cutting. When you turn the trowel edge to cut, rotate your arm so your thumb moves down. This will rotate the trowel so that the bottom of the blade turns away from you.

INTRODUCTION TO MASONRY 1.23

If you rotate only your wrist, after a while you will strain it. Use the larger muscles in your arm and shoulder to rotate the trowel.

Rotating the blade also gives you a scooping motion. Turning your thumb down will give you a forehand scoop. Turning your thumb up will rotate the bottom of the blade toward you and give you a backhand scoop. Using a forehand or backhand movement will depend on the position of the material you are trying to scoop.

10.3.2 Picking Up Mortar from a Board

After putting the mortar on the board, follow these steps:

Step 1 Work the mortar into a pile in the center of the board, and smooth it off with a backhand stroke.

Step 2 Use the trowel edge to cut off a slice of mortar from the edge.

Step 3 Pull and roll the slice of mortar to the edge of the board. Work the mortar into a long, tapered roll, as shown in *Figure 24*.

Step 4 Slide the trowel under the mortar, then lift the mortar up with a light snap of your wrist. Raising the trowel quickly will break the bond between the mortar and the board. If done correctly, the mortar will completely fill the trowel blade.

10.3.3 Picking Up Mortar from a Pan

Try this method when the mortar is in a pan:

Step 1 Cut a slice of mortar, as shown in *Figure 25*.

Step 2 Without removing the trowel from the mortar, slide the trowel under the mortar so the blade becomes parallel to the floor.

Step 3 Firmly push the trowel, with the blade parallel to the floor, toward the middle of the pan. The mortar will pile up on the blade.

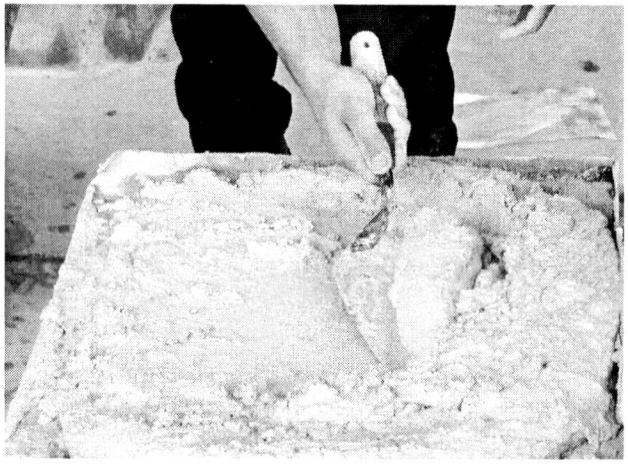

Figure 24 ◆ Picking up mortar from a board.

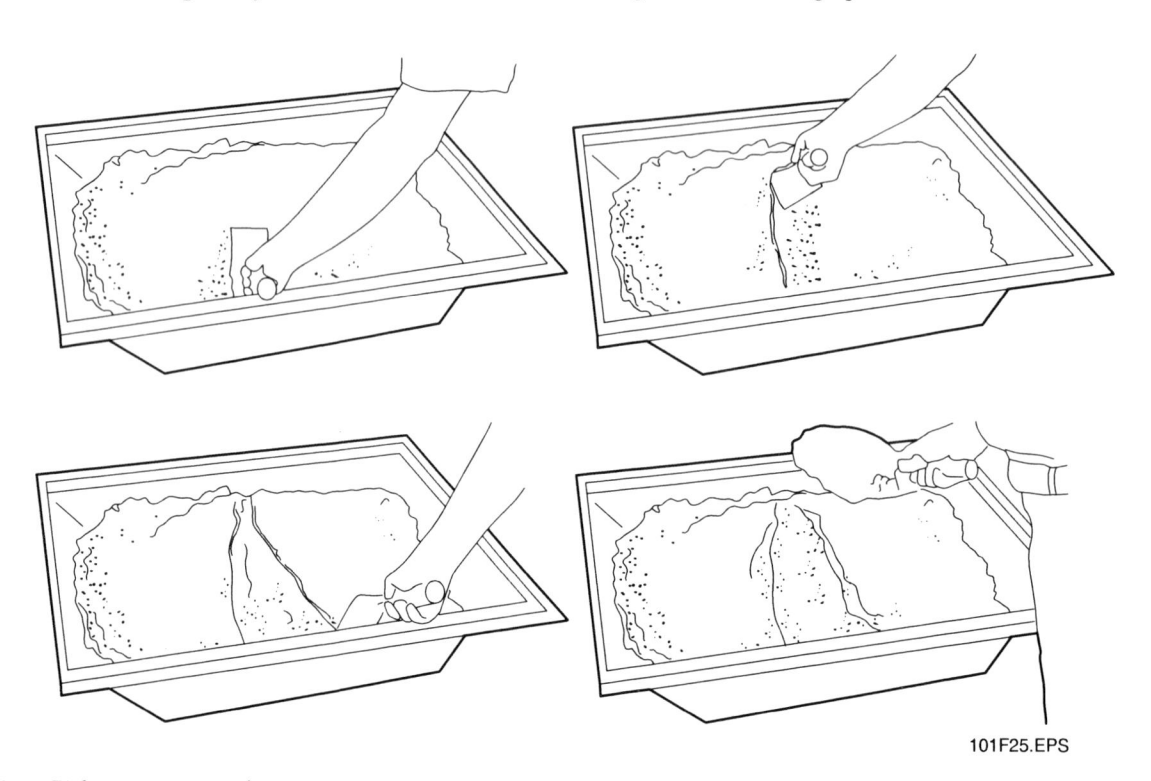

Figure 25 ◆ Picking up mortar from a pan.

Step 4 Lift the trowel from the mortar at the end of the stroke. The trowel should be fully loaded with a tapered section of mortar.

Step 5 To prevent the mortar from falling off the trowel, snap your wrist slightly to set the mortar on the trowel as you lift.

10.4.0 Spreading, Cutting, and Furrowing

The next sections describe spreading the mortar, shaping its edges, and **furrowing** it. You can practice spreading, cutting, and furrowing the mortar along a 2 × 4 board spread between two cement blocks or other props. Practice until you feel comfortable with these movements.

10.4.1 Spreading

Spreading the mortar means applying it in a desired location at a uniform thickness (see *Figure 26*). Mortar is spread for bed joints. The process of spreading the mortar for the bed joint is also called **stringing** the mortar. The spreading motion has two components to it, and they occur at the same time.

Figure 26 ♦ Spreading mortar.

Mortar application should adhere to the following guidelines:

- The joints are completely filled with no small voids for water to enter.
- The mortar is still pliable while you level and plumb the unit.
- The finished joint is the specified thickness after you level and plumb the masonry unit.
- The mortar does not smear the face of the masonry unit.

The first component of the spreading motion is a horizontal sweep from the starting point or the point where the last spread of mortar ended, back toward you. The mortar deposited is called a **spread**. Try to make the spread about two bricks long to begin with. If you are working with block, try to string the spread about one block long at first. After practice, you should be able to string the spread three to four bricks long, or two blocks long.

The second component of the motion is a vertical rotation. The trowel starts with its blade horizontal. As you move the trowel back toward you, you are also rotating it. As you rotate it, your thumb moves downward, and the back of the blade moves away from you. As the blade tilts, with the trowel traveling horizontally, the mortar is deposited along the path of the trowel.

Practice spreading until you can deposit a trail rather than a mound of mortar. Keep the trowel in the center of the wall for the length of the spread, so mortar will not get thrown on the face of the masonry. Start with a goal of 16" and work up to a spread of 24" to 32".

The joint spread should be about ¾" tall for brick and 1½" tall for block. Full joint spreads are used for all brick but not for all block. Block is usually mortared on its face shells and not its webs. However, block needs a full bed joint when it fits into any of the following categories:

- The first or starting course on a foundation, footing, or other structure
- Part of masonry columns, piers, or **pilasters** designed to carry heavy loads
- In a reinforced masonry structure, where all cores are to be grouted

Check the specifications to be sure. After the first course, the remaining block is mortared on shells, or shells and webs, according to specifications.

Whether you work with block or brick, you will need to know how to spread a full bed joint, cut it, and furrow it.

10.4.2 Cutting or Edging

After each spread, use the edge of the trowel to cut off excess mortar. To cut, hold the edge of the trowel at about a 60-degree angle, perpendicular to the edge of the mortar. Use the edge of the trowel to shave off the edge of the mortar. *Figure 27* shows the correct angle for shaving the edge of the spread.

Keep the edge of the trowel at a flat angle as shown. This will allow you to catch the mortar as you shave the edge. At this stage in your practice,

INTRODUCTION TO MASONRY 1.25

Figure 27 ◆ Cutting an edge.

learn how to catch the mortar as you cut it. The excess mortar can be returned to the mortar pan or used to fill any spaces in the bed joint.

Catching the mortar as you shave it means you do not have to go back and pick it up afterwards. On the job, having mortar stuck to the face of the masonry unit or lying in piles at the foot of a wall is unacceptable. Mortar is hard to remove when it dries, easy to clean when it is fresh. Learn to clean mortar as you lay it.

10.4.3 Furrowing

Furrowing is the act of shaping the bed joint before laying a masonry unit on it. A furrow is a shallow triangular depression, like a trough, extending the length of the bed joint. The furrow gives the mortar room to move slightly, just enough to let you adjust the masonry unit to its proper position. If the furrow is too shallow, the masonry unit will not move easily. If the furrow is too deep, it may expose the unit below and eventually cause a leak.

> **NOTE**
> Furrowing can be done with the trowel upright or upside down.

To make the furrow, hold the trowel blade at a 35-degree angle to the length of the spread, with the point into the spread. The point of the trowel should not go below the depth you want the finished furrow to stand. Tap the trowel point into the mortar at that angle and repeat the taps along the length of the spread. *Figure 28* shows the furrow. Notice the overlaid spacing for the trowel taps.

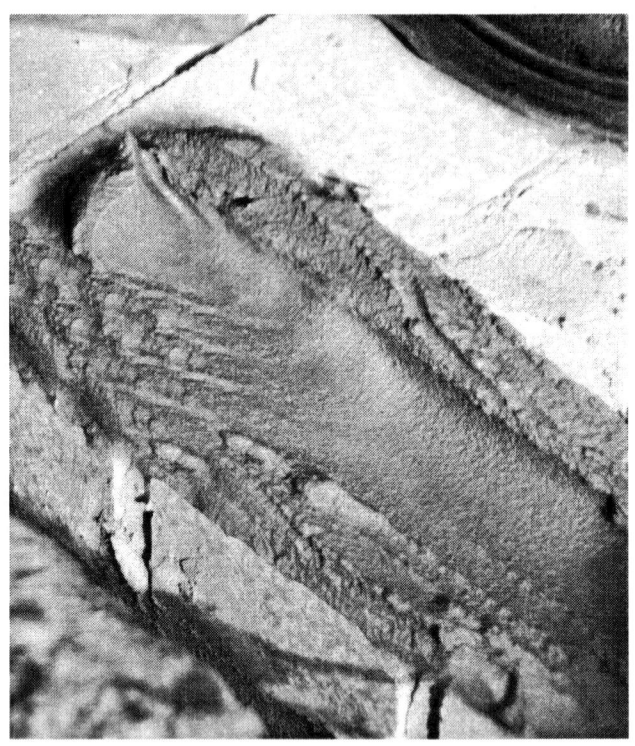

Figure 28 ◆ A furrow.

After furrowing the length of the spread, cut back the excess mortar. Use the edge of the trowel blade to shave off excess mortar hanging over the face of the wall. As you shave it, catch the excess mortar on the trowel blade. Use the excess mortar to butter the head joint on the next masonry unit to be laid.

10.5.0 Buttering Joints

Buttering the head joint is applying mortar to a header surface of a masonry unit. Buttering occurs after the bed joint is spread and the first masonry unit is laid in the bed. Buttering techniques are different for brick and block.

Buttering brick is a two-handed job. Begin by spreading the mortar on the bed joint. Keeping the trowel in your hand, pick up the first brick with your other hand. Press this brick into position in the mortar. Cut off the excess mortar on the outside face with the edge of the trowel.

Keeping the trowel in your hand, pick up a second brick in your brick hand. As you hold it, apply mortar to the header end of the brick. *Figure 29* shows a properly buttered head joint.

The buttered mortar should cover all the header surface but should not extend past the edges of the brick. Hold the trowel at an angle to the header surface to keep the mortar off the sides of the brick.

When the brick is buttered, use your brick hand to press it into position next to the first brick (*Figure 30*).

1.26   MASONRY LEVEL ONE — TRAINEE MODULE 28101-04

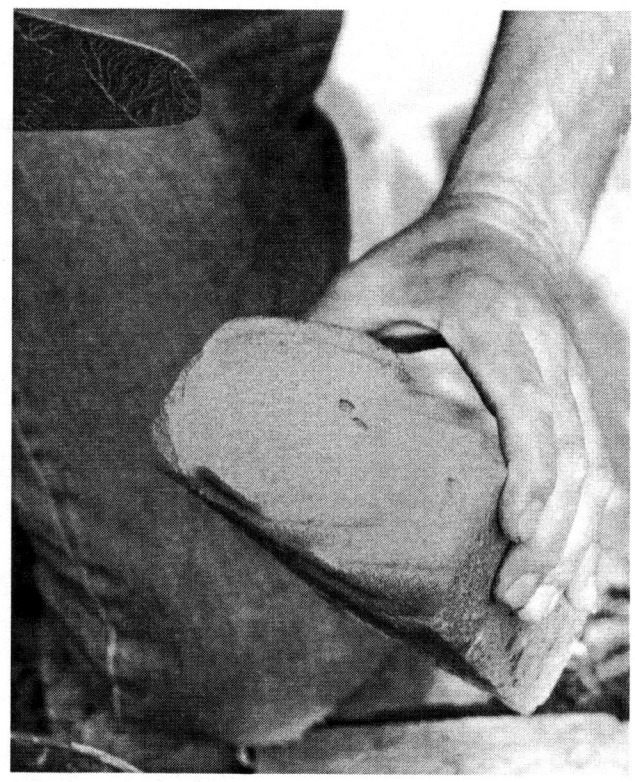

Figure 29 ◆ A buttered joint.

Figure 30 ◆ Placing the brick.

After placing the brick, cut off the excess mortar with the edge of your trowel.

Unlike blocks, you can easily hold a brick in one hand. Take advantage of this to use both hands for laying bricks. Try to develop a rhythmic set of movements. This will make the work faster and easier on you. Remember to use your shoulders and arms, not just your wrists.

After you have laid six bricks, check them for placement. Use your mason's level to check both plumb and level. If a brick is out of line, tap it gently with the handle of your trowel (*Figure 31*). Do not tap the level. Do not use the point or blade of your trowel or it will lose its edge.

10.6.0 General Rules

The way you work the mortar determines the quality of the joints between the masonry units. The mortar and the joints form a vital part of the structural strength and water resistance of the wall. Learning these general rules and applying them as you spread mortar will help you build good walls:

- Use mortar with the consistency of mud, so it will cling to the masonry unit, creating a good bond.
- Butter the head joints fully for brick and block; butter both ears of the head joints for block.

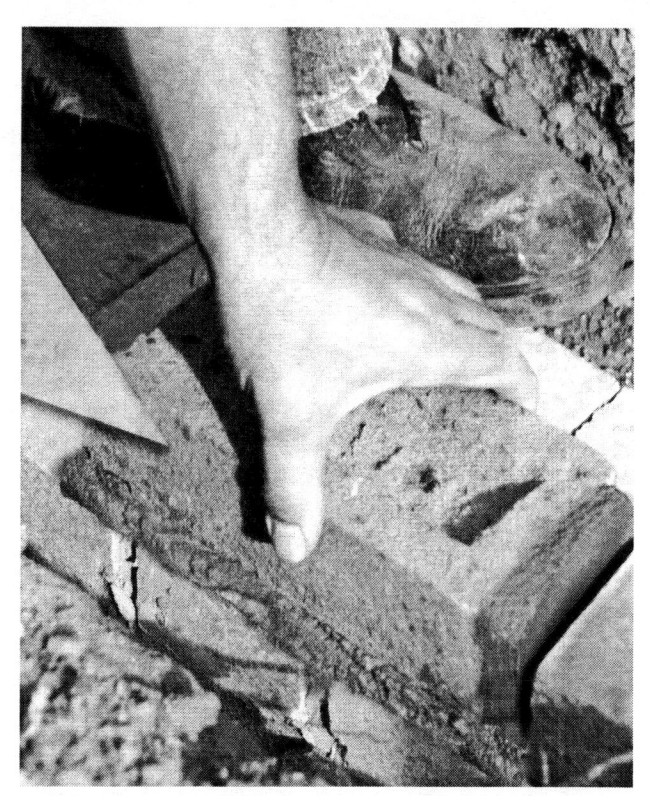

Figure 31 ◆ Checking the level.

INTRODUCTION TO MASONRY

- When laying a unit on the bed joint, press down slightly and sideways, so the unit goes against the one next to it.
- If mortar falls off a moving unit, replace the mortar before placing the unit.
- Put down more mortar than the size of the final joint; remember that placing the unit will compress the mortar.
- Do not string a spread that is more than 6 bricks or 3 blocks long; longer spreads will get too stiff to bond properly as water evaporates from them.
- Do not move a unit once it is placed, leveled, plumbed, and aligned.
- If a unit must be moved after it is placed, remove all the mortar on it and rebutter it.
- After placing the unit, cut away excess mortar with your trowel and put it back in the pan, or use it to butter the next joint.
- Throw away mortar after 2 to 2½ hours. At that point, it is beginning to set and will not give a good bond.

11.0.0 ◆ SAFETY PRACTICES

Masons operate in a high-risk environment. All around them are stacks of materials, trucks, and heavy equipment. Work sites have many possibilities for accidents. Workers themselves can cause accidents. They can drop masonry or tools off scaffolding and onto other workers. They can assemble scaffolding so poorly that it collapses under the weight of the load. They can fall off scaffolding. They can use damaged or poorly maintained tools that could injure themselves or others.

You must think and practice safety at all times. Your work must be planned so that it is safe as well as efficient.

All workers at a construction site must wear appropriate personal protective equipment (PPE) to protect their skin and eyes from mortar, grout, and flying masonry chips. They also need to protect themselves by being aware of what is happening around them. Workers need to keep track of the rest of their crew and of other crews. Unusual movements or noises can indicate something is moving that should not be. Masons need to have the knowledge, skill, and ability to do the following:

- Recognize an unsafe situation.
- Alert fellow crew members to the danger.
- Take evasive or corrective action.

11.1.0 The Cost of Job Accidents

Unsafe working conditions and practices can result in the following:

- Personal injury or death
- Injury or death of other workers
- Damage to equipment
- Damage to the work site

Insurance may cover some of the costs, but there are hidden (indirect or uninsured) costs as well (*Figure 32*).

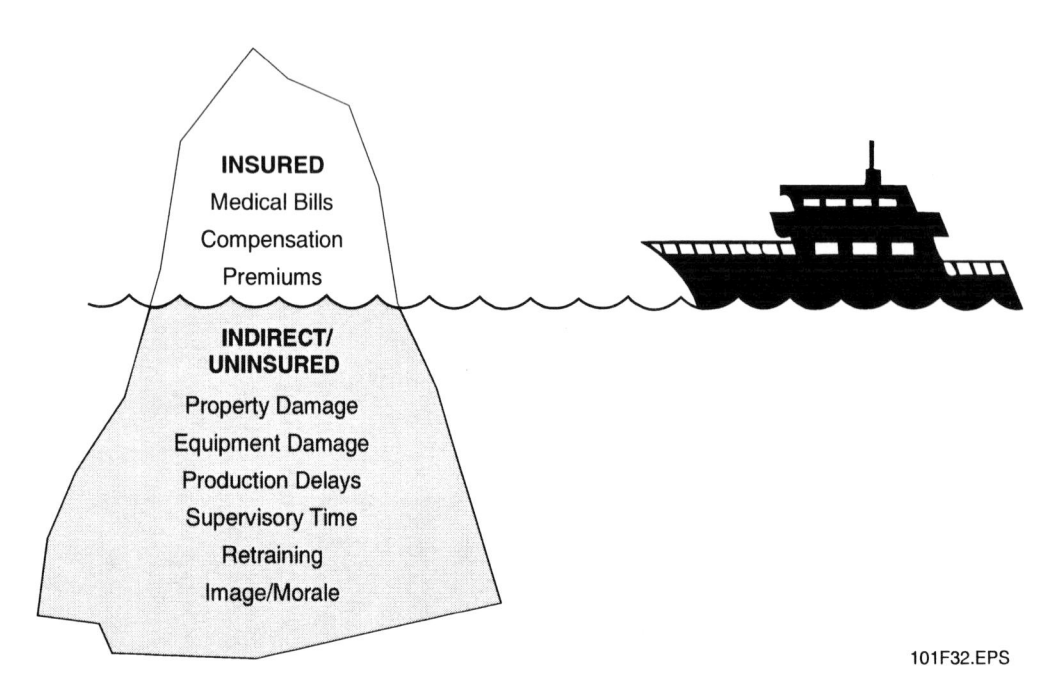

Figure 32 ◆ Hidden costs of accidents.

In addition to the pain and suffering for the individuals involved and their families, accidents can affect also others For example, accidents can slow down or stop a job, thereby putting the entire operation in jeopardy, and possibly resulting in site-wide layoffs. A high accident rate can cause an employer's insurance rates to rise, making the company less competitive with other construction companies and less likely to secure future work.

11.2.0 Wearing Safety Gear and Clothing

In general, the employer is responsible to Occupational Safety and Health Administration (OSHA) for making sure that all employees are wearing appropriate personal protective equipment whenever those employees are exposed to possible hazards to their safety. In turn, you are responsible for wearing the gear and clothing assigned to you. *Figure 33* shows a mason properly dressed and equipped for most masonry jobs.

> **DID YOU KNOW?**
> *Eye Injuries*
>
> The average cost of an eye injury is $1,463. That includes both the direct and indirect costs of accidents, not to mention the long-term effects on the health of the worker; that's priceless.
>
> Source: The Occupational Safety and Health Administration (OSHA)

It is important to take the following safety precautions when dressing for masonry work:

- Remove all jewelry, including wedding rings, bracelets, necklaces, and earrings. Jewelry can get caught on or in equipment, which could result in a lost finger, ear, or other appendage.
- Confine long hair in a ponytail or in your hard hat. Flying hair can obscure your view or get caught in machinery.

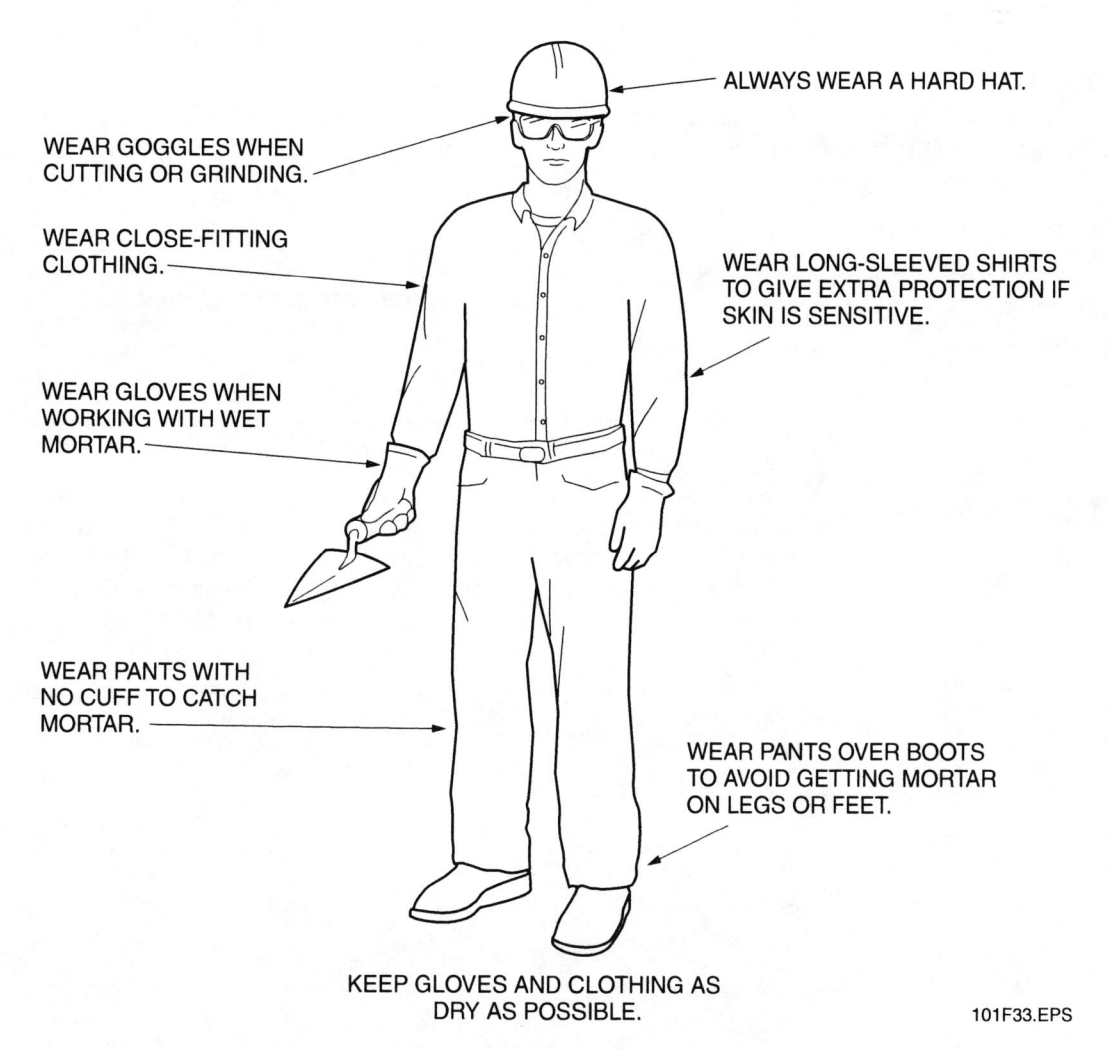

Figure 33 ◆ Dressed for masonry work.

INTRODUCTION TO MASONRY

- Wear close-fitting clothing that is appropriate for the job. Clothing should be comfortable and should not interfere with the free movement of your body. Clothing or accessories that do not fit tightly, or that are too loose or torn, may get caught in tools, materials, or scaffolding.
- Wear face and eye protection as required, especially if there is a risk from flying particles, debris, or other hazards such as brick dust or chemicals.
- Wear hearing protection as required.
- Wear respiratory protection as required.
- Wear a long-sleeved shirt to provide extra protection for your skin.
- Protect any exposed skin by applying skin cream, body lotion, or petroleum jelly.
- Wear sturdy work boots or work shoes with thick soles. Never show up for work dressed in sneakers, loafers, or sport shoes.
- Wear fall protection equipment as required.

11.3.0 Hazards on the Job

Construction sites may contain numerous hazards. You need to walk and work with all due respect for those hazards. The following list includes some of the hazardous conditions at a typical job site:

- Improper ventilation
- Inadequate lighting
- High noise levels
- Slippery floors
- Unmarked low ceilings
- Excavations, holes, and open, unguarded spaces, including open, unbarricaded elevator shafts
- Poorly constructed or poorly rigged scaffolds
- Improperly stacked materials
- Live wires, loose wires, and extension cords
- Unsafe ladders
- Unsafe crane operations
- Water and mud
- Unsafe storage of hazardous or flammable materials
- Defective or unsafe tools and equipment
- Poor housekeeping

Your safety and that of your fellow workers should be a primary consideration in your work life. Some common-sense rules and ways of doing things can make the job site safer for everyone. These safety tips should be a part of your everyday thinking. Develop a positive safety attitude. It will keep you and your co-workers safe and sound.

11.4.0 Falling Objects

Falling objects are a real danger on the job site. Follow these guidelines to stay safe when working around overhead hazards:

- Always wear a hard hat.
- Keep the working area clear by removing excess mortar, broken or scattered masonry units, and all other materials and debris on a regular basis.

BUILDING BLOCKS

Use GFCIs with Power Tools

Always plug your electrical power tools into a ground fault circuit interrupter (GFCI). The GFCI is designed to protect you from electrocution in case of a short circuit. If the work site isn't wired with GFCIs, use an extension cord with a GFCI like the one shown here.

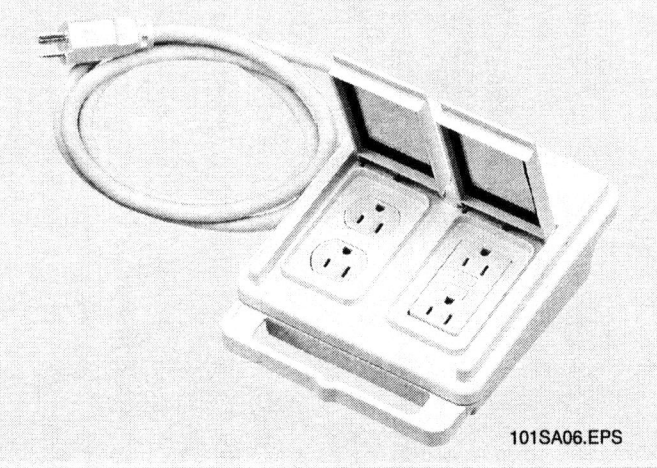

101SA06.EPS

- Keep openings in floors covered. When guardrail systems are used to prevent materials from falling from one level to another, any openings must be small enough to prevent the passage of potential falling objects.
- Do not store materials other than masonry and mortar within 4' (1.2 meters) of the working edges of a guardrail system.
- Be very careful around operating cranes. Stay clear of the crane's working area (*Figure 34*).
- Never work or walk under loads that are being hoisted by a crane.
- Learn the basic hoisting signals for cranes. Be sure to learn the stop signal and the emergency stop signal, as shown in *Figure 35*.
- Erect toeboards or guardrail systems to protect yourself from objects falling from higher levels. Toeboards should be erected along the edges of the overhead walking/working surface for a distance that is sufficient to protect the workers below.
- Erect paneling or screening from the walking/working surface or toeboard to the top of a guardrail system's top rail or mid-rail if tools, equipment, or materials are piled higher than the top edge of the toeboard.
- Raise or lower tools or materials with a rope and bucket or other lifting device. Never throw tools or materials.
- Never put tools or materials down on ladders or in other places where they can fall and injure people below. Before moving a ladder, make sure there are no tools left on it.

STOP — Extend arm, palm down, and hold. Move hand and forearm in a horizontal chopping motion.

EMERGENCY STOP — Same position as for Stop; extend and retract arms rapidly.

Figure 35 ♦ Emergency hand signals.

11.5.0 Mortar and Concrete Safety

Another hazard encountered by masonry craftworkers is exposure to mortar, grout, and concrete. These cement-based materials have ingredients that can hurt your eyes or skin. The basic ingredient of mortar, portland cement, is **alkaline** in nature and is therefore caustic. It is also **hygroscopic**, which means that it will absorb moisture from your skin. Prolonged contact between the fresh mix and skin can cause skin irritation and chemical burns to hands, feet, and exposed skin areas. It can also saturate a worker's clothes and transmit alkaline or hygroscopic effects to the skin. In addition, the sand contained in fresh mortar can cause skin abrasions through prolonged contact.

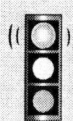

> **WARNING!**
> Those working with dry cement or wet concrete should be aware that it is harmful. Dry cement dust can enter open wounds and cause blood poisoning. The cement dust, when it comes in contact with body fluids, can cause chemical burns to the membranes of the eyes, nose, mouth, throat, or lungs. It can also cause a fatal lung disease known as silicosis.
> Wet cement or concrete can also cause chemical burns to the eyes and skin. Always wear appropriate personal protective equipment when working with dry cement or wet concrete. If wet concrete enters waterproof boots from the top, remove the boots and rinse your legs, feet, boots, and clothing with clear water as soon as possible. Repeated contact with cement or wet concrete can also cause an allergic skin reaction known as cement dermatitis.

Figure 34 ♦ Crane hazard on the job.

INTRODUCTION TO MASONRY

Avoid injuries by taking the following precautions:

- Keep your thumb on the ferrule of the trowel away from the mortar.
- Keep cement products off your skin at all times by wearing the proper protective clothing, including boots, gloves, and clothing with snug wristbands, ankle bands, and neckband. Make sure they are all in good condition.
- Prevent your skin from rubbing against cement products. Rubbing increases the chance of serious injury. If your skin does come in contact with any cement products, wash your skin promptly. If a reaction persists, seek medical attention.
- Wash thoroughly to prevent skin damage from cement dust.
- Keep cement products out of your eyes by wearing safety glasses when mixing mortar or pumping grout. If any cement or cement mixtures get in your eye, flush it immediately and repeatedly with water, and consult a physician promptly.
- Rinse off any clothing that becomes saturated from contact with fresh mortar. A prompt rinse with clean water will prevent continued contact with skin surfaces.
- Be alert! Watch for trucks backing into position and overhead equipment delivering materials. Listen for the alarms or warning bells on mixers, pavers, and ready-mix trucks.
- Never put your hands, arms, or any tools into rotating mixers.
- Be certain adequate ventilation is provided when using epoxy resins, organic solvents, brick cleaners, and other toxic substances.
- Use good work practices to reduce dust in the air when handling mortar and lime. For example, do not shake out mortar bags unnecessarily and do not use compressed air to blow mortar dust off clothing or a work surface. Stand upwind when dumping mortar bags.
- Never use solvents to clean skin.
- Immediately remove epoxy, solvents, and other toxic substances from skin using the appropriate cleansing agents.
- Know the locations of all eye wash stations and emergency showers on your job site. Be sure you know how to use them properly.

11.6.0 Flammable Liquid Safety

Flammable liquids are particularly dangerous and require additional safety precautions. Always adhere to all relevant codes, job-site rules, and the following guidelines:

- Carefully read labels on all flammable liquids, and use flammable liquids only in open, well-ventilated areas.
- Do not inhale or ignite fumes from flammable liquids.
- Be sure that all flammable liquids are marked correctly for storage.
- Store all flammable liquids properly and only in approved safety containers, such as safety cans and safety cabinets.
- Store oily rags and flammable materials in metal containers with self-sealing lids.
- When clothing is soaked by a flammable liquid, immediately change clothing, and cleanse the body with an appropriate cleaner.
- Use flammable liquids only for their intended purposes; for example, never use gasoline as a cleaner.
- Never use flammable liquids near fire or flame.
- Always be aware of the location of an appropriate fire extinguisher.
- Learn your company's emergency response procedures for fire and explosion.

Don't Get Burned

Here are some facts about gasoline:

- One gallon of gasoline contains the same explosive force as 14 sticks of dynamite.
- Gasoline vapors are heavier than air, can travel several feet to an ignition source, and can ignite at temperatures as low as 45°F. To be safe, keep open gasoline containers well removed from all potential ignition sources.
- Gasoline has a low electrical conductivity. As a result, a charge of static electricity builds up on gasoline as it flows through a pipe or hose. Getting into and out of your vehicle during refueling can build up a static charge, especially during dry weather. That charge can cause a spark that can ignite gasoline vapors if it occurs near the fuel nozzle.

Source: U.S. Department of Energy

11.7.0 Material Handling

Approximately 25 percent of all occupational injuries occur when handling or moving construction materials. Strains, sprains, fractures, and crushing injuries can be minimized with a knowledge of safe lifting and handling procedures and proper ergonomics. General guidelines for you to keep in mind when handling or moving construction materials are as follows:

- Wear steel-toe safety shoes.
- Keep floors free of water, grease, and other slippery substances so as to prevent falls.
- Inspect materials for grease, slivers, and rough or sharp edges.
- Determine the weight of the load before applying force to move it.
- Know your own limits for how much you can lift.
- Be sure that your intended pathway is free from obstacles.
- Make sure your hands are free of oil and grease.
- Take a firm grip on the object before you move it, being careful to keep your fingers from being pinched.
- When the load is too large or heavy, get help, or, if possible, simply reduce the load and make more trips.
- Whenever possible, use mechanical means of material handling.
- When stacking materials, be sure to follow OSHA regulations as to the height, shape, and stability of the pile.

Materials in the general working area and on the stockpile should be stacked safely. Masonry units that are not stacked properly and secured in some way are very likely to fall, which could cause injury to you or a fellow worker.

11.7.1 Materials Stockpiling and Storage

OSHA has several guidelines regarding the stockpiling and handling of materials. You must adhere to these and all OSHA guidelines:

- All materials stored in tiers should be stacked, racked, blocked, interlocked, or otherwise secured to prevent sliding, falling, or collapse.
- Maximum safe load limits of floors within buildings and structures should be posted in all storage areas, and maximum safe loads should not be exceeded.
- Aisles and passageways should be kept clear to provide for the free and safe movement of material handling equipment or employees.
- Materials stored inside buildings under construction should not be placed within 6' of any hoistway or inside floor openings nor within 10' of an exterior wall that does not extend above the top of the material stored.
- Each employee required to work on stored material in silos, hoppers, tanks, and similar storage areas should be equipped with personal fall-arrest equipment.
- Noncompatible materials should be segregated in storage.
- Bagged materials should be stacked by stepping back the layers and cross-keying the bags at least every ten bags high.
- Materials should not be stored on scaffolds or runways in excess of supplies needed for immediate operations.
- Stockpiles of palletized brick should not be higher than 7'. When a loose brick stockpile reaches a height of 4', it should be tapered back 2" in every foot of height above the 4' level.
- When loose masonry blocks are stockpiled higher than 6', the stack should be tapered back one-half block per tier above the 6' level.

11.7.2 Working Stacks

Working stacks of brick are used at the wall, on the ground, or on the scaffold platform. Palletized brick (*Figure 36*) is unbundled and moved to where it will be needed. When stacking materials in working piles, keep the pile neat and vertically in line to eliminate the possibility of snagging your clothes. Keep the piles about 3' high so that you can easily get to the brick. Bricks and other materials stacked too high not only pose a safety hazard, but may reduce productivity.

Figure 36 ◆ Palletized brick.

INTRODUCTION TO MASONRY

The most common way to stack bricks is to reverse the direction of every other course so that the stack is secure. Such a stack should be no more than 3' high and no closer than 2' to the wall. Wider stacks can be made by alternating a pattern of eight bricks, as shown in *Figure 37*.

11.8.0 Gasoline-Powered Tools

In masonry work, you may need to use gasoline-powered tools. Follow these safety guidelines:

- Be sure there is proper ventilation before operating gasoline-powered equipment indoors.
- Use caution to prevent contact with hot manifolds and hoses.
- Be sure the equipment is out of gear before starting it.
- Use the recommended starting fluid.
- Always keep the appropriate fire extinguishers near when filling, starting, and operating gasoline-powered equipment. OSHA requires that gasoline-powered equipment be turned off prior to filling.

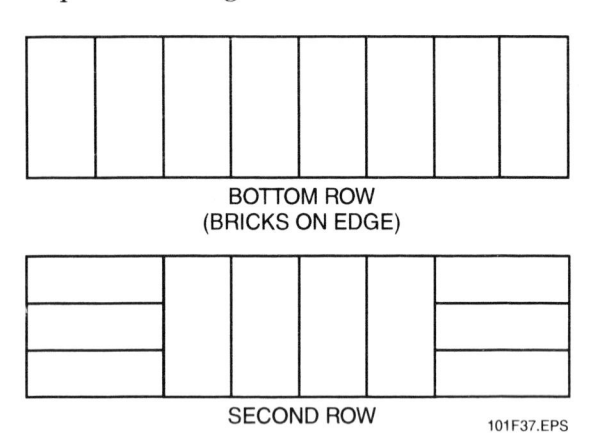

Figure 37 ◆ Stacking brick.

- Do not pour gasoline into the carburetor or cylinder head when starting the engine.
- Never pour gasoline into the fuel tank when the engine is hot or when the engine is running.
- Do not operate equipment that is leaking gasoline.

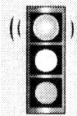

> **WARNING!**
> Never operate a tool without proper training and personal protective equipment.

11.9.0 Powder-Actuated Tools

A powder-actuated fastening tool (*Figure 38*) is a low-velocity fastening system powered by gunpowder cartridges, commonly called boosters. Powder-actuated tools are used to drive specially designed fasteners into masonry and steel.

Manufacturers use color-coding schemes to identify the strength of a powder load charge. It is extremely important to select the right charge for the job, so learn the color-coding system that applies to the tool you are using. *Table 2* shows an example of a color-coding system.

Table 2 Powder Charge Color-Coding System

Power Level*	Color
1	Gray
2	Brown
3	Green
4	Yellow
5	Red
6	Purple

*From the least powerful (1) to the most powerful (6)

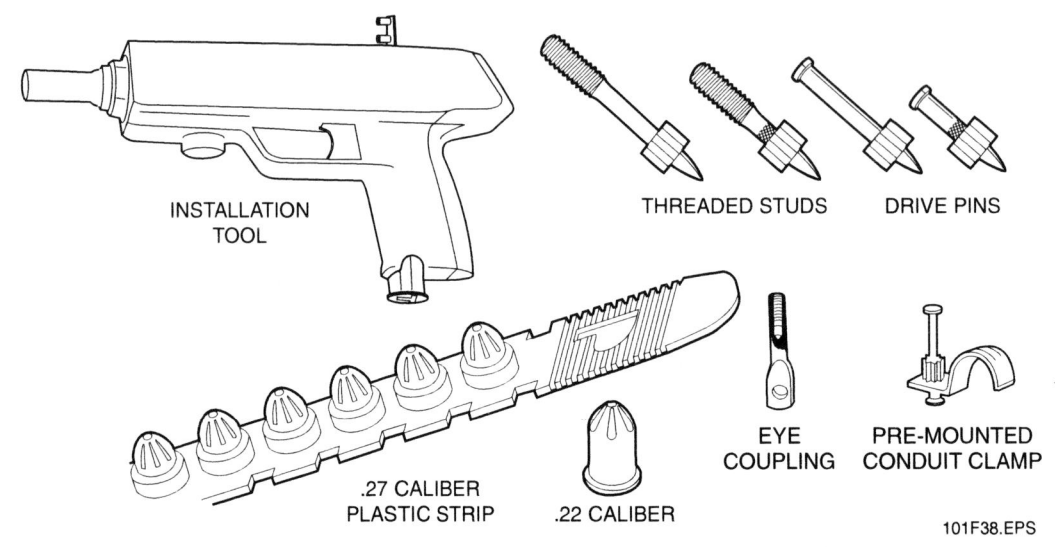

Figure 38 ◆ Powder-actuated fastening tool.

Fumes from Liquid-Fuel Tools

A worker at a large, enclosed construction site died of carbon monoxide poisoning after he and six other workers were exposed to high levels of the gas. He died because ventilation on the site was inadequate, and three machines were giving off carbon monoxide: a portable mixer and a trowel, both powered by gasoline, and a forklift powered by propane. This worker would have survived if the work area had been properly ventilated and if he were using the proper personal protective equipment, including a respirator.

The Bottom Line: Keep work areas well ventilated, and wear respirators when required. Hazardous air conditions can develop without warning.

Source: The Occupational Safety and Health Administration (OSHA)

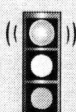

WARNING!
OSHA requires that all operators of powder-actuated tools be qualified and certified by the manufacturer of the tool. You must carry a certification card whenever using the tool.

If a gun does not fire, hold it against the work surface for at least 30 seconds. Follow the manufacturer's instructions for removing the cartridge. Do not try to pry it out because some cartridges are rim-fired and could explode.

Other rules for safely operating a powder-actuated tool are as follows:

- Do not use a powder-actuated tool unless you are certified.
- Follow all safety precautions in the manufacturer's instruction manual.
- Always wear safety goggles and a hard hat when operating a powder-actuated tool.
- Use the proper size pin for the job you are doing.
- When loading the driver, put the pin in before the charge.
- Use the correct booster (powder load) according to the manufacturer's instructions.
- Never hold your hand behind or near the material you are fastening.
- Never hold the end of the barrel against any part of your body or cock the tool against your hand.
- Do not shoot close to the edge of concrete.
- Never attempt to pry the booster out of the magazine with a sharp instrument.
- Always wear ear protection.
- Always hold the muzzle perpendicular (90 degrees) to the work.

11.10.0 Pressure Tools

You may use pressure tools on the job. These tools can be dangerous. Always use extreme caution, and never operate a pressure tool on which you have not been trained.

Pneumatic saws and grinders should have automatic overspeed controls. Runaway speeds can cause carborundum saw discs and buffers to disintegrate. Guards should also be in good condition in order to prevent flying particles.

All tools that produce dust need exhaust systems to collect the dust in order to prevent contamination of the air in the work areas. Automatic emergency valves should be installed at all compressed air sources to shut off the air immediately if the hose becomes disconnected or is severed. This will help prevent wild thrashing of the hose.

Powder-Actuated Tool

A 22-year-old worker was killed when he was struck in the head by a nail fired from a powder-actuated tool in an adjacent room. The tool operator was attempting to anchor plywood to a hollow wall and fired the gun, causing the nail to pass through the wall, where it traveled nearly 30' before striking the victim. The tool operator had never received training in the proper use of the tool, and none of the employees in the area were wearing personal protective equipment.

The Bottom Line: The use of powder-actuated tools requires special training and certification. In addition, all personnel in the area must be aware that the tool is in use and should be wearing appropriate personal protective equipment.

11.11.0 Weather Hazards

Masons usually work outdoors. Under certain environmental conditions, such as extreme hot or cold weather, work can become uncomfortable and possibly dangerous. There are specific things to be aware of when working under these adverse conditions.

11.11.1 Cold Weather

The amount of injury caused by exposure to abnormally cold temperatures depends on wind speed, length of exposure, temperature, and humidity. Freezing is increased by wind, humidity, or a combination of the two factors. Follow these guidelines to prevent injuries such as frostbite during extremely cold weather:

- Always wear the proper clothing.
- Limit your exposure as much as possible.
- Take frequent, short rest periods.
- Keep moving. Exercise fingers and toes if necessary, but do not overexert.
- Do not drink alcohol before exposure to cold. Alcohol can dull your sensitivity to cold and make you less aware of over-exposure.
- Do not expose yourself to extremely cold weather if any part of your clothing or body is wet.
- Do not smoke before exposure to cold. Breathing can be difficult in extremely cold air. Smoking can worsen the effect.
- Learn how to recognize the symptoms of over-exposure and frostbite.
- Place cold hands under dry clothing against the body, such as in the armpits.

If you live in a place with cold weather, you will most likely be exposed to it when working. Spending long periods of time in the cold can be dangerous. It's important to know the symptoms of cold weather exposure and how to treat them. Symptoms of cold exposure include the following:

- Shivering
- Numbness
- Low body temperature
- Drowsiness
- Weak muscles

Follow these steps to treat cold exposure:

Step 1 Get to a warm inside area as quickly as possible.

Step 2 Remove wet or frozen clothing and anything that is binding such as necklaces, watches, rings, and belts.

Step 3 Rewarm by adding clothing or wrapping in a blanket.

Step 4 Drink hot liquids, but do not drink alcohol.

Step 5 Check for frostbite. If you suspect frostbite, seek medical help immediately.

Frostbite is an injury resulting from exposure to cold elements. It happens when crystals form in the fluids and underlying soft tissues of the skin. The frozen area is generally small. The nose, cheeks, ears, fingers, and toes are usually affected. Affected skin may be slightly flushed just before frostbite sets in. Symptoms of frostbite include the following:

- Skin that becomes white, gray, or waxy yellow. Color indicates deep tissue damage. Victims are often not aware of frostbite until someone else recognizes the pale, glossy skin.

BUILDING BLOCKS

Cold Weather Clothing Tips

Use the following tips to prevent injury due to cold weather:

- Dress in layers.
- Wear thermal-type woolen underwear.
- Wear outer clothing that will repel wind and moisture.
- Wear a face helmet and head and ear coverings.
- Carry an extra pair of dry socks when working in snowy or wet conditions.
- Wear warm boots, and make sure that they are not so tight that circulation becomes restricted.
- Wear wool-lined mittens or gloves covered with wind- and water-repellent material.

> **NOTE**
>
> In advanced cases of frostbite, mental confusion and poor judgment occur, the victim staggers, eyesight fails, the victim falls and may pass out. Shock is evident, and breathing may cease. Death, if it occurs, is usually due to heart failure.

- Skin tingles and then becomes numb.
- Pain in the affected area starts and stops.
- Blisters show up on the area.
- The area of frostbite swells and feels hard.

Use the following steps to treat frostbite.

Step 1 Protect the frozen area from refreezing.

Step 2 Warm the frostbitten part as soon as possible.

Step 3 Get medical attention immediately.

11.11.2 Hot Weather

Hot weather can be as dangerous as cold weather. When someone is exposed to excessive amounts of heat, they run the risk of overheating. Conditions associated with overheating include the following:

- Heat cramps
- Heat exhaustion
- Heat stroke

Heat cramps can occur after an attack of heat exhaustion. Cramps are characterized by abdominal pain, nausea, and dizziness. The skin becomes pale with heavy sweating, muscular twitching, and severe muscle cramps.

If you experience heat cramps, sit or lie down in a cool area, preferably indoors. Drink a half a glass of water every 15 minutes, and gently stretch and massage any cramped muscles. Do not resume work until you feel fully recovered.

Heat exhaustion is characterized by pale, clammy skin; heavy sweating with nausea and possible vomiting; a fast, weak pulse; and possible fainting.

Treat heat exhaustion by having the victim lie down with his or her feet elevated 6 to 8 inches. If nauseous, have the victim lie on his or her side, not back. Remove any heavy clothing and loosen all other clothing. Apply cool, wet cloths, and fan the victim, but stop if chills develop. If the victim is fully conscious, give him or her a half a glass of water every 15 minutes. If the condition does not improve quickly, call emergency medical services (911).

Heat stroke is an immediate, life-threatening emergency that requires urgent medical attention. It is characterized by headache, nausea, and visual problems. Body temperature can reach as high as 106°F. This will be accompanied by hot, flushed, dry skin; slow, deep breathing; possible convulsions; and loss of consciousness.

If someone experiences heat stroke, call emergency medical services (911) immediately. Then, move the victim to a cool area and have him or her lie on his back. As with heat exhaustion, if the victim is nauseous, have the victim lie on his or her side instead. Move all nearby objects, as heat stroke may cause convulsions or seizures. Apply cool, wet cloths and/or fan the victim. If the victim is not nauseous and is conscious, give small amounts of water in 15-minute intervals. Place ice packs under the armpits and in the groin area. Remain with the victim until emergency medical assistance arrives.

Follow these guidelines when working in hot weather in order to prevent heat exhaustion, cramps, or heat stroke:

- Drink plenty of water.
- Do not overexert yourself.
- Wear lightweight clothing.
- Keep your head covered and face shaded.
- Take frequent, short work breaks.
- Rest in the shade whenever possible.

12.0.0 ◆ FALL PROTECTION

Falls are the leading cause of death in the construction industry. In fact, more than one third of all deaths in the industry are the result of a fall. Fall protection is required when workers are exposed to falls from work areas with elevations that are 6' or higher. The types of work areas that put the worker at risk include the following:

- Scaffolding
- Ladders
- Leading edges
- Ramps or runways
- Wall or floor openings
- Roofs
- Excavations, pits, and wells
- Concrete forms
- Unprotected sides and edges

Falls happen because of the inappropriate use or lack of fall-protection systems (*Figure 39*). They also happen because of worker carelessness. It is your responsibility to learn how to set up, use, and maintain fall-protection equipment. Not only will this keep you alive and uninjured, it could save the lives of your co-workers.

Falls are classified into two groups: falls from an elevation and falls on the same level. Falls from an elevation can happen when you are doing work from scaffolding, work platforms, decking, concrete forms, ladders, or excavations. Falls from elevations are almost always fatal. This is not to say that falls on the same level aren't also extremely dangerous. When a worker falls on the same level, usually from tripping or slipping, head injuries often occur. Sharp edges and pointed objects such as exposed rebar could cut or stab the worker.

The following safe practices can help prevent slips and falls:

- Wear safe, strong work boots that are in good repair.
- Watch where you step. Be sure your footing is secure.
- Install cables, extension cords, and hoses so that they will not become tripping hazards.
- Do not allow yourself to get in an awkward position. Stay in control of your movements at all times.
- Maintain clean, smooth walking and working surfaces. Fill holes, ruts, and cracks. Clean up slippery material and litter.
- Do not run on scaffolding, work platforms, decking, roofs, or other elevated work areas.

The best way to survive a fall from an elevation is to use fall-protection equipment. The three most common types of fall-protection equipment are guardrails, personal fall-arrest systems, and safety nets.

12.1.0 Guardrails

Guardrails (*Figure 40*) protect workers by providing a barrier between the work area and the ground or lower work areas. They may be made of wood, pipe, steel, or wire rope and must be able to support 200 pounds of force applied to the top rail.

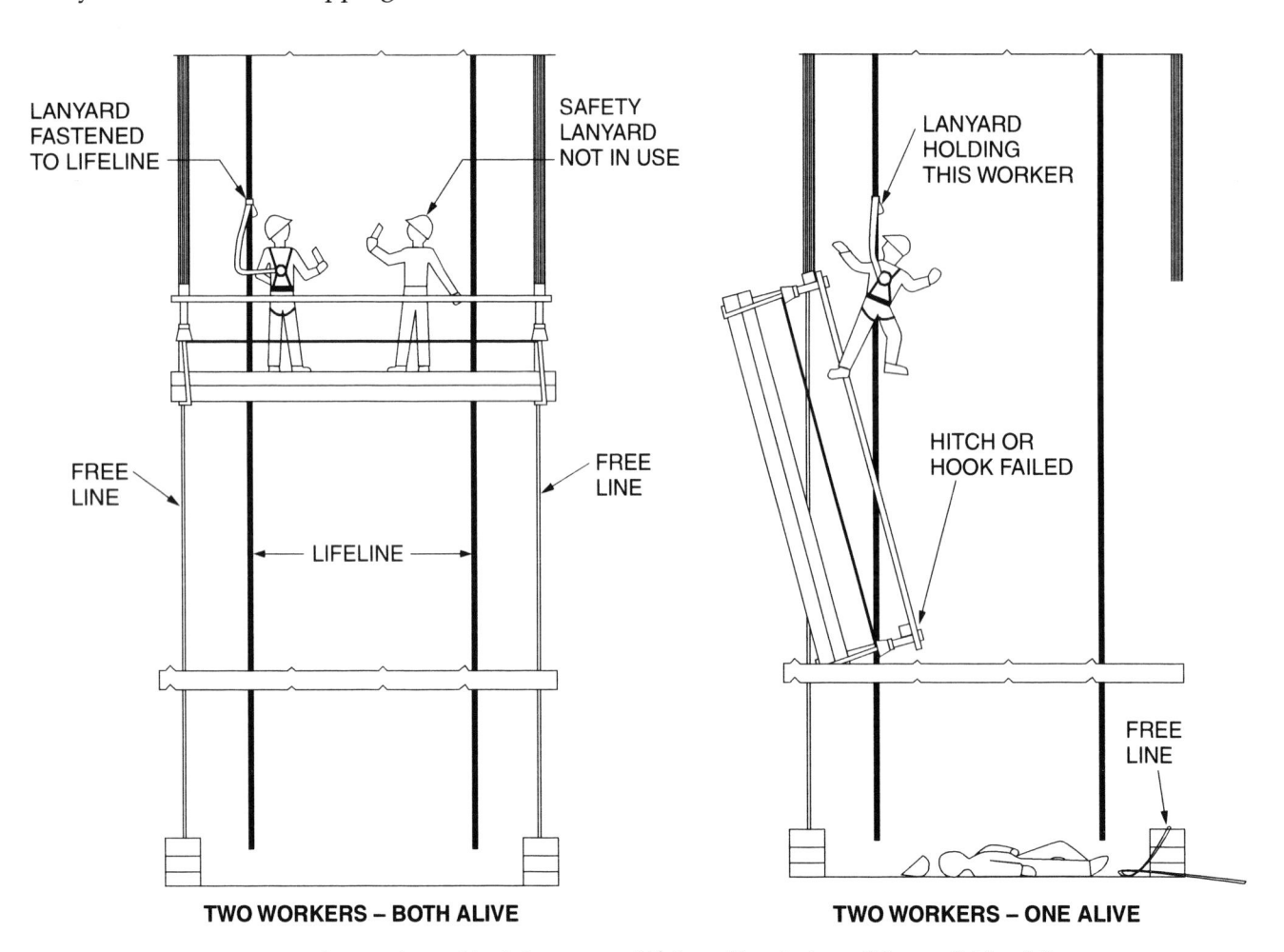

Figure 39 ◆ Proper and improper safety harness use.

Figure 40 ◆ Guardrails.

Inspect All Materials

CASE HISTORY

A crew laying bricks on the upper floor of a three-story building built a 6' platform to connect two scaffolds. The platform was correctly constructed of two 2" × 12" planks with standard guardrails. One of the planks however, was not scaffolding-grade lumber. It also had extensive dry rot in the center. When a bricklayer stepped on the plank, it disintegrated, and he fell 30' to his death.

The Bottom Line: Make sure that all planking is sound and secure. Your life depends on it.

Source: The Occupational Safety and Health Administration (OSHA)

12.2.0 Personal Fall-Arrest Systems

Personal fall-arrest systems catch workers after they have fallen. They are designed and rigged to prevent a worker from free falling a distance of more than 6' and hitting the ground or a lower work area. When describing personal fall-arrest systems, these terms must be understood:

- *Free-fall distance* – The vertical distance a worker moves after a fall before a deceleration device is activated.
- *Deceleration device* – A device such as a shock-absorbing lanyard or self-retracting lifeline that brings a falling person to a stop without injury.
- *Deceleration distance* – The distance it takes before a person comes to a stop when falling. The required deceleration distance for a fall-arrest system is a maximum of 3½'.
- *Arresting force* – The force needed to stop a person from falling. The greater the free-fall distance, the more force is needed to stop or arrest the fall.

Personal fall-arrest systems use specialized equipment. This equipment includes the following:

- Body harnesses and belts
- Lanyards
- Deceleration devices
- Lifelines
- Anchoring devices and equipment connectors

INTRODUCTION TO MASONRY

12.2.1 Body Harnesses

Full-body harnesses with sliding back D-rings (*Figure 41*) are used in personal fall-arrest systems. They are made of straps that are worn securely around the user's body. This allows the arresting force to be distributed throughout the body, including the shoulders, legs, torso, and buttocks. This distribution decreases the chance of injury. When a fall occurs, the sliding D-ring moves to the nape of the neck. This keeps the worker in an upright position and helps to distribute the arresting force. The worker then stays in a relatively comfortable position while waiting for rescue.

Selecting the right full-body harness depends on a combination of job requirements and personal preference. Harness manufacturers normally provide selection guidelines in their product literature. Some types of full-body harnesses can be equipped with front chest D-rings, side D-rings, or shoulder D-rings. Harnesses with front chest D-rings are typically used in ladder climbing and personal-positioning systems (*Figure 42*). Those with side D-rings are also used in personal-positioning systems. Personal-positioning systems allow workers to hold themselves in place, keeping their hands free to accomplish a task.

A personal-positioning system should not allow a worker to free fall more than 2'. The anchorage that it's attached to should be able to support at least twice the impact load of a worker's fall or 3,000 pounds, whichever is greater.

12.2.2 Lanyards

Lanyards are short, flexible lines with connectors on each end. They are used to connect a body harness or body belt to a lifeline, deceleration device, or anchorage point. There are many kinds of lanyards made for different uses and climbing situations. All must have a minimum breaking strength of 5,000 pounds. They come in both fixed and adjustable lengths and are made out of steel, rope, or nylon webbing. Some have a shock absorber (*Figure 43*) that absorbs up to 80% of the arresting force when a fall is being stopped. When choosing a lanyard for a particular job, always follow the manufacturer's recommendations.

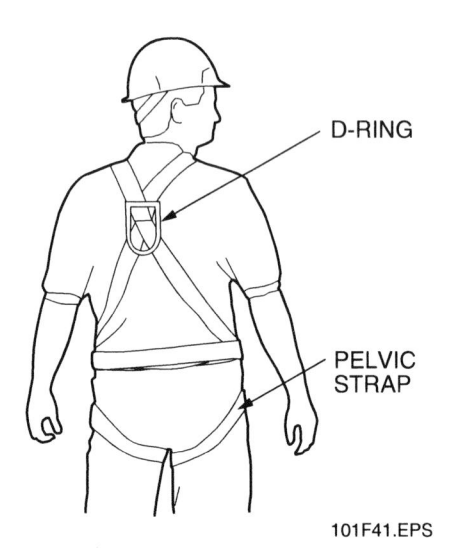

Figure 42 ◆ Harness with front chest D-ring.

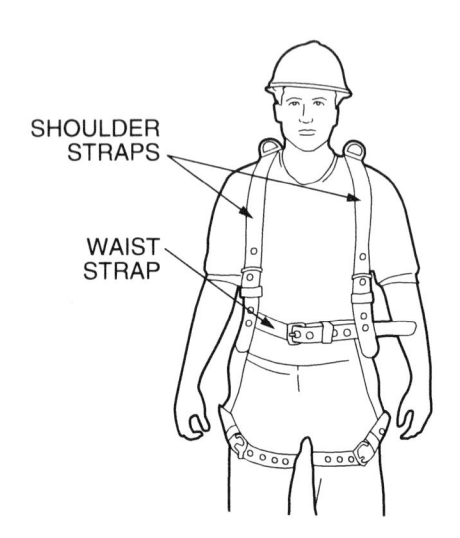

Figure 41 ◆ Full-body harnesses with sliding back D-rings.

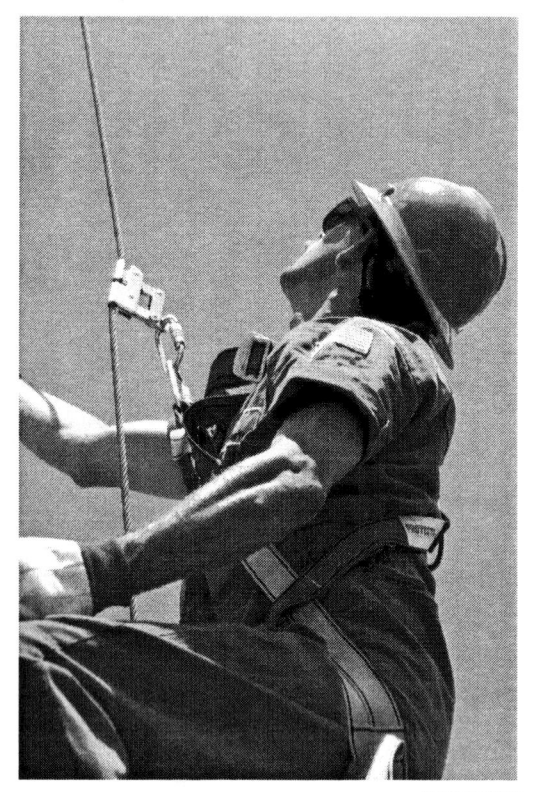

Figure 43 ◆ Lanyard with a shock absorber.

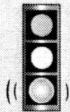

> **NOTE**
>
> In the past, body belts were often used instead of full-body harnesses as part of a fall-arrest system. As of January 1, 1998, however, they have been banned from such use. This is because body belts concentrate all of the arresting force in the abdominal area, which can cause significant injuries. It also causes the worker to hang in an uncomfortable and potentially dangerous position while awaiting rescue.

> **WARNING!**
>
> When activated during the fall-arresting process, a shock-absorbing lanyard stretches in order to reduce the arresting force. This potential increase in length must always be taken into consideration when determining the total free-fall distance from an anchor point.

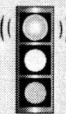

12.2.3 Deceleration Devices

Deceleration devices limit the arresting force to which a worker is subjected when the fall is stopped suddenly. Rope grabs and self-retracting lifelines are two common deceleration devices (*Figure 44*). A rope grab connects to a lanyard and attaches to a lifeline. In the event of a fall, the rope grab is pulled down by the attached lanyard, causing it to grip the lifeline and lock in place. Some rope grabs have a mechanism that allows the worker to unlock the device and slowly descend down the lifeline to the ground or surface below.

Self-retracting lifelines provide unrestricted movement and fall protection while workers are climbing and descending ladders and similar equipment or when working on multiple levels. Typically, they have a 25' to 100' galvanized-steel cable that automatically takes up the slack in the attached lanyard, keeping the lanyard out of the worker's way. In the event of a fall, a centrifugal braking mechanism engages to limit the worker's free-fall distance. Self-retracting lifelines and lanyards that limit the free-fall distance to 2' or less

> ### CASE HISTORY
>
> ### Icy Scaffolding Can Be Deadly
>
> A laborer was working on the third level of a tubular welded-frame scaffold. It was covered with ice and snow. Planking on the scaffold was weak, and a guardrail had not been set up. The worker slipped and fell head first approximately 20' to the pavement below.
>
> **The Bottom Line:** Never work on a wet or icy scaffold. Make sure all scaffolding is sturdy and includes the proper guardrails.
>
> *Source: The Occupational Safety and Health Administration (OSHA)*

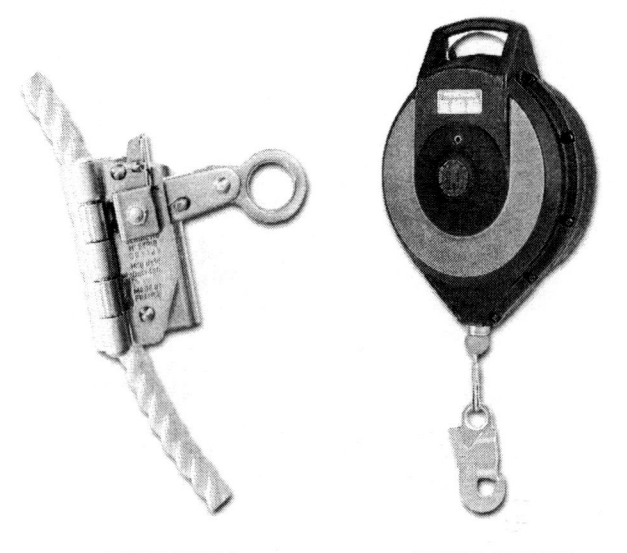

Figure 44 ◆ Rope grab and retractable lifeline.

must be able to support a minimum tensile load of 3,000 pounds. Those that do not limit the free fall to 2' or less must be able to hold a tensile load of at least 5,000 pounds.

12.2.4 Lifelines

Lifelines are ropes or flexible steel cables that are attached to an anchorage. They provide a means for tying off personal fall-protection equipment. Vertical lifelines (*Figure 45*) are suspended vertically from a fixed anchorage. A fall-arrest device such as a rope grab is attached to the lifeline. Vertical lifelines must have a minimum breaking strength of 5,000 pounds. Each worker must use his or her own line. This is because if one worker falls, the movement of the

---

INTRODUCTION TO MASONRY 1.41

lifeline during the fall arrest may also cause the other workers to fall. A vertical lifeline must be connected in a way that will keep the worker from moving past its end, or it must extend to the ground or the next lower working level.

Horizontal lifelines (*Figure 46*) are connected horizontally between two fixed anchorages. These lifelines must be designed, installed, and used under the supervision of a qualified, competent person. The more workers who are tied off to a single horizontal line, the stronger the line and anchors must be.

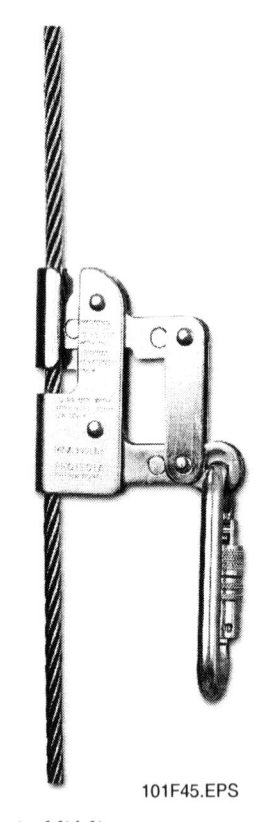

Figure 45 ◆ Vertical lifeline.

Figure 46 ◆ Horizontal lifeline.

12.2.5 Anchoring Devices and Equipment Connectors

Anchoring devices, commonly called tie-off points, support the entire weight of the fall-arrest system. The anchorage must be capable of supporting 5,000 pounds for each worker attached. Eye bolts (*Figure 47*) and overhead beams are considered anchorage points.

The D-rings, buckles, carabiners, and snaphooks (*Figure 48*) that fasten and/or connect the parts of a personal fall-arrest system are called connectors. There are regulations that specify how they are to be made and that require D-rings and snaphooks to have a minimum tensile strength of 5,000 pounds.

> **NOTE**
> As of January 1, 1998, only locking-type snaphooks are permitted for use in personal fall-arrest systems.

12.2.6 Selecting an Anchor Point and Tying Off

Connecting the body harness either directly or indirectly to a secure anchor point is called tying off. Tying off is always done before you get into a position from which you can fall. Follow the manufacturer's instructions on the best tie-off methods for your equipment.

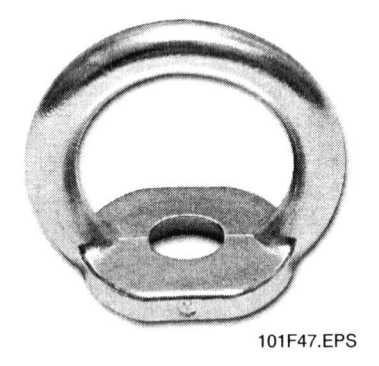

Figure 47 ◆ Eye bolt.

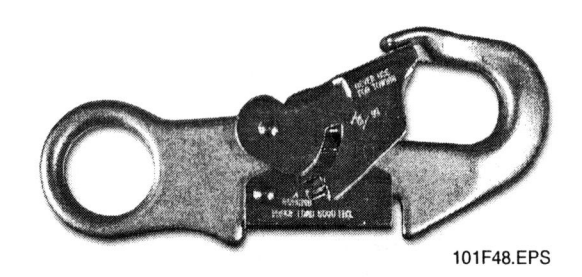

Figure 48 ◆ Double locking snaphook.

Building Blocks

Myths and Facts About Falls in Construction

Myth 1: In the construction industry, falls are not a leading cause of death.
Fact: One third of all deaths in the construction industry are caused by falls.

Myth 2: You have to fall a long distance to kill yourself.
Fact: Half of the construction workers who die in falls fall from a height of 21' or less. If you hit your head hard enough, you can die at any height. Even if you survive a fall, you may be laid up for some time with an injury.

Myth 3: Experienced workers don't fall.
Fact: The average age of construction workers who have fallen to their death is 47. That's not exactly young and inexperienced.

> "It just happens so fast. It's when you think you're safe that you need to be more careful." – *Gene, Builder*

Myth 4: Working safely is costly.
Fact: Some fall-protection equipment is inexpensive, such as ladder stabilizers, guardrail holders, and fall-protection kits. Other items such as harnesses, lifelines, and safe scaffolding are more costly. Injury and death, however, are much more expensive in the end.

> "I fell three stories and was out of work for 8 weeks. I was subcontracting and didn't have comp [workman's compensation insurance]. This was a long time ago, but I probably lost around $5,000. A harness would have cost me $50 back then." – *Dan, General Contractor*

Myth 5: Fall-protection equipment is more of a hindrance than a help.
Fact: Nothing is more of a hindrance than a lifelong disability you may experience due to a fall.

Source: Electronic Library of Construction Occupational Safety and Health

---

In addition to the manufacturer's instructions, an anchorage point should be as follows:

- Directly above the worker
- Easily accessible
- Damage-free and capable of supporting 5,000 pounds per worker
- High enough so that no lower level is struck should a fall occur
- Separate from work basket tie offs

Be sure to check the manufacturer's equipment labels, and allow for any equipment stretch and deceleration distance.

12.2.7 Using Personal Fall-Arrest Equipment

Before using fall-protection equipment on the job, you must know the basics and proper usage of fall protection equipment. All equipment supplied by your employer must meet established standards for strength. Before each use, always read the instructions and warnings on any fall-protection equipment. Inspect the equipment using the following guidelines:

- Examine harnesses and lanyards for mildew, wear, damage, and deterioration.
- Ensure no straps are cut, broken, torn, or scraped.
- Check for damage from fire, chemicals, or corrosives.
- Make sure all hardware is free of cracks, sharp edges, and burrs.
- Check that snaphooks close and lock tightly and that buckles work properly.
- Check ropes for wear, broken fibers, pulled stitches, and discoloration.
- Make sure lifeline anchors and mountings are not loose or damaged.

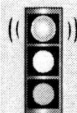

> **WARNING!**
> Never use fall-protection equipment that shows signs of wear or damage.

Do not mix or match equipment from different manufacturers. All substitutions must be approved by your supervisor. All damaged or defective parts must be taken out of service immediately and tagged as unusable or destroyed. If the equipment was used in a previous fall, remove it from service until it can be inspected by a qualified person.

12.3.0 Safety Net Systems

Safety nets are used for fall protection on bridges and similar projects. They must be installed as close as possible, not more than 30', beneath the work area. There must be enough clearance under a safety net to prevent a worker who falls into it

> ### Death Due to Unguarded Protruding Steel Bar
>
> A laborer fell approximately 8' through a roof opening to a foundation that had about 20 half-inch rebars protruding straight up. The laborer was impaled on one of the bars and died.
>
> **The Bottom Line:** Even a short-distance fall can be fatal. Use a personal fall-protection system, and check the area for potential hazards.
>
> *Source: The Occupational Safety and Health Administration (OSHA)*

from hitting the surface below. There must also be no obstruction between the work area and the net.

Depending on the actual vertical distance between the net and the work area, the net must extend 8' to 13' beyond the edge of the work area. Mesh openings in the net must be limited to 36 square inches and 6" on the side. The border rope must have a 5,000-pound minimum breaking strength, and connections between net panels must be as strong as the nets themselves. Safety nets must be inspected at least once a week and after any event that might have damaged or weakened them. Worn or damaged nets must be removed from service.

12.4.0 Rescue After a Fall

Every elevated job site should have an established rescue and retrieval plan. Planning is especially important in remote areas where help is not readily available. Before beginning work, make sure that you know what your employer's rescue plan calls for you to do in the event of a fall. Find out what rescue equipment is available and where it is located. Learn how to use equipment for self-rescue and the rescue of others.

If a fall occurs, any employee hanging from the fall-arrest system must be rescued safely and quickly. Your employer should have previously determined the method of rescue for fall victims, which may include equipment that lets the victim rescue himself or herself, a system of rescue by co-workers, or a way to alert a trained rescue squad. If a rescue depends on calling for outside help such as the fire department or rescue squad, all the needed phone numbers must be posted in plain view at the work site. In the event a co-worker falls, follow your employer's rescue plan. Call any special rescue service needed. Communicate with the victim, and monitor him or her constantly during the rescue.

13.0.0 ♦ FORKLIFT SAFETY

Forklifts are common on many masonry work sites. They are useful for lifting and moving heavy or awkward loads of materials, supplies, and equipment (*Figure 49*). While extremely useful and relatively easy to operate, these machines can also be very dangerous. They present several risks, including hitting other workers, dropping loads, tipping over, and causing fires and explosions.

Mechanical and hydraulic problems can cause accidents, but the most common cause of forklift

Figure 49 ♦ Forklift.

Drop Testing Safety Nets

Safety nets should be drop-tested at the job site after the initial installation, whenever relocated, after a repair, and at least every six months if left in one place. The drop test consists of a 400-pound bag of sand of 29" to 31" in diameter that is dropped into the net from at least 42" above the highest walking/working surface at which workers are exposed to fall hazards. If the net is still intact after the bag of sand is dropped, it passed the test.

> ### Case History
>
> ### Location is Everything
> A worker was placing metal bridge decking onto the stringers of a bridge deck to be welded. After the first decking was placed down on stringers, the employee stepped onto it in order to put down the next decking. The decking was not secured in place and shifted.
> Although safety nets were being used under another section of the bridge, they had not been moved forward as the crew moved to another area. The worker fell approximately 80' into the river and was killed.
> **The Bottom Line:** Make sure that all safety equipment is in place before beginning any job.

accidents is human error. That means most forklift accidents can be avoided if the operator and other workers in the area stay alert and use caution and common sense. In fact, research by Liberty Mutual Insurance Company shows that drivers with more than a year of experience operating a forklift are more likely to have an accident than someone with little experience. This is because operators tend to become too comfortable and less attentive after they gain experience on the equipment. The same study showed that the most common type of forklift accident is one in which a pedestrian is hit by the truck.

13.1.0 Before You Operate a Forklift

Before you can begin operating a forklift, you must be trained and certified on that particular piece of equipment. Once you are trained and certified, you must thoroughly inspect your forklift before you begin each shift.

13.1.1 Training and Certification

It is a common misconception that if you can drive a car, truck, or piece of heavy equipment, you can operate a forklift. However, OSHA requires forklift operators to be trained and certified on each piece of equipment before they operate it on the job site. The operator's card only applies to the specific piece of equipment on which they are trained. Powered forklift operators must have the visual, hearing, physical, and mental abilities necessary to safely operate the equipment. Personnel who have not been trained in forklift operation may only operate them for the purpose of training. The training must be conducted under the direct supervision of a qualified trainer.

13.1.2 Pre-Shift Inspection

You must perform a pre-shift inspection before operating a forklift. The more thorough you are when inspecting the forklift, the safer and more productive you will be during your shift. *Figure 50* shows a sample checklist covering the basic items that need to be checked during a pre-shift inspection.

Your company's checklist and your supervisor can provide you with specific information about what you should check before you begin each shift on a forklift. Your training and the forklift operator's manual will help you understand what to look for when you inspect the forklift. If you find any problems during your pre-shift inspection, notify your supervisor or maintenance manager immediately. The forklift should be locked out and tagged. It cannot be used until all problems are corrected.

13.1.3 General Safety Precautions

Safe operation is the operator's responsibility. Operators must develop safe working habits and be able to recognize hazardous conditions in order to avoid equipment and property damage and to protect themselves and others from death or injury. They must always be aware of unsafe conditions, so they can protect the load and the forklift from damage. They must also understand the operation and function of all controls and instruments before operating any forklift. Operators must read and fully understand the operator's manual for each piece of equipment being used.

The following safety rules are specific to forklift operation:

- Always check the capacity chart mounted on the machine before operating any forklift.
- Never put any part of the body into the mast structure or between the mast and the forklift.
- Never put any part of the body within the reach mechanism.
- Understand the limitations of the forklift.
- Do not permit passengers to ride in the forklift unless a safe place to ride has been provided by the manufacturer.
- Never leave the forklift running unattended.
- Never carry passengers on the forks.

INTRODUCTION TO MASONRY

OPERATOR'S DAILY CHECKLIST

Check Each Item Before Start of Each Shift Date: _____

Check One: Gas/LGP/Diesel Truck ☐   Electric Sit-Down ☐   Electric Stand-Up ☐   Electric Pallet ☐

Truck Serial Number: _____   Operator: _____   Supervisor's OK: _____

Hour Meter Reading: _____

Check each of the following items before the start of each shift. Let your supervisor and/or maintenance department know of any problem. DO NOT OPERATE A FAULTY TRUCK. Your safety is at risk.

After checking, mark each item accordingly. Explain below as necessary.

Check boxes as follows: ☐ OK   ☐ NG, needs attention or repair. Circle problem and explain below.

OK	NG	Visual Checks	OK	NG	Visual Checks
		Tires/Wheels: wear, damage, nuts tight			Steering: loose/binding, leaks, operation
		Head/Tail/Working Lights: damage, mounting, operation			Service Brake: linkage loose/binding, stops OK, grab
		Gauges/Instruments: damage, operation			Parking Brake: loose/binding, operational, adjustment
		Operator Restraint: damage, mounting, operation oily, dirty			Seat Brake (if equipped): loose/binding, operational, adjustment
		Warning Decals/Operator's Manual: missing, not readable			Horn: operation
		Data Plate: not readable, missing adjustment			Backup Alarm (if equipped): mounting, operation
		Overhead Guard: bent, cracked, loose, missing			Warning Lights (if equipped): mounting, operation
		Load Back Rest: bent, cracked, loose, missing			Lift/Lower: loose/binding, excessive drift, leaks
		Forks: bent, worn, stops OK			Tilt: loose/binding, excessive drift, "chatters," leaks
		Engine Oil: level, dirty, leaks			Attachments: mounting, damaged operation, leaks
		Hydraulic Oil: level, dirty, leaks			Battery Test (electric trucks only): indicator in green
		Radiator: level, dirty, leaks			Battery: connections loose, charge, electrolyte low while holding full forward tilt
		Fuel: level, leaks			Control Levers: loose/binding, freely return to neutral
		Covers/Sheet Metal: damaged, missing			Directional Controls: loose/binding, find neutral OK
		Brakes: linkage, reservoir fluid level, leaks			
		Engine: runs rough, noisy, leaks			

Explanation of problems marked above: _____
_____
_____
_____
_____

Figure 50 ◆ Forklift operator's daily checklist.

Forklift operators must pay special attention to the safety of any pedestrians on the job site. Safeguard pedestrians at all times by observing the following rules:

- Always look in the direction of travel.
- Do not drive the forklift up to anyone standing in front of an object or load.
- Make sure that personnel stand clear of the rear swing area before turning.
- Exercise particular care at cross aisles, doorways, and other locations where pedestrians may step into the travel path.
- Always use a spotter or signal person when moving an elevated load with a telescoping-boom forklift.

13.2.0 Traveling

Traveling refers to driving your forklift both with and without a load. To move either the forklift or a load of materials, you must travel. Sometimes the job requires only short travel distances, such as from a flatbed truck or rail car to a storage or staging area (*Figure 51*). Other times you must travel longer distances on the forklift. For example, you may need to move a load of bricks from a storage area all the way across the work site to a building under construction.

13.2.1 Stay Inside

You are safest when traveling on a forklift if you keep your whole body inside the vehicle. Many experienced drivers get into the unsafe habit of hanging an elbow outside of the truck, sliding a foot off the platform, or resting one hand with the fingers hanging over the edge of the truck. This often results in crushing injuries and amputation. The operator's compartment, along with the use of seat belts, is designed to protect the operator from falling objects, impact from collisions, contact with electrical utilities, and tipping accidents. For example, if you allow your elbow to hang over the edge of the truck, and then accidentally back into a support beam, you could easily crush your arm between the forklift and the beam.

> **WARNING!**
> Keep all parts of your body inside the forklift cab during operation or travel.

Figure 51 ◆ Forklift working in a storage area.

13.2.2 Pedestrians

Always yield the right-of-way to pedestrians. Forklifts are heavy machines that typically require a distance equal to the length of the forklift in order to stop. Pedestrians are usually not aware of this and walk around the site expecting these large and cumbersome machines to be able to stop quickly. Because of this, it is important to look out for any pedestrians.

13.2.3 Passengers

Forklifts are designed to carry one person: the operator. No one else should ever ride in or stand on a forklift. There are a few specially designed and certified attachments that allow forklifts to be converted to personnel lifts. Other than those few situations, no one other than the operator should be on the forklift.

13.2.4 Blind Corners and Intersections

As you approach a blind corner or intersection, always assume that a pedestrian or another piece of equipment is coming the other way. Stop at the intersection. Sound your horn. Then proceed slowly through the intersection or around the corner. Be prepared to stop if necessary.

INTRODUCTION TO MASONRY

13.2.5 Keeping the Forks Low

Whether traveling with or without a load, keep the forks as low as possible (*Figure 52*). As a general rule of thumb, the forks should never be higher than 6" from the travel surface while traveling, unless you are moving over an extremely rough surface. The forks are strong and pointed. Ramming into something or someone can cause serious damage or injury. If the forks are low, the chance of critically or fatally injuring someone is greatly reduced.

Figure 52 ◆ Forklift with forks in low position.

13.2.6 Horseplay

Driving a forklift is a serious operation that requires maturity and attention. Never drive a forklift toward another person as a joke, particularly if they are in front of a solid object or another piece of equipment. Doing so could easily lead to a very serious crushing accident. Accidents caused by horseplay are the most avoidable problem on the job site. Working with heavy equipment is dangerous, and your behavior on the job should reflect that. Your safety and that of your co-workers depends on it.

13.2.7 Travel Surface

Whenever possible, take the smoothest and driest route when traveling. Rough or bumpy surfaces can cause a lot of bouncing, which may destabilize a forklift and make the forklift or its load tip. Wet or slippery surfaces can cause the tires to lose traction, resulting in loss of control of the forklift.

13.3.0 Handling Loads

A forklift's main use is to transport large, heavy, or awkward loads. If not handled correctly, loads can fall from the forks, obstruct the operator's view, or cause the forklift to tip. The most important factor to consider when using any forklift is its capacity. Each forklift is designed with an intended capacity, and this capacity must never be exceeded. Exceeding the capacity jeopardizes not only the machine and the load, but also the safety of everyone on or near the forklift. Every manufacturer supplies a capacity chart for each forklift. The operator must be aware of the capacity of the machine before being allowed to operate the forklift.

13.3.1 Picking Up Loads

Some forklifts are equipped with a sideshift device that allows the operator to shift the load sideways several inches in either direction with respect to the mast. A sideshift device enables more precise placing of loads, but it also changes the forklift's center of gravity and must be used with caution. If the forklift being used is equipped with a sideshift device, be sure to return the fork carriage to the center position before attempting to pick up a load.

13.3.2 Traveling with Loads

Always travel at a safe rate of speed with a load. Never travel with a raised load. Keep the load as low as possible, and be sure the mast is tilted rearward to cradle the load.

As you travel, stay alert, and pay attention. Watch the load and the conditions ahead of you, and alert others to your presence. Avoid sudden stops and abrupt changes in direction. Be careful when downshifting because sudden deceleration can cause the load to shift or topple. Watch the machine's rear clearance when turning.

If you are traveling with a telescoping-boom forklift, be sure the boom is fully retracted. If you have to drive on a slope, keep the load as low as possible. Do not drive across steep slopes. If you have to turn on an incline, make the turn wide and slow.

13.3.3 Traveling with Long Loads

Traveling with long loads presents special hazards, particularly if the load is flexible and subject to damage. Traveling multiplies the effect of bumps over the length of the load. A stiffener may be added to the load to give it extra rigidity.

To prevent slippage, secure long loads to the forks. A field-fabricated cradle may be used to support the load. While this is an effective method, it requires that the load be jacked up.

The forklift may be used to carry pieces of rigging equipment. This method requires the use of slings and a spreader bar.

In some cases, long loads may be snaked through openings that are narrower than the load itself. This is done by approaching the opening at an angle and carefully maneuvering one end of the load through the opening first. Avoid making quick turns because abrupt maneuvers will cause the load to shift.

13.3.4 Placing Loads

Position the forklift at the landing point so that the load can be placed where you want it. Be sure everyone is clear of the load. The area under the load must be clear of obstructions and able to support the weight of the load. If you cannot see the placement, use a signaler to guide you.

With the forklift in the unloading position, lower the load and tilt the forks to the horizontal position. When the load has been placed and the forks are clear from the underside of the load, back away carefully to disengage the forks.

13.3.5 Placing Elevated Loads

When placing elevated loads, you must be especially careful. Some forklifts are equipped with a leveling device that allows the operator to rotate the fork carriage to keep the load level during travel. When placing elevated loads, it is extremely important to level the machine before lifting the load.

One of the biggest potential safety hazards during elevated load placement is poor visibility. There may be workers in the immediate area who cannot be seen. The landing point itself may not be visible. Your depth perception decreases as the height of the lift increases. To be safe, use a signal person to help you position the load.

Use tag lines to tie off long loads to the mast of the forklift. Drive the forklift as closely as possible to the landing point with the load kept low. Set the parking brake, and then raise the load slowly and carefully while maintaining a slight rearward tilt to keep the load cradled. Under no circumstances should the load be tilted forward until the load is over the landing point and ready to be set down.

If the forks start to move, sway, or lean, stop immediately but not abruptly. Lower the load slowly. Reposition it, or break it down into smaller components if necessary. If ground conditions are poor at the unloading site, you may need to reinforce the ground with planks to provide greater stability. As the load approaches the landing point, slow the lift speed to a minimum. Continue lifting until the load is slightly higher than the landing point.

13.3.6 Tipping

There are three main causes for a forklift tipping:

- The load is too heavy.
- The load is placed too far forward on the forks.
- The operator is not driving safely.

To avoid tipping, you need to understand what the center of gravity is and how it applies to forklifts. The center of gravity is the point around which all of an object's weight is evenly distributed (*Figure 53*). Your forklift has a center of gravity, and the load you're moving will have its own center of gravity. When the forklift picks up the load, the center of gravity shifts to the combined center of gravity (*Figure 54*).

The forklift will tip if the center of gravity moves too far forward, backward, right, or left. Putting too heavy a load on the forklift or placing the load too far forward on the forks will cause the combined center of gravity to be too far forward, causing the forklift to tip forward. Turning too sharply or quickly can cause the forklift to sway or swing to the left or right, causing the combined center of gravity to veer far enough off center to tip the forklift over. These types of accidents can easily result in the operator or a bystander being crushed by the forklift.

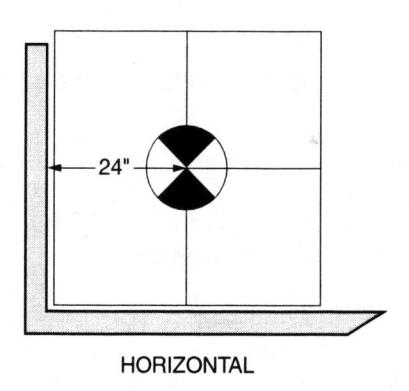

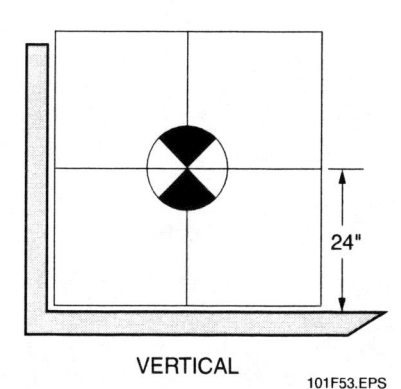

Figure 53 ◆ Center of gravity.

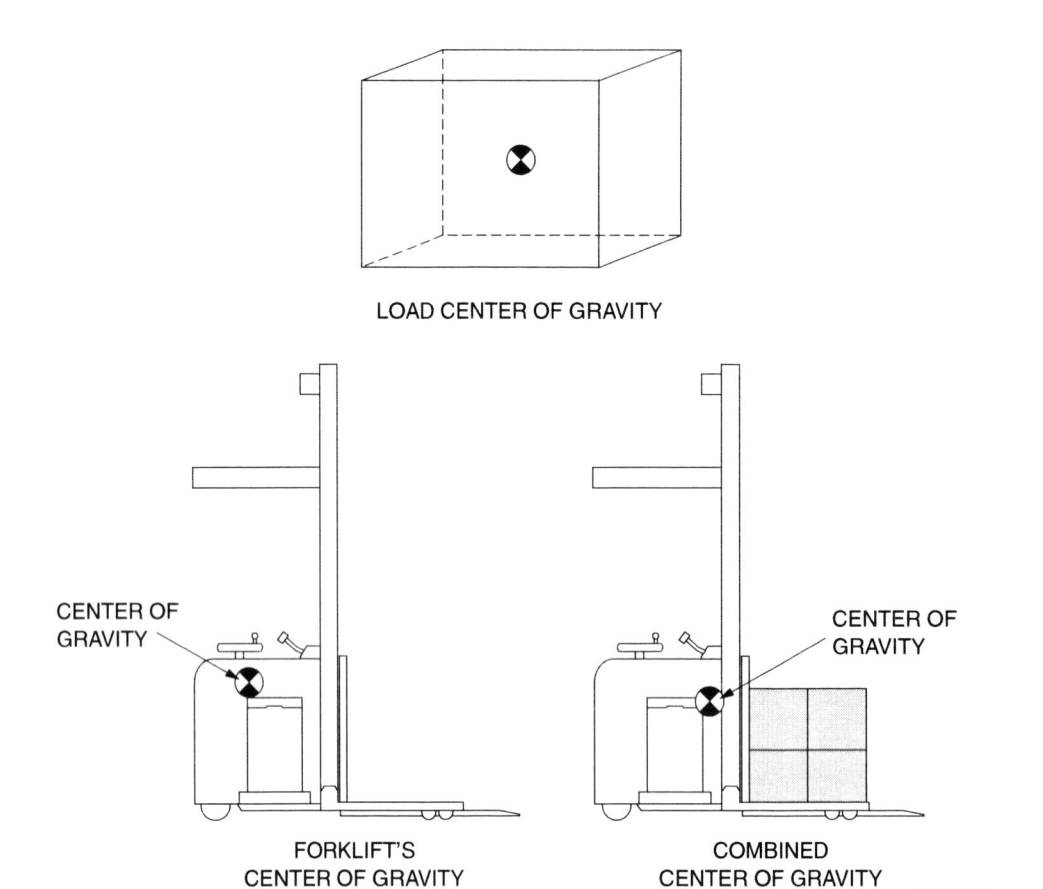

Figure 54 ◆ Combined center of gravity.

13.3.7 Using a Forklift to Rig Loads

A forklift can be a very useful piece of rigging equipment if it is properly and safely used. Loads can be suspended from the forks with slings, moved around the job site, and placed. All the rules of careful and safe rigging apply when using a forklift to rig loads. Never drag the load or let it swing freely. Use tag lines to control the load.

Never attempt to rig an unstable load with a forklift. Be especially mindful of the load's center of gravity when rigging loads with a forklift.

When carrying cylindrical objects, such as oil drums, keep the mast tilted rearward to cradle the load. If necessary, secure the load to keep it from rolling off the forks.

13.3.8 Dropping Loads

Momentum is a physical force that makes objects in motion tend to stay in motion, even if the mode of transportation stops. Have you ever carried a meal on a cafeteria tray? In your experience, what happened to the items on your tray if you had to stop suddenly or if you turned too quickly? The items on your tray probably started to topple or slide. They might have fallen right off the tray, or they may have shifted the load, so that the tray tipped, spilling everything.

The same principle applies to the load on your forklift. If you turn or stop too quickly, the load will keep going in the direction you were originally headed, causing it to slide off the forks. At the very least, this may be inconvenient, as you have to stop everything to restack your load. It can be expensive if you drop fragile materials or equipment. It can be deadly if the load falls on a co-worker or causes your truck to tip over. Avoid sudden maneuvers when operating a forklift.

13.3.9 Obstructing the View

Forklift operators commonly try to move as much as possible in the fewest number of trips in order to save time. Sometimes this causes them to stack loads too high. Doing so may cause them to exceed the safe weight limit of the forklift, and it may also block the operator's view. To operate a forklift safely, the operator must be able to clearly see what is in front of and behind the forklift without leaning outside of the operator's compartment. Leaning outside the operator's compartment can result in serious injuries.

> ### Deadly Overload
>
> A forklift operator was carrying a load that was stacked too high. The load obstructed his view. To make up for it, he stuck his head out the side of the operator's compartment to see around the load. Unfortunately, as he was preparing to drive forward, another forklift was backing up and sideswiped his machine, decapitating him.
>
> **The Bottom Line:** Never carry a load that obstructs your view. Always keep all of your body parts inside the operator's compartment.
>
> *Source: The Occupational Safety and Health Administration (OSHA)*

13.4.0 Working on Ramps and Docks

Ramps and docks have special working conditions with specific safety requirements. Ramps allow wheeled vehicles to move easily from one level to the next. However, going up and down an angled surface has an impact on the forklift's center of gravity, making it easier for the forklift to tip. Operators can be crushed by or thrown from the forklift if this happens. Docks elevate the driving surface to a convenient height for the loading and unloading of over-the-road (OTR) trucks and rail cars. However, they also create a risk that the forklift might fall off the edge of the dock.

13.4.1 Ramps

The ramp's grade increases the tipping hazard for a forklift. Follow these rules when working on ramps:

- Keep the load as low as possible.
- Always keep the load uphill.
- When working on any graded surface, make sure your load is pushed as far back onto the forks as possible.
- If possible, tilt the forks so that the load is level with the graded surface.
- Do not turn or make quick starts or stops on a ramp.

13.4.2 Docks

It is possible to drive a forklift off the edge of a dock if you are not careful and attentive. Not only can this damage the equipment, it can also kill or injure the operator and anyone near the area of the falling load.

Forklift operators must be aware of the edge of the dock. The edge is normally painted a bright color, such as yellow. Besides wheel chocks for the truck, devices called dock plates may be used to smoothly bridge the gap between a dock and the floor of a truck.

13.5.0 Fire and Explosion Hazards

Forklifts use fuels such as gasoline, liquid propane (LP) gas, and diesel fuel. All of these fuels are capable of causing a fire or explosion if not handled properly. In addition, LP gas is stored in cylinders under pressure, creating an explosion hazard if the

> ### Chock the Wheels
>
> As a forklift passes from a dock to an over-the-road truck and back, the force can slowly move the truck forward if the wheels are not properly chocked. In one case, a forklift operator incorrectly assumed that the truck driver had chocked the wheels of the truck. Every time the forklift passed from the dock onto the truck, the truck moved away from the dock. Because the operator was concentrating on his job, he did not notice that the gap between the dock and the truck was growing. After several trips in and out of the truck, the gap was wide enough for the forklift's front wheels to fall into the gap. This caused the forklift to tip forward into the truck and tip the truck trailer backward, crushing the operator between the truck and the forklift.
>
> **The Bottom Line:** Never assume that the wheels are chocked properly. Always chock the wheels yourself, or check to make sure it was done correctly.
>
> *Source: The Occupational Safety and Health Administration (OSHA)*

INTRODUCTION TO MASONRY

cylinder is exposed to extreme heat or fire. It is very important to keep these fuels away from any source of fire and to keep the areas in which the forklift is used free of any flammable materials. There are specific precautions that must be taken to avoid the possibility of a fire or explosion.

The best way to prevent a fire is to make sure that the three elements needed for fire (fuel, heat, and oxygen) are never present in the same place at the same time. Here are some basic safety guidelines for fire prevention:

- Always work in a well-ventilated area, especially when you are using flammable materials.
- Never smoke or light matches when you are working with flammable materials.
- Keep oily rags in approved, self-closing metal containers.
- Store combustible materials only in approved containers.
- Know where to find fire extinguishers, what kind of extinguisher to use for different kinds of fires, and how to use the extinguishers.
- Keep open fuel containers away from any sources of sparks, fire, or extreme heat.
- Make sure all extinguishers are fully charged. Never remove the tag from an extinguisher; it shows the date the extinguisher was last serviced and inspected.
- Don't fill a gasoline or diesel fuel container while it is resting on a truck bed liner or other ungrounded surface. The flow of fuel creates static electricity that can ignite the fuel if the container is not grounded.
- Always use approved containers, such as safety cans, for flammable liquids.

13.5.1 Flammable and Combustible Liquids

Liquids can be flammable or combustible. Flammable liquids have a flash point below 100°F. Combustible liquids have a flash point at or above 100°F. Fires can be prevented by doing the following things:

- *Removing the fuel* – Liquid does not burn. What burns are the gases (vapors) given off as the liquid evaporates. Keeping the liquid in an approved, sealed container prevents evaporation. If there is no evaporation, there is no fuel to burn.
- *Removing the heat* – If the liquid is stored or used away from a heat source, it cannot ignite.
- *Removing the oxygen* – The vapor from a liquid will not burn if oxygen is not present. Keeping safety containers tightly sealed prevents oxygen from coming into contact with the fuel.

WARNING!
Never transfer flammable or combustible liquids between containers without proper training. Doing so can result in fire or explosion.

13.5.2 Flammable Gases

Flammable gases used on construction sites include acetylene, hydrogen, ethane, and LP gas. To save space, these gases are compressed so that a large amount can be stored in a small cylinder or bottle. As long as the gas is kept in the cylinder, oxygen cannot get to it and start a fire. The cylinders must be handled carefully and stored away from sources of heat.

Oxygen is also classified as a flammable gas. If it is allowed to escape and mix with another flammable gas, the resulting mixture can explode.

13.5.3 Fire Fighting

You are not expected to be an expert firefighter, but you may have to deal with a fire to protect your safety and the safety of others. You need to know the location of fire-fighting equipment on your job site. You also need to know which equipment to use on different types of fires. However, only qualified personnel are authorized to fight fires.

Most companies tell new employees where fire extinguishers are kept. If you have not been told, be sure to ask. Also ask how to report fires. The telephone number of the nearest fire department should be clearly posted in your work area. If your company has a company fire brigade, learn how to contact them. Learn your company's fire-safety procedures.

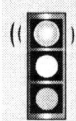

WARNING!
Before working with any flammable materials, be sure you know the closest escape route and the appropriate measures to take in case of emergency.

13.6.0 Pedestrian Safety

You may not be a forklift operator, but you will probably work on a site with forklifts. Here are some guidelines for working safely around forklifts.

Remember that it may be difficult for the operator to see you. The operator may be concentrating on the load and may not be paying attention to pedestrians. Therefore, always assume that the forklift operator doesn't see you. Remember, the most common type of forklift accident is hitting a pedestrian.

BUILDING BLOCKS

Static Electricity Can Kill

In recent incidents reported to the National Institute for Occupational Safety and Health (NIOSH), fires spontaneously ignited when workers or others attempted to fill portable gasoline containers (gas cans) in the backs of pickup trucks equipped with plastic bed liners or in cars with carpeted surfaces. Serious skin burns and other injuries resulted. Similar incidents in the last few years have resulted in warning bulletins from several private and government organizations.

These fires result from the buildup of static electricity. The insulating effect of the bed liner or carpet prevents the static charge generated by gasoline flowing into the container or other sources from grounding. The discharge of this buildup to the grounded gasoline dispenser nozzle may cause a spark and ignite the gasoline. Both ungrounded metal (most hazardous) and plastic gas containers have been involved in these incidents.

Source: The National Institute for Occupational Safety and Health (NIOSH)

Forklifts are heavy and usually carry large loads. This results in a large amount of momentum, which means that it will probably take a distance equal to the length of the forklift to stop it. Never risk your safety on a forklift's ability to stop in time. Make sure that you're never positioned between a forklift and an immovable object.

Sometimes forklifts carry heavy loads high overhead. Never stand under the raised forks of a forklift. Objects may fall from the forks or, if there is a sudden loss of hydraulic pressure, the forks and load may drop suddenly, crushing any people or objects beneath them.

When working in a storage room or warehouse with racking, never work in the aisle on the other side of a racking unit where a forklift is working. Occasionally, the operator may push a load into the rack, causing it to fall off the other side. If you are unlucky enough to be on the other side, there is a good chance that you will be struck or crushed by a falling object.

Never hitch a ride on a forklift. Forklifts are designed for one operator and a load. Riding on the forks creates a high risk of falling and being run over by the forklift. Riding on the tractor of the forklift is also not permitted. It is easy to fall and/or be crushed between the forklift and other objects.

Summary

Masonry is a craft that has existed for thousands of years. Over the centuries, hand-formed dried mud brick has been replaced by molded mud brick, then by fired, molded clay brick. Today, fired clay brick is available in a variety of shapes, sizes, colors, and textures.

Modern clay products are categorized as solid and hollow brick, structural and nonstructural tile, and made-to-order architectural terra-cotta. Clay has been joined by concrete as a modern masonry material. Concrete masonry units outnumber clay units in their variety. Concrete products are block, cement brick, and special-purpose block. Stone, the oldest recovered building material, is still laid by masons. Mortars and grouts have evolved from mud to special-purpose, high-strength cements.

Science and engineering have brought masonry into the modern age. With modern construction techniques, masonry is now used in high-rise buildings as well as residential and commercial projects. This allows for widespread use of masonry construction throughout North America.

Masonry as a career dates back to the ancient kingdoms of the Middle East. Masons were among the first craftworkers to organize into guilds. The guilds protected the secrets of the craft and maintained standards for the work. The legacy of these guilds includes the apprenticeship program available today.

Masonry offers advancement through the recognized career steps of apprentice, mason, layout person, supervisor, superintendent, and masonry contractor. Your success as a mason requires the willingness to keep on learning.

Safety is a particularly important issue for you as a masonry worker. You will often work at heights where falls and falling objects are major hazards. You must also be careful when working with mortar and concrete products because they can cause skin irritations and lung ailments if not handled properly. You will often work on sites where heavy equipment is used. For that reason, you must be especially vigilant. Forklift safety is of particular concern because forklifts are commonly used to transport and lift masonry materials at a work site.

Review Questions

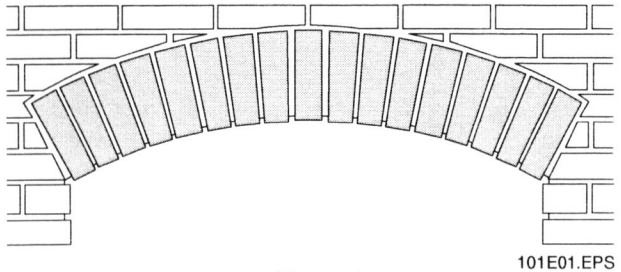

Figure 1
101E01.EPS

1. The arch shown in *Figure 1* is a _____ arch.
   a. Roman
   b. radial
   c. jack
   d. Babylonian

2. Structural clay products include _____.
   a. brick, block, and tile
   b. brick, tile, and terra-cotta
   c. tile, terra-cotta, and pre-faced units
   d. block, grout, and tile

3. A cube of bricks contains _____ standard bricks.
   a. 90
   b. 300
   c. 400
   d. 500

4. Solid masonry units have voids occupying _____.
   a. less than 25 percent of their surface
   b. more than 25 percent of their mass
   c. less than 10 percent of their mass
   d. more than 75 percent of their surface

5. Rubble stone is _____.
   a. called slump brick
   b. made of concrete
   c. cut with smooth bedding surfaces
   d. irregular in size and shape

6. Modern masonry mortar is made of _____.
   a. cement, lime, and sand
   b. cement, sand, admixtures, and water
   c. sand, lime, ceramic, and water
   d. masonry cement, sand, and water

7. Veneer walls are _____.
   a. built 6½" away from the weight-bearing wall
   b. commonly used in industrial construction
   c. not loadbearing
   d. not designed for appearance

8. The National Registry for construction craft training is maintained by _____.
   a. the U.S. Department of Labor
   b. local apprenticeship councils
   c. NCCER
   d. state labor departments

9. Using mortar with the consistency of mud will *not* enable you to create strong bonds.
   a. True
   b. False

10. The process of spreading a bed of mortar is called _____.
    a. snapping
    b. stringing
    c. leading
    d. furrowing

11. The process of carving a trough in a bed of mortar is called _____.
    a. furrowing
    b. buttering
    c. tailing
    d. tuckpointing

12. Portland cement is hygroscopic, which means that it will absorb moisture from your skin, resulting in skin irritation.
    a. True
    b. False

13. Fall protection is required when working at heights _____ or higher.
    a. 6'
    b. 8'
    c. 10'
    d. 12'

14. Short, flexible lines with connectors on each end are called _____.
    a. anchorages
    b. deceleration devices
    c. lanyards
    d. lifelines

15. The most common type of forklift accident is the _____.
    a. forklift tipping over
    b. load falling
    c. forklift falling off a dock
    d. forklift hitting a pedestrian

PROFILE IN SUCCESS

Arnold Shueck, Field Superintendent

Nester Brothers Masonry
Pennsburg, PA

Arnold Shueck grew up on a farm and, in doing so, learned the value of hard work that led to his success in the masonry trade. He decided early on that he wanted to work outdoors and work with his hands. This led him to enter the high school masonry program that was the beginning of a long and rewarding career.

How did you get started in the masonry trade?
I learned early on that I wanted to work outdoors. I had heard that masons make good money, so I enrolled in the masonry program at my high school. The training I received in high school helped me land a job with Nester Brothers. I started out as a laborer apprentice, but my training helped me move up pretty quickly to laying brick and block. I was promoted to foreman, and was eventually moved up to field superintendent.

What does your present job entail?
As a field superintendent I go from project to project checking on the work and helping to solve problems. It's my job to make sure all the projects flow smoothly. I estimate new projects, plan the projects, and order the materials. Nester Brothers works mostly on commercial masonry projects, so I get to work on some very challenging jobs.

What do you think it takes to become a success?
You have to be willing to work hard and stick with it. Growing up on a farm taught me about the value of hard work and gave me an appreciation for working with my hands.

What do you like most about your job?
First, I like the company I work for. I've been with the same company for my entire career and they have been very good to me. Second, I like working outdoors. I never saw myself in an indoor job. I get a lot of satisfaction from seeing the work I've accomplished over the years as I drive around the area.

What would you say to someone just entering the trade?
Be sure this is what you want to do. If it is, then give it everything you have. Don't be afraid to take advice from experienced people. Take advantage of every training and educational opportunity that comes your way. Materials and techniques are constantly changing. It's important to keep up with the industry.

GLOSSARY

Trade Terms Introduced in This Module

Admixture: A chemical or mineral other than water, cement, or aggregate added to mortar immediately before or during mixing to change its setting time or curing time, to reduce water, or to change the overall properties of the mortar.

Aggregate: Materials such as crushed stone or gravel used as a filler in concrete and concrete block.

Adobe: Sun-dried, molded clay brick.

Alkaline: Bitter, slippery, or caustic.

American Society for Testing and Materials (ASTM) International: The publisher of masonry standards.

Ashlar: A squared or rectangular cut stone masonry unit; or, a flat-faced surface having sawed or dressed bed and joint surfaces.

Butter: Apply mortar to the end of a masonry unit.

Capital: The top part of an architectural column.

Concrete masonry unit (CMU): A hollow or solid block made from portland cement and aggregates.

Cornice: The horizontal projection crowning the wall of a building.

Course: A row or horizontal layer of masonry units.

Cube: A strapped bundle of approximately 500 standard bricks, or 90 standard blocks, usually palletized. The number of units in a cube will vary according to the manufacturer.

Facing: That part of a masonry unit or wall that shows after construction; the finished side of a masonry unit.

Footing: The base for a masonry unit wall, or concrete foundation, that distributes the weight of the structural member resting on it.

Furrowing: Making an indentation with a trowel point along the center of the mortar bed joint.

Grout: A mixture of portland cement, lime, and water, with or without fine aggregate, with a high enough water content that it can be poured into spaces between masonry units and voids in a wall.

Head joint: A vertical joint between two masonry units.

Hygroscopic: The tendency of a substance to absorb moisture.

Joints: The area between each brick or block that is filled with mortar.

Mason: A person who assembles masonry units by hand, using mortar, dry stacking, or mechanical connectors.

Masonry unit: Any building block made of brick, cement, ashlar, clay, adobe, rubble, glass, tile, or any other material, that can be assembled into a structural unit.

Mortar: A mixture of portland cement, lime, fine aggregate, and water, plastic or stiff enough to hold its shape between masonry units.

Nonstructural: Not bearing weight other than its own.

Parapet: A low wall or railing.

Pilaster: A square or rectangular pillar projecting from a wall.

Spread: A row of mortar placed into a bed joint.

Stringing: Spreading mortar with a trowel on a wall or footing for a bed joint.

Structural: Bearing weight in addition to its own.

Tuckpointing: Filling fresh mortar into cutout or defective joints in masonry.

Weephole: A small opening in mortar joints or faces to allow the escape of moisture.

Wythe: A continuous section of masonry wall, one masonry unit in thickness, or that part of a wall which is one masonry unit in thickness.

REFERENCES

Additional Resources

This module is intended to be a thorough resource for task training. The following reference works are suggested for further study. These are optional materials for continued education rather than for task training.

Building Block Walls—A Basic Guide, 1988. Herndon, VA: National Concrete Masonry Association.

Bricklaying: Brick and Block Masonry. Reston, VA: Brick Industry Association.

Concrete Masonry Handbook. Skokie, IL: Portland Cement Association.

ACKNOWLEDGMENTS

Figure Credits

Copyright © 2004, Corel Corporation	101F01, 101SA01
Topaz Publications, Inc.	101SA02–101SA04, 101F15, 101F16, 101F21, 101F24, 101F26, 101F27, 101F36
Hanson Brick	101F03–101F05, 101F07
Used with permission of the Brick Industry Association, Reston, Virginia, www.gobrick.com	101F06, 101F08, 101F10, 101F11
Portland Cement Association	101F12–101F14, 101T01
Associated General Contractors	101F18, 101F33
Easi-Set Industries	101SA05
Bon Tool Company	101F19
www.freefoto.com	101F28–101F31
Coleman Cable, Inc.	101SA06
Construction Safety Association of Ontario	101F40A
Fall Protection Systems, Inc.	101F40B, 101F40C
Protecta International, Inc.	101F42–101F48
Sellick Equipment Limited	101F49, 101F51
Manitou North America, Inc.	101F52

CONTREN® LEARNING SERIES — USER UPDATES

The NCCER makes every effort to keep these textbooks up-to-date and free of technical errors. We appreciate your help in this process. If you have an idea for improving this textbook, or if you find an error, a typographical mistake, or an inaccuracy in NCCER's Contren® textbooks, please write us, using this form or a photocopy. Be sure to include the exact module number, page number, a detailed description, and the correction, if applicable. Your input will be brought to the attention of the Technical Review Committee. Thank you for your assistance.

Instructors – If you found that additional materials were necessary in order to teach this module effectively, please let us know so that we may include them in the Equipment/Materials list in the Instructor's Guide.

Write: Product Development and Revision
National Center for Construction Education and Research
P.O. Box 141104, Gainesville, FL 32614-1104

Fax: 352-334-0932

E-mail: curriculum@nccer.org

Craft _____ Module Name _____

Copyright Date _____ Module Number _____ Page Number(s) _____

Description _____

_____

_____

_____

_____

(Optional) Correction _____

_____

_____

_____

(Optional) Your Name and Address _____

_____

_____

_____

Module 28102-04

Masonry Tools and Equipment

COURSE MAP

This course map shows all of the modules in the first level of the *Masonry* curriculum. The suggested training order begins at the bottom and proceeds up. Skill levels increase as you advance on the course map. The local Training Program Sponsor may adjust the training order.

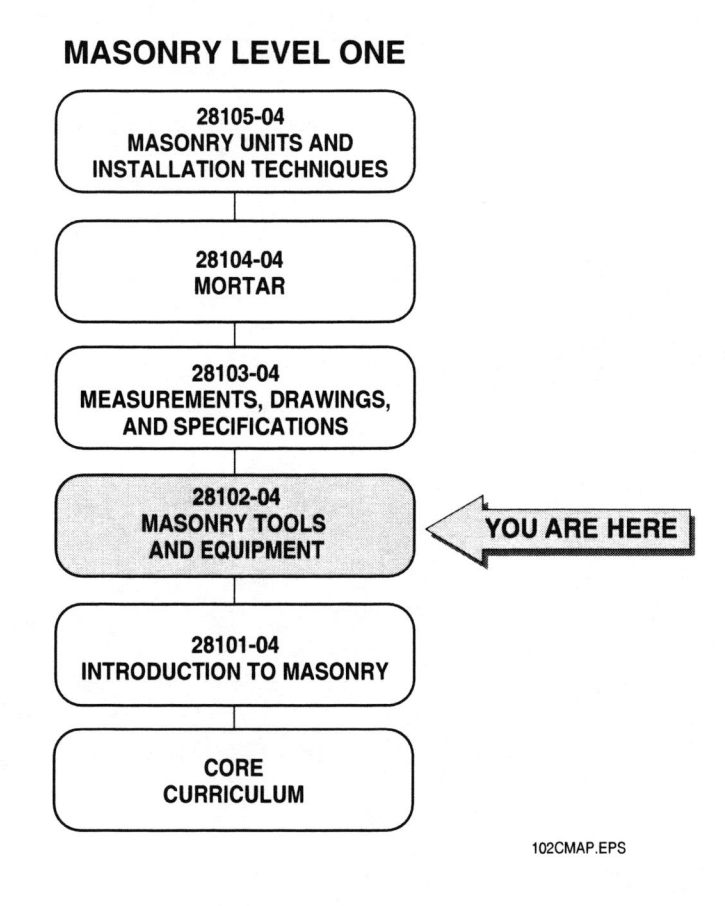

MASONRY LEVEL ONE

- 28105-04 MASONRY UNITS AND INSTALLATION TECHNIQUES
- 28104-04 MORTAR
- 28103-04 MEASUREMENTS, DRAWINGS, AND SPECIFICATIONS
- 28102-04 MASONRY TOOLS AND EQUIPMENT ⇐ YOU ARE HERE
- 28101-04 INTRODUCTION TO MASONRY
- CORE CURRICULUM

102CMAP.EPS

Copyright © 2004 National Center for Construction Education and Research, Gainesville, FL 32614-1104. All rights reserved. No part of this work may be reproduced in any form or by any means, including photocopying, without written permission of the publisher.

MODULE 28102-04 CONTENTS

1.0.0 INTRODUCTION .2.1
    1.1.0   Tool Safety .2.1
    1.2.0   Tool Maintenance .2.2

2.0.0 HAND TOOLS .2.2
    2.1.0   Trowels .2.3
    2.2.0   Hammers .2.4
    2.2.1   Cutting Hammers .2.4
    2.2.2   Mashes and Mauls .2.5
    2.3.0   Chisels .2.6
    2.4.0   Jointers .2.8
    2.5.0   Brushes .2.9
    2.6.0   Brick Tongs .2.10
    2.7.0   Other Hand Tools .2.10
    2.8.0   Tool Bag .2.10

3.0.0 MEASURES AND MEASURING TOOLS .2.11
    3.1.0   Levels .2.11
    3.2.0   Rules .2.12
    3.3.0   Squares .2.14
    3.4.0   Mason's Line and Fasteners .2.14
    3.5.0   Corner Pole .2.16
    3.6.0   Chalkline and Chalk Box .2.16
    3.7.0   Plumb Bob .2.17

4.0.0 HAND-POWERED MORTAR EQUIPMENT .2.18
    4.1.0   Mortar Boxes .2.18
    4.2.0   Mixing Accessories .2.19
    4.3.0   Water Bucket and Barrel .2.19
    4.4.0   Barrows .2.20

5.0.0 POWER TOOLS .2.20
    5.1.0   Saws .2.21
    5.2.0   Splitters .2.22
    5.3.0   Grinders .2.22
    5.4.0   Power Drills and Powder-Actuated Drivers .2.23

6.0.0 POWER EQUIPMENT .2.23
    6.1.0   Mortar Mixer .2.24
    6.2.0   Masonry Pump and Vibrator .2.25
    6.3.0   Pressurized Cleaning Equipment .2.25
    6.3.1   Pressure Washing .2.26
    6.3.2   Steam Cleaning .2.26
    6.3.3   Sandblasting .2.26

MODULE 28102-04 CONTENTS (CONTINUED)

7.0.0 LIFTING EQUIPMENT 2.26
    7.1.0 Mounted Hoists 2.27
    7.2.0 Portable Materials Hoists 2.28
    7.3.0 Hydraulic Lift Materials Truck 2.28
    7.4.0 Cranes and Derricks 2.28
    7.4.1 Tower Cranes 2.29
    7.4.2 Mobile Cranes 2.29
    7.5.0 Forklifts, Pallet Jacks, and Buggies 2.29
    7.6.0 Conveyors 2.30

8.0.0 SCAFFOLDING 2.31
    8.1.0 Tubular Steel Sectional Scaffolding 2.31
    8.1.1 Putlog 2.33
    8.1.2 Steel Tower 2.34
    8.2.0 Swing Stage 2.34
    8.3.0 Hydraulic Personnel Lift 2.35
    8.4.0 Scaffold Safety 2.35

SUMMARY 2.36

REVIEW QUESTIONS 2.36

PROFILE IN SUCCESS 2.38

GLOSSARY 2.39

REFERENCES 2.40

ACKNOWLEDGMENTS 2.41

Figures

Figure 1	Parts of a trowel	2.3
Figure 2	Brick trowel shapes	2.3
Figure 3	Different types of trowels	2.4
Figure 4	Brick hammers	2.5
Figure 5	Tile hammer	2.5
Figure 6	Stonemason's hammer	2.5
Figure 7	Mash	2.5
Figure 8	Toothed bushhammer	2.6
Figure 9	Rubber mallets	2.6
Figure 10	Mason's chisel	2.6
Figure 11	Brick set chisel	2.7
Figure 12	Rubber-grip mason's chisel	2.7
Figure 13	Tooth chisel and pitching tool	2.7
Figure 14	Plugging or joint chisel	2.8
Figure 15	Tooled mortar joints	2.8
Figure 16	Jointers	2.9
Figure 17	Runner jointers	2.9
Figure 18	Rakers	2.9
Figure 19	Mason's brushes	2.10
Figure 20	Brick tongs	2.10
Figure 21	Common mason's hand tools	2.11
Figure 22	Mason's tool bag	2.11
Figure 23	An air bubble shows level or plumb	2.11
Figure 24	Common levels	2.12
Figure 25	Spacing rules	2.13
Figure 26	Folding rule and steel tape	2.14
Figure 27	Framing, or steel, square	2.14
Figure 28	Combination square and sliding T-bevel	2.14
Figure 29	Line pin	2.15
Figure 30	Corner block	2.15
Figure 31	Line stretchers	2.16
Figure 32	Line trig or twig	2.16
Figure 33	Corner pole or deadman	2.17
Figure 34	Chalk box	2.17
Figure 35	Plumb bob	2.18
Figure 36	Mortar pan and mortar box with wheels	2.18
Figure 37	A hawk and a hod	2.18
Figure 38	Cubic-foot measuring box	2.19

Figure 39	Mixing aids	2.19
Figure 40	Water bucket and barrel	2.19
Figure 41	Barrows for masonry units	2.20
Figure 42	Large masonry saw	2.21
Figure 43	Small masonry saw and stand	2.21
Figure 44	Handheld masonry saw	2.22
Figure 45	Hand-operated masonry splitter	2.22
Figure 46	Large foot-operated hydraulic splitter	2.23
Figure 47	Tuckpoint grinder	2.23
Figure 48	Hammer drill	2.23
Figure 49	Components of powder-actuated driver	2.24
Figure 50	Mortar mixer	2.25
Figure 51	Masonry pump	2.25
Figure 52	Handheld vibrator	2.25
Figure 53	Typical steam-cleaning operation	2.26
Figure 54	Typical raised work station	2.26
Figure 55	Materials hoist	2.27
Figure 56	Ladder hoist	2.27
Figure 57	Materials, or brick, hoist	2.28
Figure 58	Tower crane	2.29
Figure 59	Mobile crane performing a lift	2.30
Figure 60	Mason's reach-type forklift	2.30
Figure 61	Motorized buggy	2.30
Figure 62	Conveyor	2.31
Figure 63	Scaffold frame types.	2.31
Figure 64	Connecting scaffold sections	2.32
Figure 65	Hook-mounted scaffold plank	2.33
Figure 66	Completed tubular steel scaffolding	2.33
Figure 67	Putlog at base of scaffold	2.34
Figure 68	Adjustable steel tower scaffold	2.34
Figure 69	Swing stage scaffolding	2.34
Figure 70	Hydraulic personnel lift	2.35

MODULE 28102-04

Masonry Tools and Equipment

Objectives

After completing this module, you will be able to do the following:

1. Identify and name the tools used in performing masonry work.
2. Identify and name the equipment used in performing masonry work.
3. Describe how each tool is used.
4. Describe how the equipment is used.
5. Associate trade terms with the appropriate tools and equipment.
6. Demonstrate the correct procedures for assembling and disassembling scaffolding according to federal safety regulations, under the supervision of a competent person.

Recommended Prerequisites

Core Curriculum; Masonry Level One, Module 28101-04

Required Trainee Materials

1. Pencil and paper
2. Appropriate personal protective equipment

1.0.0 ◆ INTRODUCTION

Why are tools so important for masonry work? Brick cannot be cut with bare hands, and mortar cannot be laid with fingers. Masonry tools are the interface between you and the work. In fact, the quality of the tool—especially the trowel—affects the quality of the work. Because of the various things that masons do, there are many special-purpose tools. Some hand tools are virtually unchanged since ancient times, while some power tools use the most up-to-date technology.

This module introduces the tools and equipment that you will need to lay masonry units. Specifically, this module covers hand tools, measuring tools, power tools, and power equipment. By the end of this module, you will be able to identify each item and what it does. Some tools are used more than others. You may not use every tool on every job.

1.1.0 Tool Safety

Using your tools and equipment safely will help ensure you have a long, productive career. Tool and equipment safety falls into the following categories:

- Following safe work practices and procedures
- Inspecting tools and equipment before use
- Using tools and equipment correctly
- Keeping tools and equipment clean and properly maintained

Tool safety means using tools properly, keeping them clean, and being careful not to damage them. Damaged tools and equipment can break, cause injuries, and slow down or stop work. To avoid damaging tools, always use the right tool for the job. The following are some rules for avoiding the misuse of tools:

- Do not use a trowel as a hammer
- Do not hammer on your level.
- Do not use your level or trowel as a pry bar.
- Do not use broken or defective tools.
- Do not cut reinforcement with a hammer, chisel, or trowel.

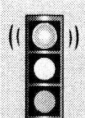

> **WARNING!**
> Always wear safety goggles when cutting or chipping masonry units.

Building Blocks

Stack Bricks Safely

A neat brick stack is a safe stack. Stack materials by reversing direction on every other layer, so they will be less prone to tip. Keep the pile neat and vertical to avoid snagging clothes.

102SA01.EPS

Accidents also happen when you leave tools and equipment in the way of other workers. Store tools safely, where other people cannot trip over them and where the tools will not get damaged. Clean, well-kept tools work better and are safer.

Do not drop or temporarily store tools or masonry units in pathways or around other workers. Stack masonry units neatly, reversing direction with each layer, to avoid the risk of them toppling over. As a general rule, do not stack masonry units higher than your chest. A higher stack is more likely to tip over.

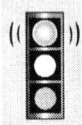

> **WARNING!**
> Mortar and grout can be very caustic. You must always protect your skin and eyes when working with mortar and grout. Wear gloves and other appropriate personal protective equipment (PPE) as required for the specific job.

1.2.0 Tool Maintenance

Tools and equipment must be in good repair to be safe to use. Repair or replace defective tools and equipment immediately.

Clean all tools and equipment that touch mortar or grout immediately after use. Use a bucket of water to soak tools that you will use again in a few minutes. Keep wooden handles out of the water. If mortar dries on a tool, it will harden and make the tool unusable.

Wash tools thoroughly with water and a wire brush. Be sure that you remove all mortar or grout completely. If you do not use tools regularly, coat them lightly with oil, grease, or other approved coating to prevent rust and corrosion. After cleaning, check that tool handles are secure and free from cracks or splinters. Oil the pivot joints and wooden parts of the tools. Sharpen blades and cutting edges when they become dull or nicked.

Clean all motorized tools and equipment after use. Always turn off and disconnect the power before cleaning or fixing power tools and equipment. When in doubt about the care of specific items, check with your supervisor.

Here are some specific hand-tool care tips:

- Keep your mason's hammer sharp and square.
- Keep folding rules oiled.
- Keep chisel blades sharp.
- Keep chisel heads and blades free of burrs.
- Keep all tool handles tight.

2.0.0 ◆ HAND TOOLS

The quality and use of masonry hand tools greatly affect the quality of the final work. Hand tools are used to coat, cut, carry, clean, align, and level masonry units. The following sections describe masonry hand tools you will typically use to perform masonry construction.

> **BUILDING BLOCKS**
>
> *Check Tools Daily*
>
> Cleaning and checking your tools at the end of each day is a good habit. This will keep them in good working order and prevent loss.

2.1.0 Trowels

The trowel is the mason's most commonly used tool. Trowels are used for placing mortar, an activity commonly known as buttering joints. Masons use trowels to move and shape mortar between masonry units. Trowels are also used to mix, scrape, and shape mortar and clean mortar from masonry units and tools. The trowel's handle is often used to tap units into place.

The mason's trowel comes in many sizes and shapes. The basic trowel (*Figure 1*) consists of a steel blade ground to the proper balance, taper, and shape. The narrow end of the blade is the point, and the wide end is the heel. The blade is connected to the handle by a shank. The handle is made of wood or plastic, and it can be covered with leather or foam for a more comfortable grip. The wood handle has a band, or ferrule, on the shank end to prevent splitting.

The blade can come with a sharply angled heel (London pattern) or a square heel (Philadelphia pattern), as shown in *Figure 2*. Some masons prefer a wider heel, which can make buttering masonry units easier.

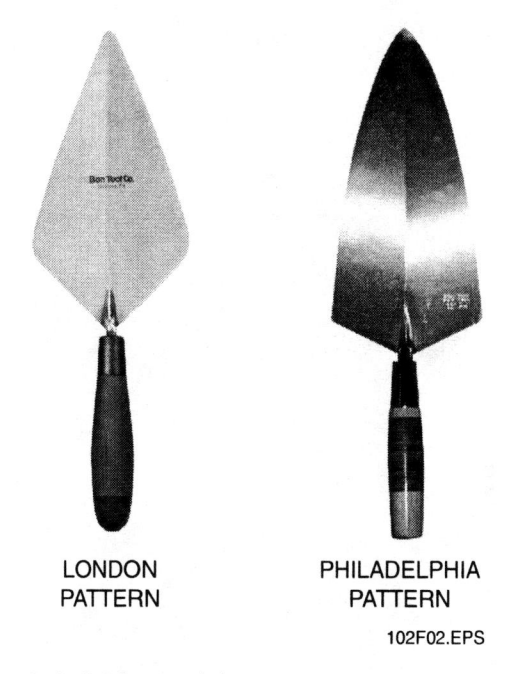

Figure 2 ◆ Brick trowel shapes.

Trowels come in different shapes and sizes for different purposes. Trowels can range in width from about 4" to 7" and can be up to 13" long. *Figure 3* shows different types of trowels. Some trowels have specialized uses and will not be used as often as a standard brick trowel. Some trowels are particular to certain parts of the country.

The standard, or brick, trowel is the all-purpose model, used for cutting, buttering, and adjusting units. The small **pointing** trowel fits into tight spaces. It is used to point, grout, and tool mortar joints. The margin trowel, with a small square blade, is used to smooth joints and mix grout or other compounds. The tuckpointing trowel is used to shape mortar between joints. It comes in

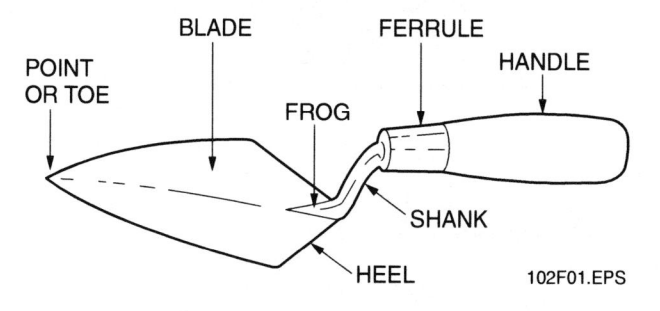

Figure 1 ◆ Parts of a trowel.

> **BUILDING BLOCKS**
>
> *Hold the Trowel Properly*
>
> When holding a trowel, keep your thumb along the top of the handle, not the shank. If your thumb is on the shank, it can get coated with mortar as you work, causing skin irritation.

MASONRY TOOLS AND EQUIPMENT

2.3

BRICK   POINTING   MARGIN   TUCKPOINTER

PARGING   DUCK BILL   BUCKET   TILE-SETTING

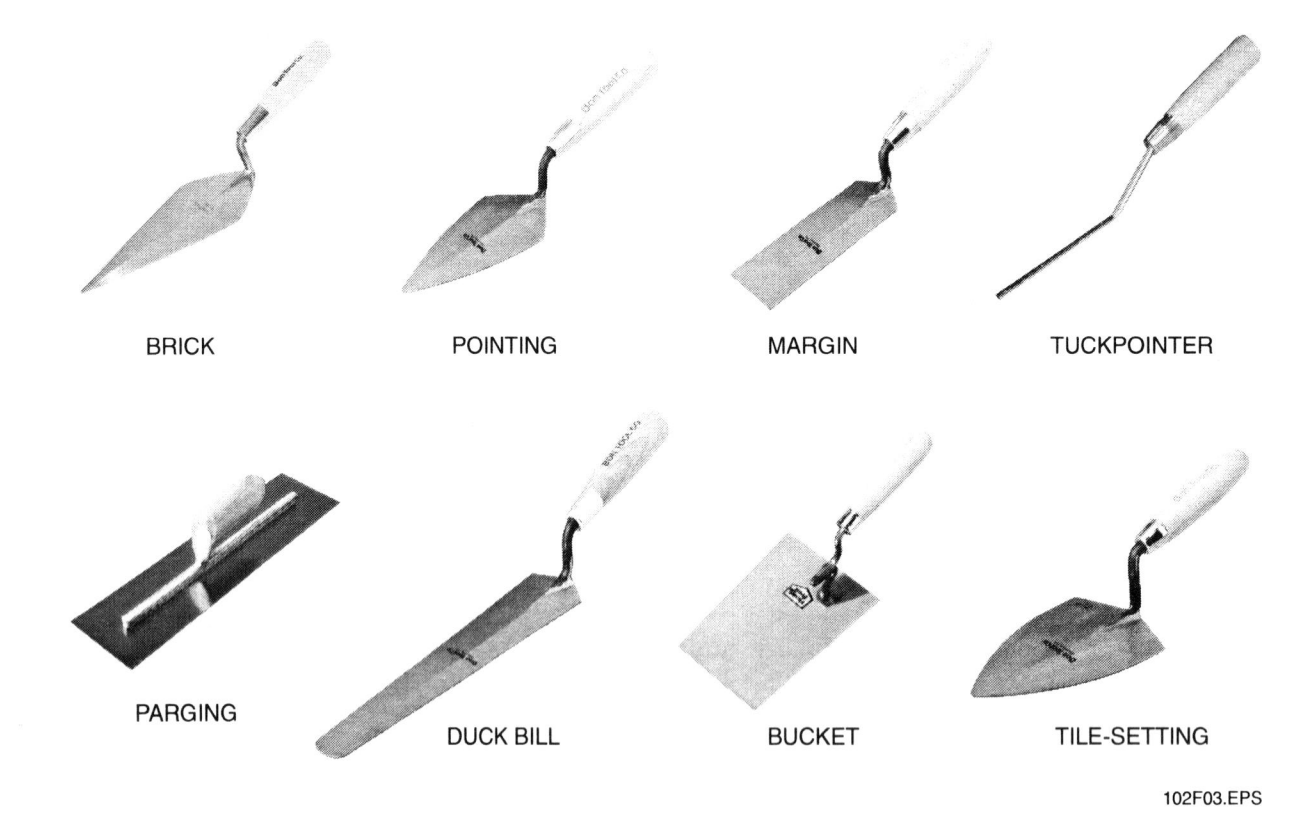

102F03.EPS

Figure 3 ◆ Different types of trowels.

widths ranging from ⅛" to 1", to fit commonly used joint sizes.

The plasterer's or parging trowel is a flat piece of steel with the handle mounted on one side. It is used to apply a thin coat, or **parge**, of mortar. The parge coat will hold tile or veneer to the masonry units. The duck bill trowel is narrow, with a long blade up to 13" long. It is primarily used for cleanup. The bucket trowel has straight sides and a square toe, which aid the mason in scraping the sides of the bucket. Tile-setting trowels have a wide blade to handle large amounts of mortar or grout.

2.2.0 Hammers

Mason's hammers fall into two categories: cutting hammers and mauls. Using the right hammer for each job will ensure the most efficient use of your energy. Many types of hammers come with wooden handles. Inspect these handles every day to make sure they are not loose, splintered, or cracked. Chisel edges on hammers need to be inspected and sharpened, just as chisels do.

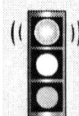

WARNING!
If the wooden handle on a hammer becomes loose, replace it immediately. If the handle breaks, the head could fly off, causing serious damage or injury.

2.2.1 Cutting Hammers

The brick hammer is a double-headed hammer with a chisel head on one side, as shown in *Figure 4*. This is the most commonly used, everyday mason's hammer. The brick hammer drives nails and strikes chisels with one head. It can break, cut, or chip masonry material with the other head. Use

BUILDING BLOCKS

Choose Wisely

A trowel should be sturdy, light, and balanced. It should have a comfortable grip and a flexible steel blade. There are many options to consider, so choose a tool that is right for you.

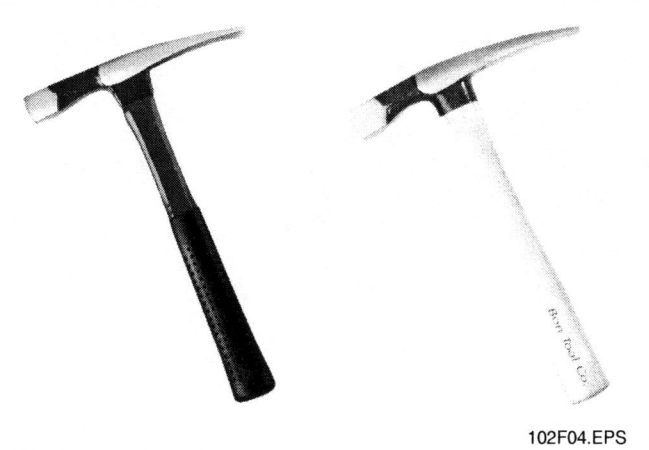

Figure 4 ◆ Brick hammers.

it for cutting masonry units, setting line pins, and nailing wall ties.

Brick hammers are one-piece, drop-forged steel tools, or they can be steel heads on wooden or fiberglass shafts. Steel hammers need a comfortable grip sleeve over the shaft. The weight of this hammer usually ranges from 12 to 24 ounces. If you are buying a hammer, pick a weight that is comfortable for your hand.

The tile hammer is smaller than a standard brick hammer with a smaller, thinner cutting head. *Figure 5* shows this hammer, which usually weighs about 9 ounces. Use it for cutting and trimming tile where you need more precision than you can get with a brick hammer. Do not drive nails, strike chisels, or do other heavy work with a tile hammer.

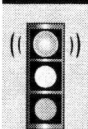

> **WARNING!**
> Never strike the heads of two hammers together. A chip could fly off one of the heads.

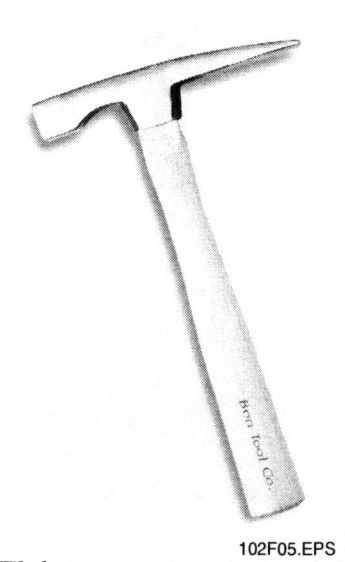

Figure 5 ◆ Tile hammer.

The stonemason's hammer, shown in *Figure 6*, resembles an ax. This heavy hammer has a cutting blade on one end of the head. Use this special-purpose hammer for dressing, cutting, splitting, or trimming stone.

2.2.2 Mashes and Mauls

Mashes and mauls are heavy club-headed hammers used for striking. They usually have two striking heads.

The mason's mash is a drop-forged, one-piece tool with a grip sleeve over the shaft, as shown in *Figure 7*. It can also have a steel head mounted on a wooden handle. It resembles a short-handled, double-headed sledgehammer. The mash is most commonly used to strike chisels to cut masonry units. This hammer is too heavy to use for setting line pins or nailing wall ties.

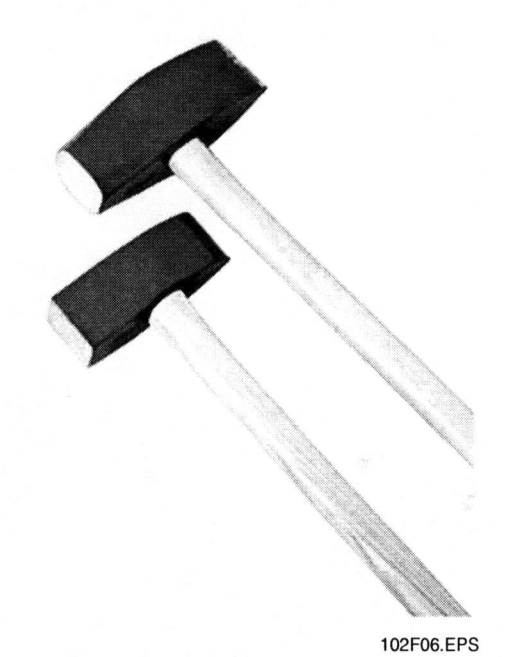

Figure 6 ◆ Stonemason's hammer.

Figure 7 ◆ Mash.

MASONRY TOOLS AND EQUIPMENT

The maul, or bushhammer, is a square-ended, rectangular, double-headed sledgehammer. It has a heavy head with toothed ends and a long wooden handle, as shown in *Figure 8*. It is a stonemason's tool used to face block, stone, or concrete.

The rubber mallet (*Figure 9*) has a rubber head with two flat ends and a wooden handle. This hammer can tap or drive something without leaving marks. Use the rubber mallet for setting stone, marble, tile, or other finely finished units into place.

When working with wooden-handled hammers, be sure to inspect them daily. Check that the handle is securely seated in the head. If the handle is loose, replace it before using the hammer.

2.3.0 Chisels

Masons use chisels to cut masonry units. For everyday work, masons use a hammer to cut brick or block, but for precise, sharp edges, they use a chisel. There are several types of chisels.

Chisels become damaged after prolonged use. Cutting edges get notches or burrs from striking rough surfaces or metal. Chisels are cutting tools, so keep them sharp. The best way to keep chisels sharp is to take them to a blacksmith. Like the mason's hammer, tempered steel, such as that used for chisels, must be sharpened carefully so it does not shatter. Blacksmiths are trained to do this properly.

Heads flatten after long use. The striking head mushrooms out, and metal burrs form at the edges of the head. These deformed edges can fly off and may cause injury. Inspect your chisels every day for dullness or deformation. To prevent injury, grind off the deformed part of the chisel head on a grinding wheel. As you grind, cool the chisel with water to keep it from overheating. If it gets overheated, the steel loses its temper and may unexpectedly shatter. Do not repair or sharpen the blade yourself; take it to a blacksmith.

> **WARNING!**
> Do not use a chisel with a deformed blade or a mushroomed head. Repair it or replace it immediately.

Figure 10 shows a mason's steel chisel, used for general cutting. It has a narrow blade of 1½" to 3" for a neat, clean cut. Use it to cut out masonry and make repairs. Use it for brick, block, and veined stone.

Figure 11 shows a brick set chisel. This is a wider chisel, up to 7" wide, with a thicker, beveled cutting blade. Traditionally, the brick set is as wide as the bricks it cuts. A wider version of the brick set is called the blocking chisel, or bolster chisel. This is usually 8" wide and used for cutting block.

Figure 8 ◆ Toothed maul, or bushhammer.

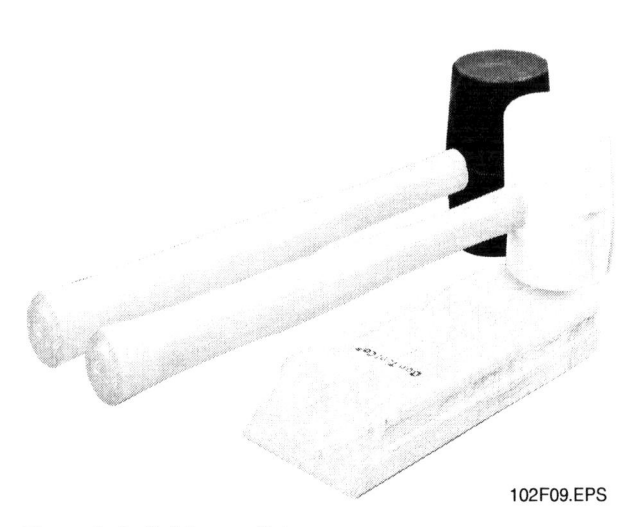

Figure 9 ◆ Rubber mallets.

Figure 10 ◆ Mason's chisel.

Building Blocks: Protect Your Hands

Some brick chisels have a protective grip. The grip is made of rubber and has a rubber hood to protect your knuckles from misplaced hammer strikes.

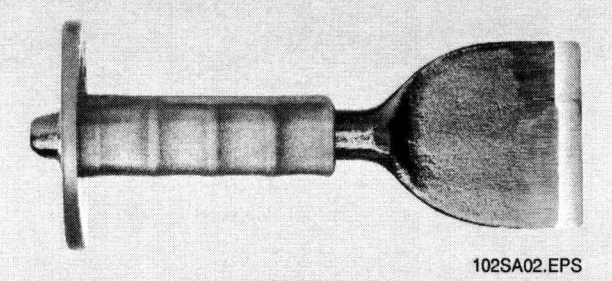

Figure 11 ◆ Brick set chisel.

Figure 12 ◆ Rubber-grip mason's chisel.

The rubber-grip mason's chisel is shown in *Figure 12*. This chisel has a rubber cushion on the handle to soften the impact of the hammer blow. It also has a wider cutting blade than the standard mason's chisel and is used for cutting brick, block, or stone.

The tooth chisel (*Figure 13*) has a toothed edge. This chisel is designed to cut soft stone and shape it to fit. It should not be used for hard stone. The pitching tool (also shown in *Figure 13*) is used for sizing, trimming, and facing hard stone.

The plugging, or joint, chisel (*Figure 14*) has a sharply tapered blade. It is also called a tuck-pointer's chisel. Use this chisel for cleaning mortar joints, for chipping mortar out of a joint, or for removing a brick or block from a wall. Each chisel is made for a specific job. Using the right tool for the job will make your work easier and more professional looking.

Figure 13 ◆ Tooth chisel and pitching tool.

MASONRY TOOLS AND EQUIPMENT

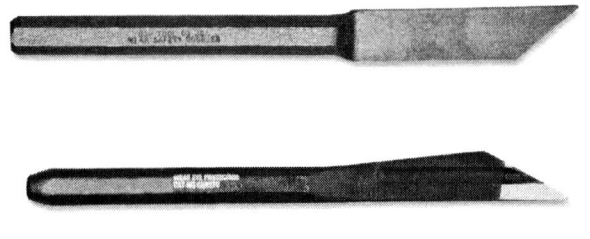

Figure 14 ◆ Plugging, or joint, chisel.

2.4.0 Jointers

Before the mortar is set, it can be troweled or tooled. Special tools are used to shape the mortar joint. Jointers, slickers, rakers, beaders, and sled runners are used for finishing or pointing the surface of mortar joints. Also called joint tools or finishing tools, the standard size jointer will fit into the standard joint between bricks or blocks. Jointers are available to fit a range of joint sizes. They come in various shapes to give different effects to the finished joint. *Figure 15* shows how the final surface appearance is determined by the shape of the jointer used.

Jointing compresses the mortar and decreases moisture absorption at the surface, so it adds water protection. Struck and raked joints are not recommended for exterior walls as they do not offer good water protection.

The jointers are usually cast or forged shaped steel rods with or without wooden handles. *Figure 16* shows some simple jointers for compressing and waterproofing joints. These convex, round, flat, and V-jointers are used on the short vertical or head joints. You can tell from the profile of the jointer what the tooled mortar joint will look like.

The **bed joints**, or the long horizontal joints, are tooled after the head joints. Longer jointers with wooden handles are used for the horizontal joints. As shown in *Figure 17*, these sled runner jointers come in a variety of shapes also, to match the head joints.

Another type of jointer is the raker or rake-out jointer; these also come mounted on skate wheels. *Figure 18* shows a skatewheel and a hand raker. The skatewheel raker is used for bed joints. The

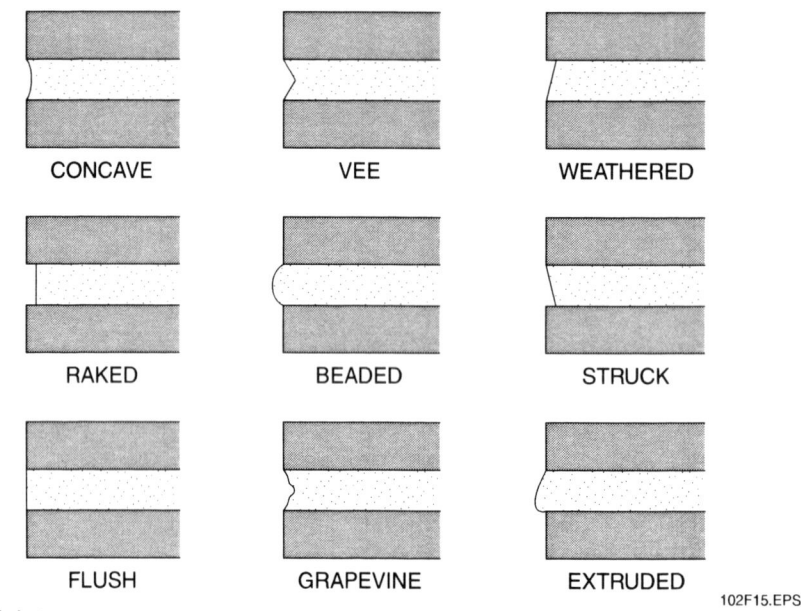

CONCAVE   VEE   WEATHERED

RAKED   BEADED   STRUCK

FLUSH   GRAPEVINE   EXTRUDED

Figure 15 ◆ Tooled mortar joints.

BUILDING BLOCKS

Multipurpose Jointers

Some jointers have different heads on each end; for example, a round head on one end and a V-head on the other end. Using double-headed jointers allows you to do more joints with fewer tools.

The flat jointer is sometimes called a slicker. Either end of a slicker can be used to shape mortar; usually the ends are different sizes. The slicker is handy for working on inside corners or other tight places where a neat joint is needed. A bull horn is a common round jointer that is larger on one end and tapered on the other.

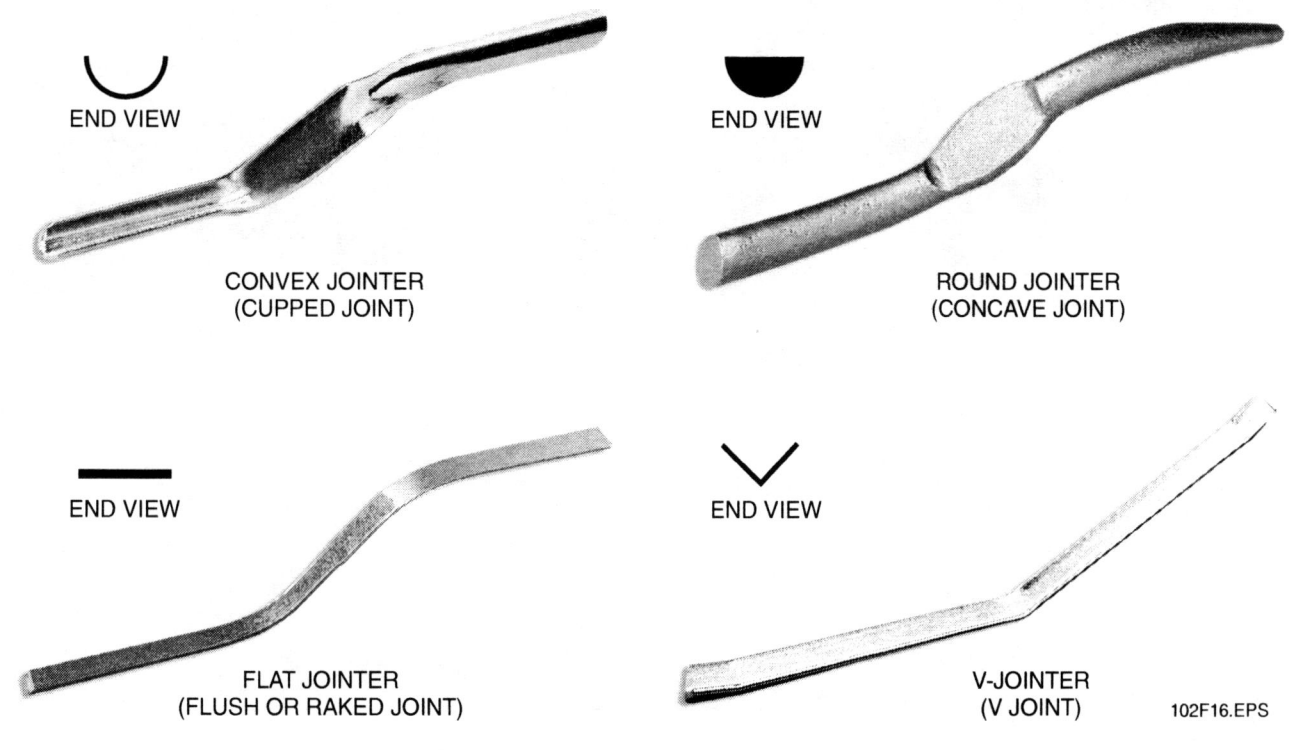

Figure 16 ◆ Jointers.

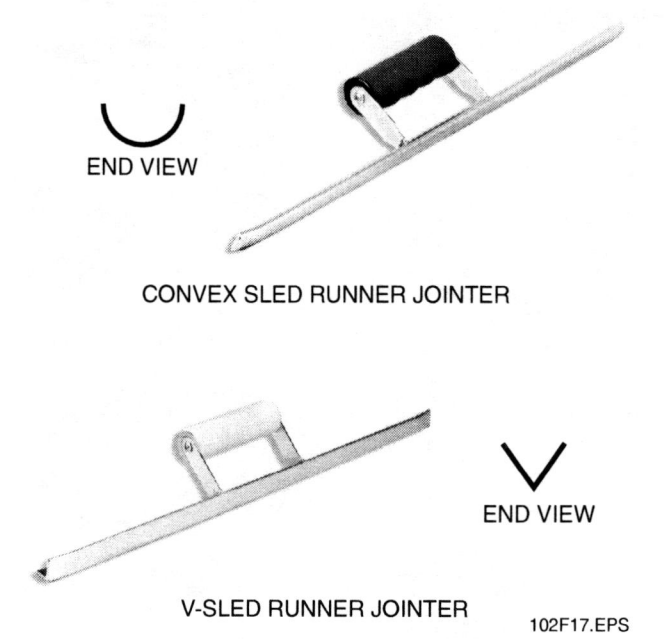

Figure 17 ◆ Runner jointers.

Figure 18 ◆ Rakers.

2.5.0 Brushes

Brushing masonry work removes any burrs or excess mortar. This is the finishing process for the wall or floor. The brush should have stiff plastic or wire bristles. *Figure 19* shows several brushes commonly used by masons.

A stove brush has a longer handle to keep fingers clear of the work area. Brushes are also useful for brushing off footings before laying masonry units and for cleaning the work area. For cleaning stains, brushes are used to apply acid to the face of masonry units. A cleaning brush may also have a

hand raker is used for joints the skatewheel raker will not reach. Note that the raking mechanism has an adjustable setscrew to rake different depths. The advantage of the skatewheel raker is the speed with which it forms a neat, hole-free joint.

All jointers need to be cleaned after they are used. If mortar hardens on them, they are not usable for smoothing.

MASONRY TOOLS AND EQUIPMENT 2.9

Figure 19 ♦ Mason's brushes.

scraper on one edge. Some brushes are designed to fit a long extension handle. Acid brushes are made of stiff plastic or fiber and have longer handles.

> **WARNING!**
> When applying acid, use a long-handled brush, and wear gloves and eye protection.

2.6.0 Brick Tongs

Brick tongs are designed to carry bricks without chipping or breaking them, as shown in *Figure 20*. Most are made of an adjustable steel clamp with a locking nut and a handle. They can be adjusted for different sizes and will hold 6 to 11 masonry units. Some brick carriers are steel rods that fit into the holes in the bricks. These carriers do not have a clamp and must be held upright. Smaller carriers are designed to carry larger individual blocks. With practice, the mason can carry loaded brick tongs in one hand, freeing the other hand for other activities.

2.7.0 Other Hand Tools

In addition to their other tools, masons use caulking guns, grout bags, pinch bars, and bolt cutters (*Figure 21*). The caulking gun is used to add caulking to expansion joints. The grout bag, with a metal or plastic tip, is squeezed to apply grout between masonry units. The pinch bar is used to pry out masonry units. The bolt cutter is used to cut rebar or ties. These and other hand tools will fit into the mason's tool bag.

Figure 20 ♦ Brick tongs.

2.8.0 Tool Bag

Figure 22 shows a mason's canvas tool bag with leather straps, handles, and reinforced bottom. The tool bag keeps tools together and within reach as the mason moves around the job site. Buckets

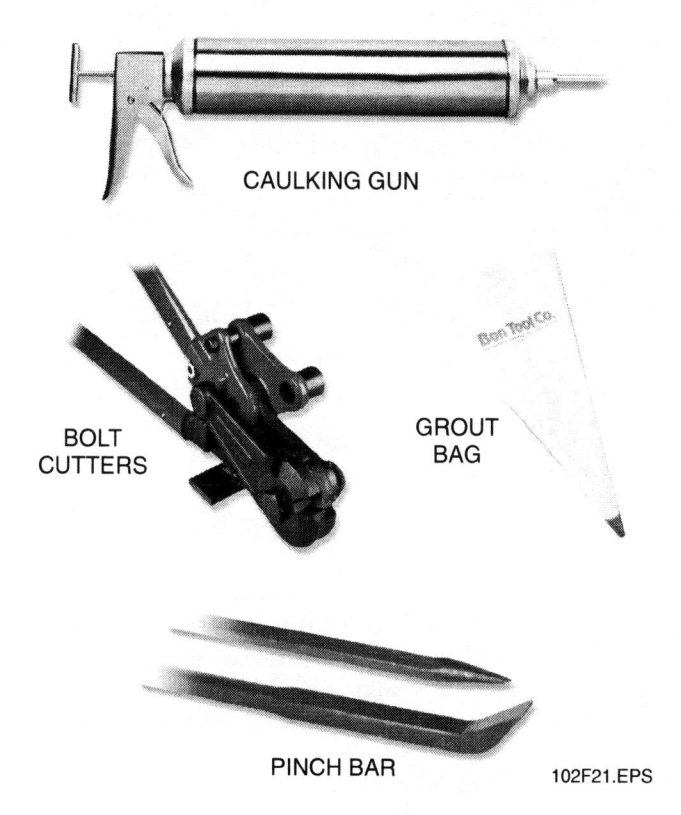

Figure 21 ♦ Common mason's hand tools.

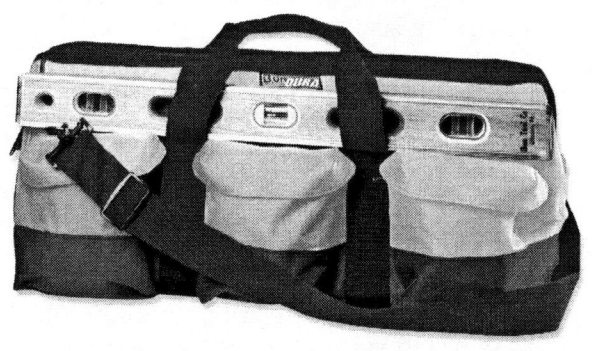

Figure 22 ♦ Mason's tool bag.

or other open containers will not keep tools dry. It is important to keep your tools dry and protect them from damage. Levels, rules, and squares must be properly maintained if they are to remain accurate.

Typically, a tool bag measures 14" to 18" across and has inside pockets for small items. The steel square and long level will fit under the leather straps on the outside of the bag. Some tool bag handles convert to shoulder straps, so masons will have hands free for climbing ladders or scaffolding. A good tool belt can also keep tools handy and your hands free.

3.0.0 ♦ MEASURES AND MEASURING TOOLS

Measures and measuring tools are key to good masonry. They are used on every course to check that a wall is level and true. If they are not used constantly and consistently, it is easy to end up with a wall that is out of plumb or full of bulges and hollows.

Because measures and measuring tools are so important to the final product, you must take the time to learn how to read them accurately and quickly. Keep them clean and free of dust or grit. Buy the sturdiest, easiest-to-read, most accurate measures and measuring tools you can find.

Levels are the most important tools in checking course alignment. Other measuring and aligning tools include rules, squares, lines, corner poles, chalklines, and plumb bobs. These course alignment tools are discussed in the following sections.

3.1.0 Levels

Levels were introduced in *Core Curriculum, Introduction to Hand Tools*. The level, sometimes called a plumb rule, establishes two measures:

- A plumb line that is vertical to the surface of the Earth
- A level line that is horizontal to the surface of the Earth

The level contains air bubbles in sealed vials filled with oil or alcohol, as shown in *Figure 23*. When the bubble in the vial is centered between the vial's two markers, the object being checked is level or plumb.

Levels can come with single or double sets of bubble vials. Two common levels are shown in *Figure 24*. They are made of hardwood, metal, or plastic to make them as light as possible without sacrificing strength. Levels come in a wide range of sizes. A smaller 8" size, often called a torpedo level,

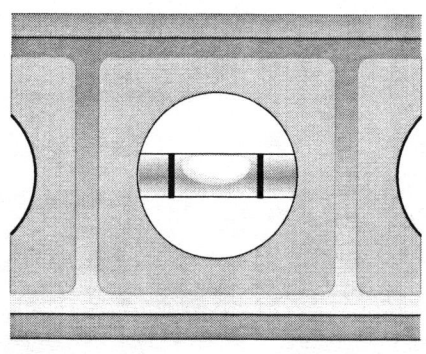

Figure 23 ♦ An air bubble shows level or plumb.

MASONRY TOOLS AND EQUIPMENT 2.11

BUILDING BLOCKS

String Line level

A line level is only a few inches long and usually made of plastic. They are designed to hang on a string line to check the accuracy of guidelines.

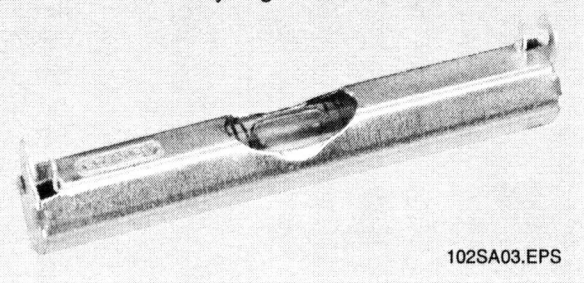

---

is used for individual blocks or smaller spaces. Longer levels can be up to 72" for leveling wall sections or spanning distances. Masons use levels continually to check the plumb and level of individual masonry units and the entire length of a wall.

The level is usually the most expensive of a mason's tools. Always check the action of the bubble vials against another level before purchasing a level. Check the accuracy of your level daily, especially if it is dropped or jarred. Do not tap a level with a hammer to level a brick. Some levels can be adjusted if they are no longer accurate.

Clean your level carefully at the end of the day to keep any mortar from hardening on it. Wipe off wooden levels with a rag dampened with linseed oil to preserve the wood. Clean a metal level with a dry cloth, so dust and grit do not stick to it. You may want to oil it occasionally as well.

3.2.0 Rules

Masons use two kinds of rules, a 6' folding rule, and a 10' retractable tape. Mason's rules have special marking scales on them.

The brick spacing rule, as shown in *Figure 25A*, has markings for different sizes of bricks. It shows the course spacing for each brick size. This measure is also called the course counter rule. Use it to lay out and space standard brick courses to dimensions that are not modular. It is useful for spacing mortar joints around door tops, window tops, or other uneven spaces.

The modular spacing rule, as shown in *Figure 25B*, is based on a module of 4". It has 6 scales ranging from 8 to 2, with scale 6 representing the size of a standard brick. The modular rule can be used for block as well as brick. There is a slight disparity between spacing rules and the modular spacing rule.

Figure 24 ◆ Common levels.

TORPEDO LEVEL

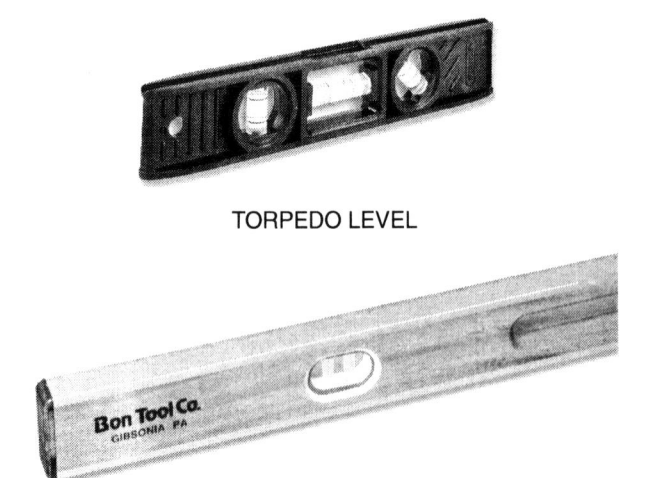

STANDARD LEVEL

BUILDING BLOCKS

Take Care of Your Tape

Steel tapes are wound with a spring, so they will retract into the housing. Never let a retractable steel tape snap back into the housing. The end tab can easily break off, making the tape useless. Always guide the tape back into the housing.

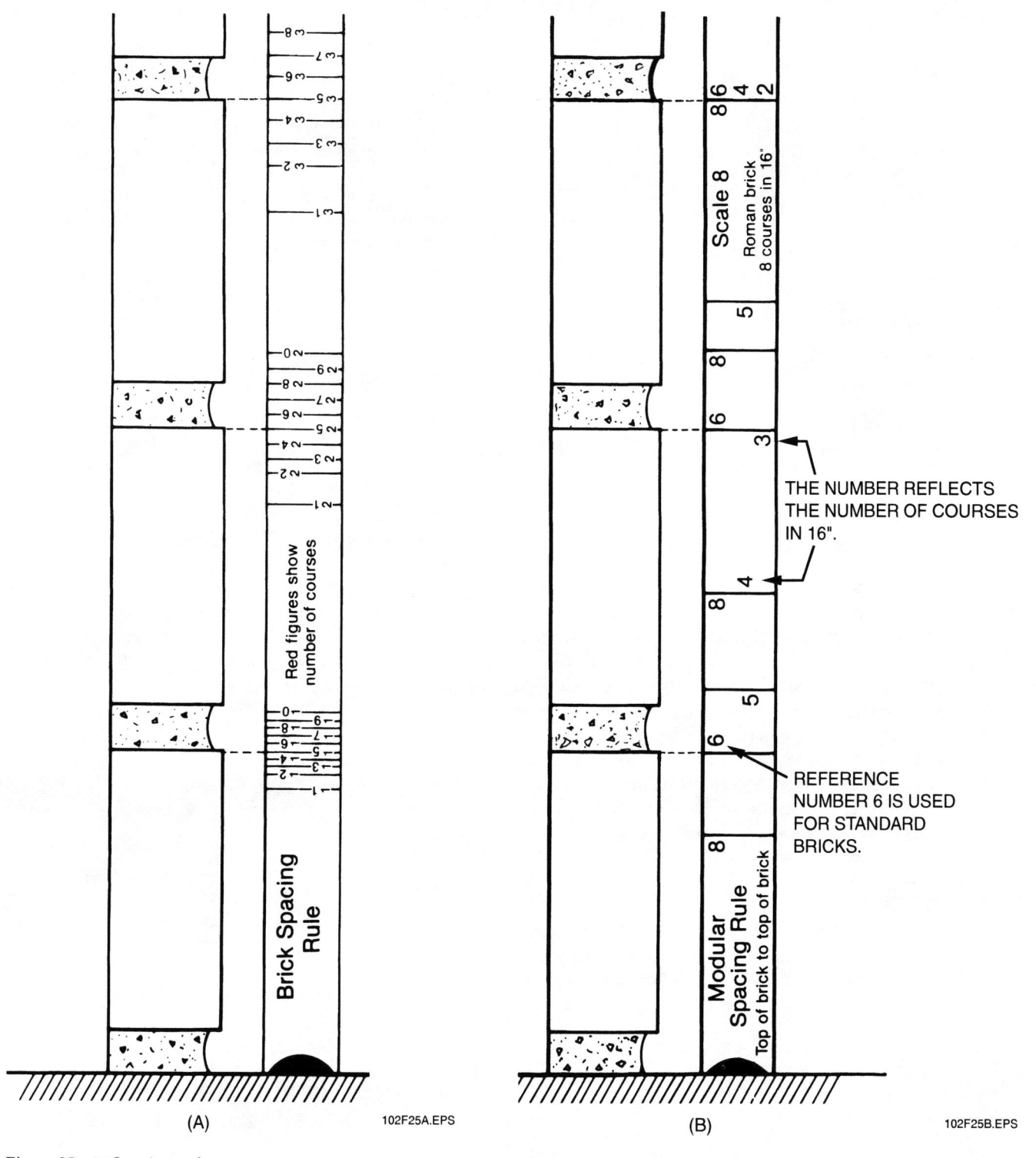

Figure 25 ♦ Spacing rules.

The mason's steel tape and folding rule are both marked with spacing measures. They are available with either the modular or the course markings. *Figure 26* shows both these rules, which also have inch measurements. Folding rules are usually made of wood, with brass joints and pivots between the sections. An oversized brick ruler has eleven scales for queen or king-size brick spacing, ranging from 2⅞" to 3½" on one side, and inches, eighths, and sixteenths on the other side. These rules are usually marked with letters and/or numerals to show the number of courses.

MASONRY TOOLS AND EQUIPMENT

2.13

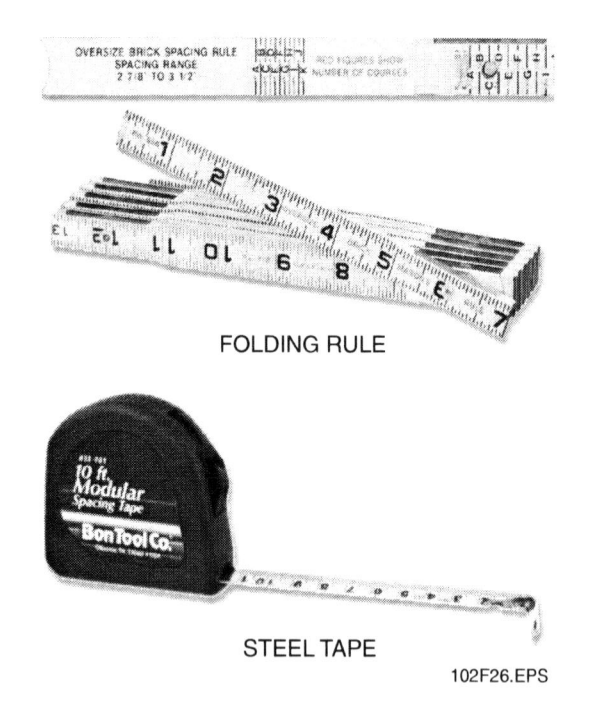

Figure 26 ◆ Folding rule and steel tape.

Masons prefer a folding rule because one person can easily and accurately extend it to the needed length. Tapes must be stretched without bowing, which may need an extra hand over longer distances. Both rules and tapes should be inspected and cleaned daily. Use a very light touch of linseed oil on wood rules, and be sure to clean dust and grit from the pivots. A belt pouch or clip will keep your rule handy and clean.

3.3.0 Squares

Masons use several types of squares. The framing, or steel, square (*Figure 27*), looks like a carpenter's square. Use this for laying out **leads** and other corners and for checking that corners are square.

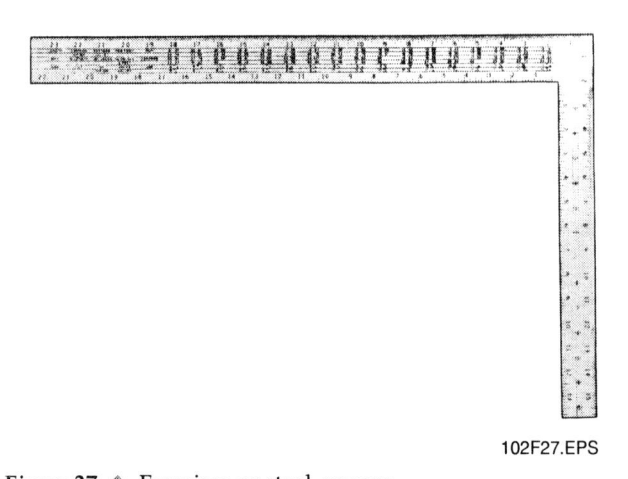

Figure 27 ◆ Framing, or steel, square.

As shown in *Figure 28A*, the combination square, or T-square, has a movable crosspiece and a built-in 45-degree angle. Use this for marking right-angle and 45-degree angled cuts and for checking these angles on cuts. The sliding T-bevel, as shown in *Figure 28B*, has a setscrew in the crosspiece. The screw fixes the angle of the crosspiece so that the same angle can be used for marking many cuts. The bevel is useful when the job calls for an odd angle other than 90 degrees or 45 degrees. Use the bevel for marking these kinds of cuts, for marking **skewbacks**, and for checking the angles on cuts.

3.4.0 Mason's Line and Fasteners

The level is used to lay out structural masonry elements 4' in length or shorter. The mason's line is used to lay out masonry structures longer than 4'. The line itself is twisted or braided nylon cord. The typical line is about 150 pounds test and should be used at 20 percent of its test. This means it is good for a pull of up to 30 pounds. Braided line is preferred because it will not sag as much when pulled tight and will last longer.

The line is stretched tight as it is strung between two fixed points. For most jobs, the fixed points are on the wall corners, or leads, which have

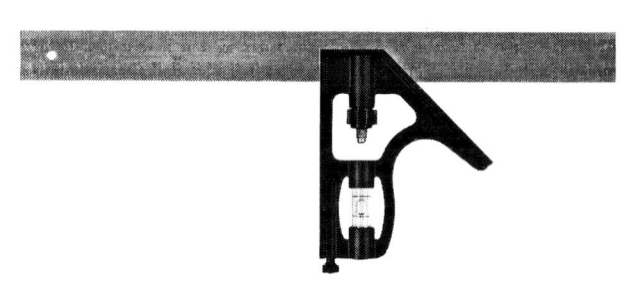

(A) COMBINATION SQUARE

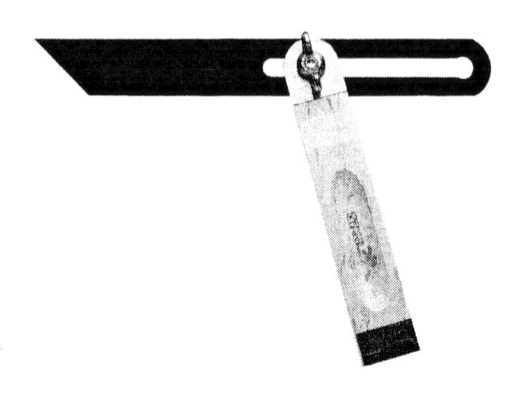

(B) SLIDING T-BEVEL

Figure 28 ◆ Combination square and sliding T-bevel.

Check Your Squares

BUILDING BLOCKS

Squares may be damaged accidentally from time to time. Make sure that you periodically check your squares with a known straightedge. This will ensure accuracy and square walls. A small error is compounded over the length of a wall. This can result in serious error in the finished structure.

---

already been laid. The line becomes the guide for laying the course of masonry units between the leads, or fixed points. Using the line properly will result in a wall without bulges or hollows. The masonry units are laid under the line, and the line is moved up for each course.

There are three methods of fastening the line in place: pins, blocks, and stretchers. Line pins (*Figure 29*) are about 4" long and made of steel. Drive them into the wall or structure at the marked point, and string the line tightly between them. Line pins leave holes, and the string must be remeasured or retightened for each course.

However, line pins are harder to accidentally knock off than line blocks.

Line blocks are also called corner blocks (*Figure 30*). They are made of wood, plastic, or metal; however, wood grips the corner better. The knotted line passes through the slit in the back of the block. The line tension between the two blocks holds the blocks in place.

Line stretchers are shown in *Figure 31*. They have a flatter profile and also use the line tension to hold them in place. They come in standard and adjustable sizes and fit more snugly to the lead than corner block.

Figure 29 ◆ Line pin.

Figure 30 ◆ Corner block.

Don't Tangle Your Line

BUILDING BLOCKS

Usually, the line is bought wound around a core. If it is rewound on a shuttle or holder, it will not tangle as it is used and re-used. Lines are made in several bright colors, so they can be easily seen.

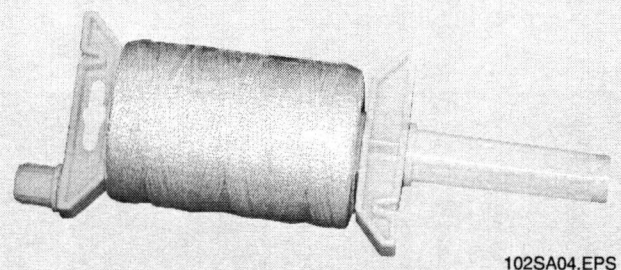

MASONRY TOOLS AND EQUIPMENT 2.15

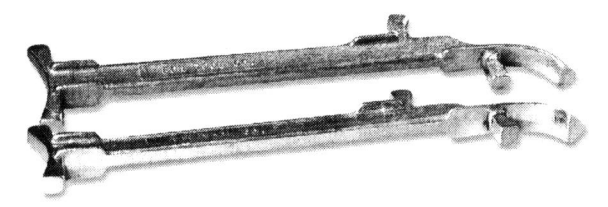

STANDARD LINE STRETCHERS

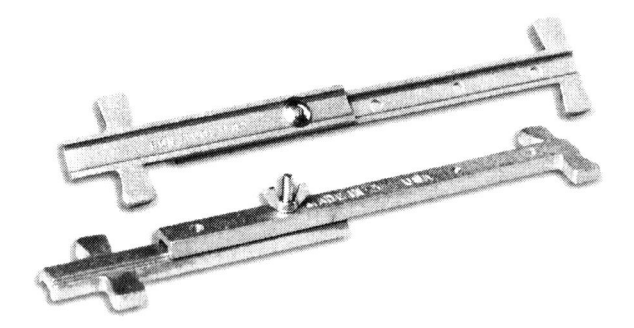

ADJUSTABLE LINE STRETCHERS

102F31.EPS

Figure 31 ♦ Line stretchers.

LINE TRIG

102F32.EPS

Figure 32 ♦ Line trig or twig.

Unlike pins, blocks and stretchers must have a corner to be held snugly against. They do not leave holes in bricks or walls, but they do project slightly from the corners of the wall. They can be accidentally knocked off. Because of the tension on the line, a flying block or stretcher can severely injury someone in its path.

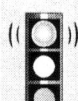

> **WARNING!**
> Use extra care working with corner blocks or stretchers to avoid knocking them off.

Once the line is in place, it may need more support. Line trigs (*Figure 32*), sometimes called twigs, are steel fasteners. They hold the line in position and keep it from sagging. A very long line may need several trigs, even if it is stretched tight. The trig slips over the line and rests on a masonry unit that has been put in place to support it. A half brick or piece of block can anchor the trig to ensure the line is not disturbed.

3.5.0 Corner Pole

The next in the series of course alignment measures is the **corner pole**, or **deadman**. The corner pole is any type of post braced into a plumb position so that a line can be fastened to it. This allows a wall to be built without building the corners first. This tool is useful when building a veneer wall. The corner pole can be braced against an existing part of a structure or the block foundation.

Corner poles can be made from dimensional lumber and braces. Commercial corner poles (*Figure 33*) are metal with proprietary line block and supporting hardware. Commercial corner poles have masonry units marked on them. A handmade corner pole can be marked off in masonry units with a grease pencil, marker, or carpenter's pencil. If you make a deadman from scrap wood, make sure that the wood is straight and not badly warped. A corner pole with markings for different stories of a building is called a story pole.

3.6.0 Chalkline and Chalk Box

A chalk box (*Figure 34*) is a metal or plastic case with finely ground chalk and about 50' of twisted cotton line wound on a spool inside. As the line is drawn out of the box, it picks up powdered chalk.

BUILDING BLOCKS

Check with Your Supplier

Masonry materials suppliers sometimes give away trigs and wooden line blocks with larger orders.

Chalk Box Maintenance

Before using a chalkline, check that there is enough chalk in the reservoir and that the string is not frayed. Both of these parts should be replaced or refreshed periodically.

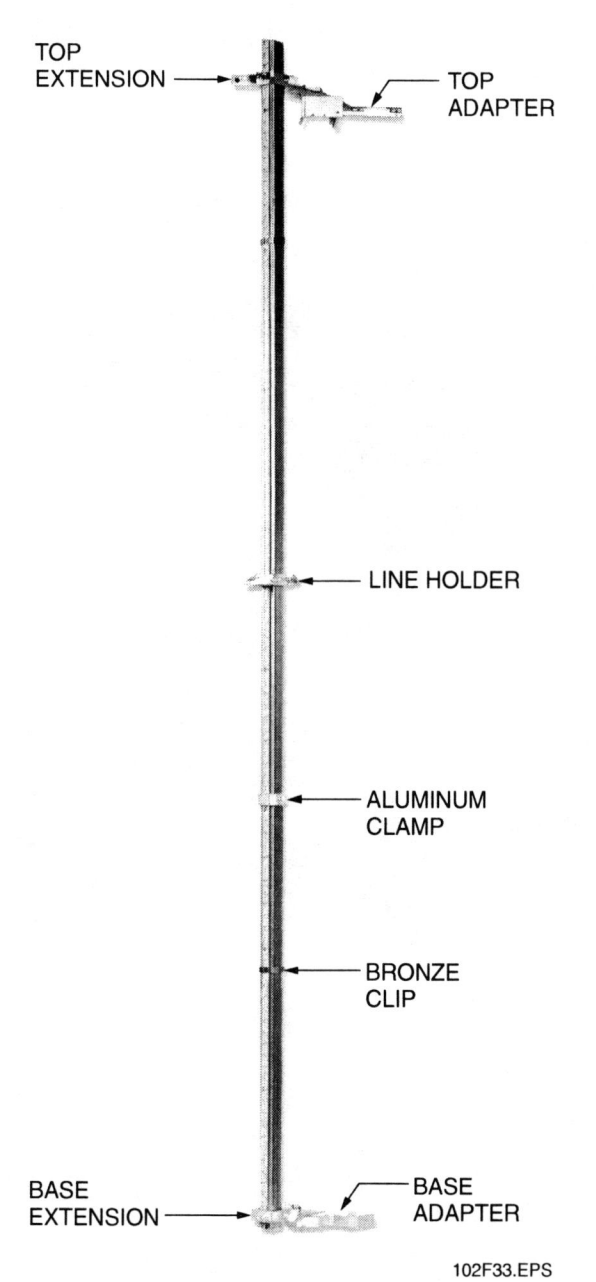

Figure 33 ◆ Corner pole or deadman.

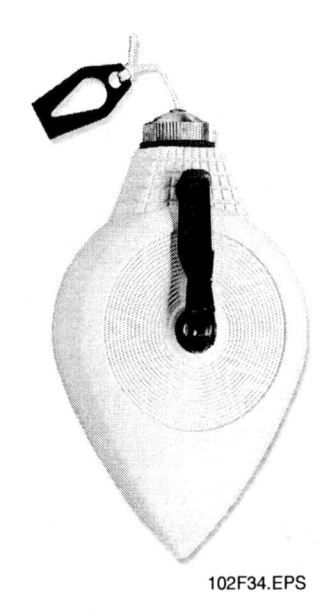

Figure 34 ◆ Chalk box.

The line can be stretched between two points and snapped. This will leave a chalk mark exactly under the line. Masons use chalklines to establish straight lines for the first course of bricks in a wall. The chalkline can be snapped against the foundation slab or footing. Some chalk boxes can also be used as a plumb bob.

3.7.0 Plumb Bob

Sometimes plumb is marked with a plumb bob (*Figure 35*). The plumb bob is a pointed weight attached to a length of mason's line. The length of the line is easily changed. A plumb bob can establish vertical plumb points. Use it to mark a point directly under another measured point. A plumb bob is also useful for checking the plumb of a story pole or corner pole.

Blowing in the Wind

When using a plumb bob outside, you must be aware that the wind may blow it out of true vertical. The longer the drop line, the more likely that wind will affect your measurement.

MASONRY TOOLS AND EQUIPMENT

Figure 35 ◆ Plumb bob.

4.0.0 ◆ HAND-POWERED MORTAR EQUIPMENT

Hand-powered mortar equipment is used on both small and large jobs. It complements the hand tools and completes the tool set needed to build a structural masonry unit.

4.1.0 Mortar Boxes

Mortar is mixed in a wheelbarrow or mortar box. A mortar box can be steel or plastic and is about 32" × 60". Some mortar boxes have wheels for easy movement (*Figure 36*). Make sure the box is set level so that water does not collect in one end. After it is mixed, the mortar may be moved to a mortar pan (*Figure 36*) or smaller mortar board.

The mortar pan and mortar board can be located next to the working mason. They can be placed on a stand, so they are convenient for working higher courses. The metal or plastic mortar pan fits more securely on the stand, but mortar tends to stick in the corners. Mortar boards are available in plastic or plywood and average about 3' on a side. A plywood board needs to be wetted thoroughly before mortar is put on it to keep the board from absorbing moisture and drying the mortar.

A smaller version of the mortar board is the hawk (*Figure 37*). This small board has a pole-grip handle underneath and is used to carry small amounts of mortar. A close relative of the plasterer's hawk, it is useful for tasks such as pointing. A wooden hawk needs to be wetted thoroughly before mortar is put on it.

Figure 36 ◆ Mortar pan and mortar box with wheels.

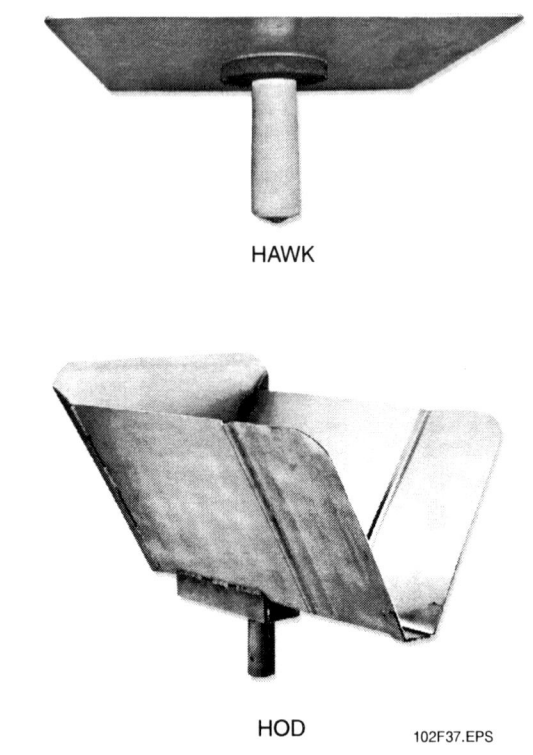

Figure 37 ◆ Hawk and hod.

The hod, also shown in *Figure 37*, carries mortar from the mixer to the mason. It is practical for moving material in tight spaces or on scaffolding. It is an aluminum trough with a long pole handle. The mortar can be lifted up to a mason on a scaffold by using a hod.

A cubic-foot measuring box (*Figure 38*) is sometimes used to prepare mortar. This box measures one cubic foot of sand or cement. It is used to measure ingredients by volume in order to proportion the mortar correctly. As pre-packaged mortar has become more popular, use of the cubic-foot box is uncommon, but sometimes it is specified by architects for measuring sand.

Mortar containers and carriers should be cleaned out with water between loads of mortar. The mixing pan should be cleaned out after each batch and cleaned thoroughly at the end of each day.

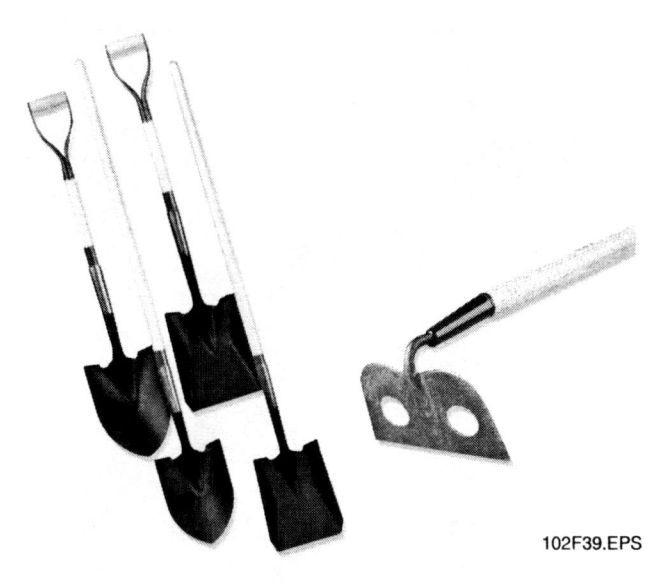

Figure 39 ◆ Mixing aids.

Figure 38 ◆ Cubic-foot measuring box.

4.2.0 Mixing Accessories

Several of the tools used in mixing mortar are shown in *Figure 39*. A long-handled shovel is used to measure, shovel, and mix dry ingredients in the mortar box. A square-end, short-handled shovel is used to move mortar from the mixing box to a pan, stand, or hod. The square end helps you to scrape the bottom of the mixing pan.

A large hoe is used to mix wet mortar and to **temper** mortar. The mortar hoe has a 10" blade with two holes in the blade to make it easier to pull the hoe through the mix. Some hoes have shorter handles for use in smaller spaces. Remember to clean mixing tools with a stiff brush and water immediately after use.

4.3.0 Water Bucket and Barrel

The water bucket and barrel (*Figure 40*) are essential in masonry. Made of steel, plastic, or galvanized metal, both are necessary equipment. Half-filled with water, the bucket provides a ready bath to clean mortar off hand tools. Masons frequently use five-gallon buckets to measure water for mixing mortar. The bucket can also serve as a carrier for small amounts of mortar or grout. The barrel provides a bath for larger tools and a ready supply of water for **retempering** mortar.

Be sure to wet the inside surface of the bucket before filling it with mortar or grout. Always wash empty buckets out immediately to prevent mortar from hardening inside. Wash the barrel out at the end of the day.

FIVE-GALLON WATER BUCKET   WATER BARREL

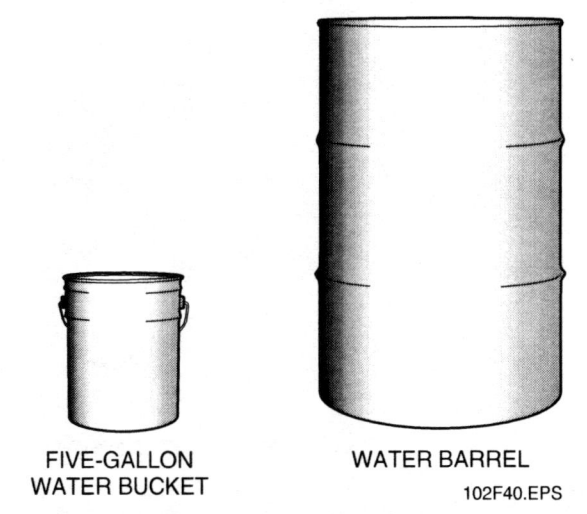

Figure 40 ◆ Water bucket and barrel.

MASONRY TOOLS AND EQUIPMENT

Building Blocks: Brick Carts

A commercial brick or block cart is also very useful for moving packaged brick or blocks. Most carts have air-filled tires and can carry up to 100 standard bricks.

4.4.0 Barrows

Masons use two types of barrows (*Figure 41*). The standard contractor's wheelbarrow is made of steel with wooden handles. It can carry about 5 cubic feet of masonry or mortar. The second type of barrow is a pallet on wheels. Its wooden body has no sides, so it is convenient to unload masonry units from either side. However, it must be unloaded so that it does not become unbalanced and tip over. This brick barrow works well for moving bags of cement, piles of block, and other bulky materials.

Both types of barrows need large, air-filled tires to move over bumpy ground without tipping. Check the wheels before using any barrow.

5.0.0 ◆ POWER TOOLS

Power tools bring time and labor savings to masonry work. Mechanical and power tools, most of which were developed after the 1920s, have become an integral part of masonry work.

When using power tools and equipment, always follow power-tool safety rules. Inspect items before using them to make sure they are clean and functional. Disconnect power cords, and turn off engines before inspecting or repairing power equipment.

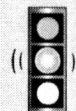

> **CAUTION**
> Always be sure the power is off before the final cleaning at the end of the day.

CONTRACTOR'S WHEELBARROW

BRICK AND TILE BARROW

Figure 41 ◆ Barrows for masonry units.

BUILDING BLOCKS

Know Your Cutting Depth

A 14" blade has approximately a 5" cutting depth. To cut all the way through an 8" block, use a 20" blade.

5.1.0 Saws

Masonry saws can make more accurate cuts than a mason's hammer or brick set. The saw does not weaken or fracture the material as hand tools do. Masonry saws are available in handheld or table-mounted models. Some larger table saws can be operated with a foot control, which leaves both hands free for guiding the piece to be cut.

Figure 42 shows a large masonry saw. Note that the saw is used with a carrier or conveyor tray on tracks. The masonry unit to be cut is placed on the carrier tray, aligned, and carried against the blade.

A smaller version of the masonry saw (*Figure 43*) is portable. It is easily moved about and quickly set up at different areas of the job site. This saw has a carrier or conveyor tray that runs on tracks. It also has hand controls. The blade on some smaller saws is mounted on a rotating arm. They are called chop saws because the blade is lowered (chopped) onto the brick.

Masonry table saws use diamond, carborundum, or other abrasive blades. Diamond blades must be irrigated to prevent them from overheating and burning up. The irrigating water wets the masonry unit, cools the blade, and controls dust.

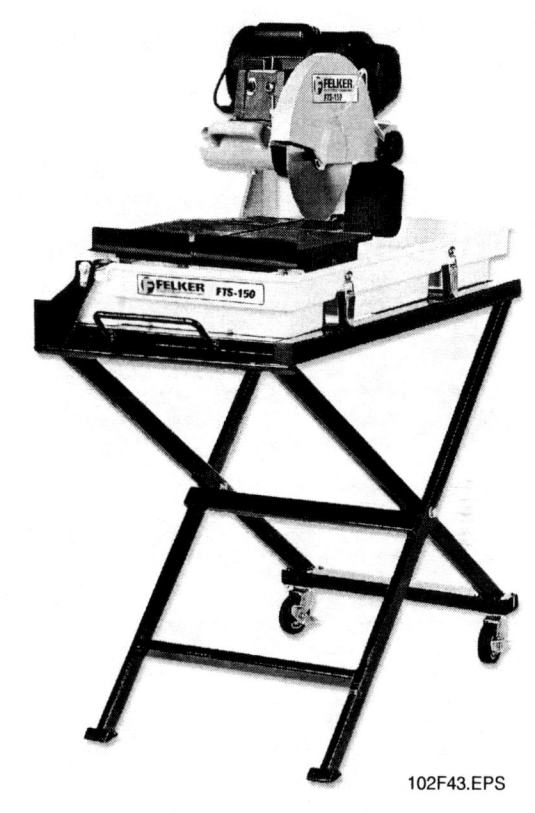

Figure 43 ♦ Small masonry saw and stand.

The wet units must be allowed to dry before they can be laid. The dry abrasive blades cut slowly and cannot cut as thinly, but the masonry can go directly to the mortar bed. Dry cutting produces dust that must be vented away from workers and the work site.

Handheld masonry saws with 14" blades (*Figure 44*) are also known as rapid-cut saws. Handheld saws with 12" blades are sometimes called cutoff or target saws. These smaller saws can be powered by gasoline, electricity, or hydraulics. The diameter of a blade on a handheld saw can be up to 14". Using a handheld saw calls for extra caution.

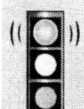

> **WARNING!**
> Make sure electrical saws are grounded, especially if you are using water to cool the blade and control dust.

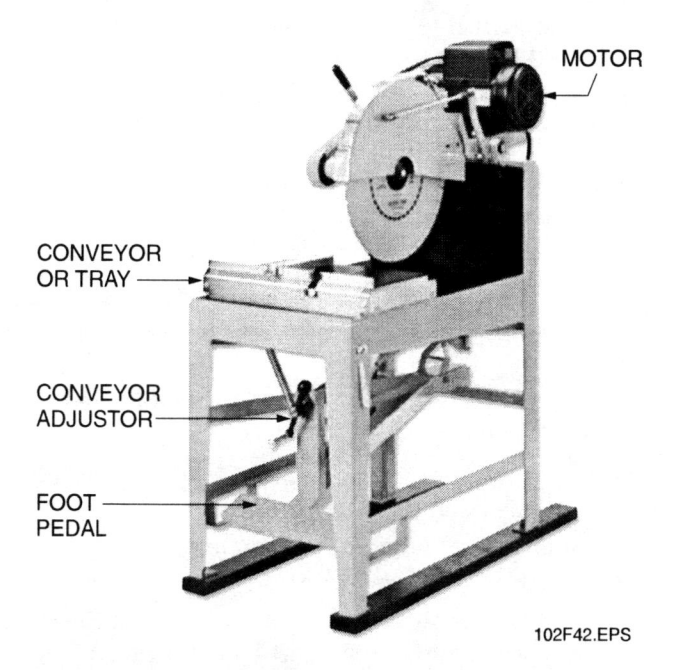

Figure 42 ♦ Large masonry saw.

MASONRY TOOLS AND EQUIPMENT 2.21

BUILDING BLOCKS

Jigs

Jigs are often used with handheld saws. A jig is a template used to keep masonry units in place as they are cut. You can make a simple jig from scrap wood. The jig can be marked if a specific cut needs to be repeated.

Figure 44 ♦ Handheld masonry saw.

Figure 45 ♦ Hand-operated masonry splitter.

When using power saws of any size, review saw control operations before starting to cut. Rules for the safe use of saws include the following:

- Wear a hard hat and eye protection to guard against flying chips.
- Wear rubber boots and gloves to reduce the chance of electric shock.
- Wear ear plugs and/or other ear protection.
- Wear a respirator when using a dry-cut saw.
- Check that the blade is properly mounted and tightened before starting to cut.

5.2.0 Splitters

Masonry splitters are mechanical cutters for all types of masonry units. They do not have engines, but some use hydraulic power. Splitters range from small hand-operated units, such as the one shown in *Figure 45*, to massive foot-operated hydraulic splitters. These mechanical units offer more precision than a brick hammer or brick set. They are also faster, especially when many bricks need to be cut.

Unlike saws, splitters do not create as much dust and do not have high-speed blades. Splitters do make cuts as neatly as saws. Larger units (*Figure 46*) can deliver as much as 20 tons of cutting pressure but are precise enough to shave ¼" off a brick or stone. Large hydraulic splitters can accommodate larger masonry units than most large saws.

5.3.0 Grinders

The tuckpoint grinder (*Figure 47*) is a handheld, electric grinding tool designed to grind out old bed and head joints in a masonry wall. It has shatterproof blades and a safety guard on the part of the machine facing the mason.

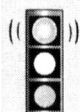

> **WARNING!**
> Wear a hard hat and eye protection when using a tuckpoint grinder.

The following are rules for the safe use of a grinder:

- Be sure the grinding head is in good condition and has been properly secured.
- Wear gloves when appropriate.
- Use the proper grinding head for the material being ground.
- Be sure the grinder head has stopped turning before putting the grinder down.

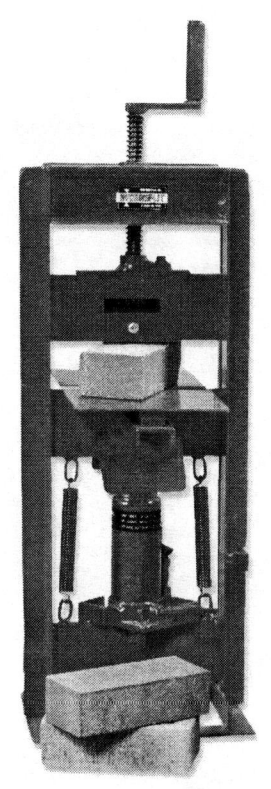

Figure 46 ♦ Large foot-operated hydraulic splitter.

Figure 47 ♦ Tuckpoint grinder.

5.4.0 Power Drills and Powder-Actuated Drivers

Usually, masons set bolts to anchor metalwork or wood when the mortar is still pliable. When the mortar is hardened, a power drill or powder-actuated driver is needed. Drilling in masonry takes a combination of power, speed, and hammering, so a ¼" drill is not adequate. A ⅜" or ½" hammer drill, such as the one shown in *Figure 48*, delivers enough power. Carbide-tipped drill bits are available in several standard sizes. Drills are used to attach wall ties or other anchors to a concrete wall.

Figure 48 ♦ Hammer drill.

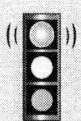

> **WARNING!**
> Most hammer drills have enough torque to break your wrist. Make sure that you have a firm grip on the side handle when using a hammer drill. You should never hold on to just the main handle. Use both hands to equalize the rotation of the drill.

Powder-actuated drivers are usually designed to drive a specific line of bolts, anchors, and other fasteners. These drivers use a small explosive charge to drive a pin or stud into masonry. Specially hardened pins or studs are used. Different charge sizes are available. Manufacturers color code the powder load charges to identify the strength of the charge. Be sure to learn the color code specific to your tool's manufacturer.

Some models of powder-actuated drivers use compressed air. *Figure 49* shows a powder-actuated driver and components needed to operate it. Do not operate a powder driver without the proper training and documentation. Always wear appropriate personal protective equipment when working with or around powder-actuated tools.

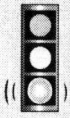

> **NOTE**
> OSHA requires that all operators of powder-actuated tools must be qualified or certified by the tool manufacturer. Operators must carry certification cards whenever using the tool.

6.0.0 ♦ POWER EQUIPMENT

Power equipment, like power tools, bring speed and economy to the masonry building process. When using power equipment, follow the general rules for power-tool safety.

MASONRY TOOLS AND EQUIPMENT 2.23

Building Blocks

Powder-Actuated Drivers

A powder-actuated fastening tool is a low-velocity fastening system powered by gunpowder cartridges, commonly called boosters. Boosters often come in a strip called a magazine. This cutaway diagram of a powder-actuated fastening tool shows how the components work together.

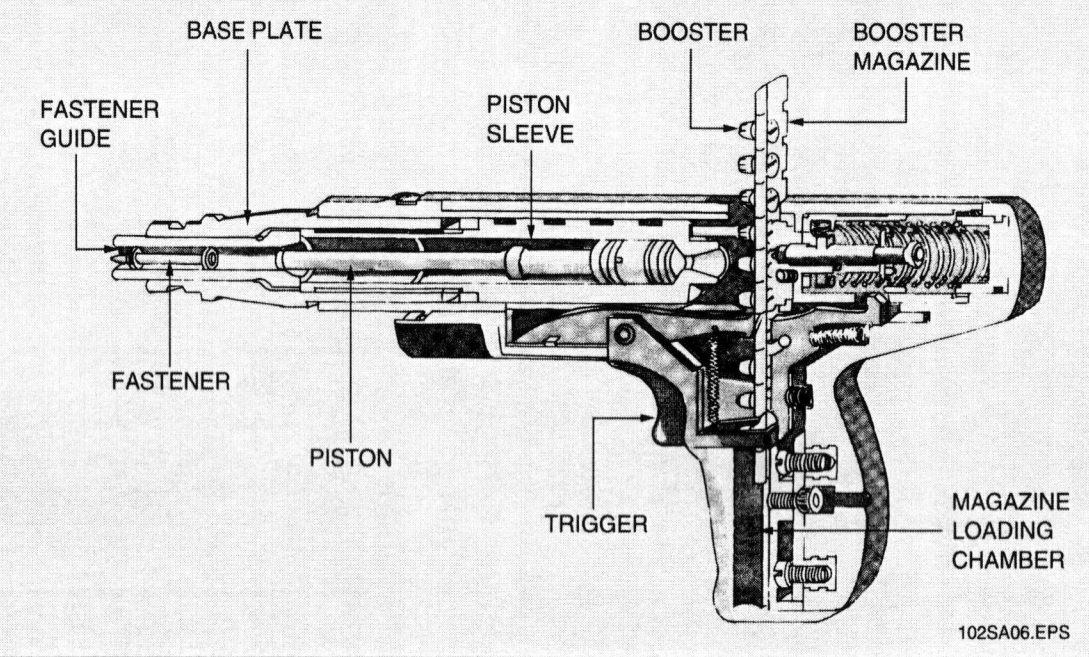

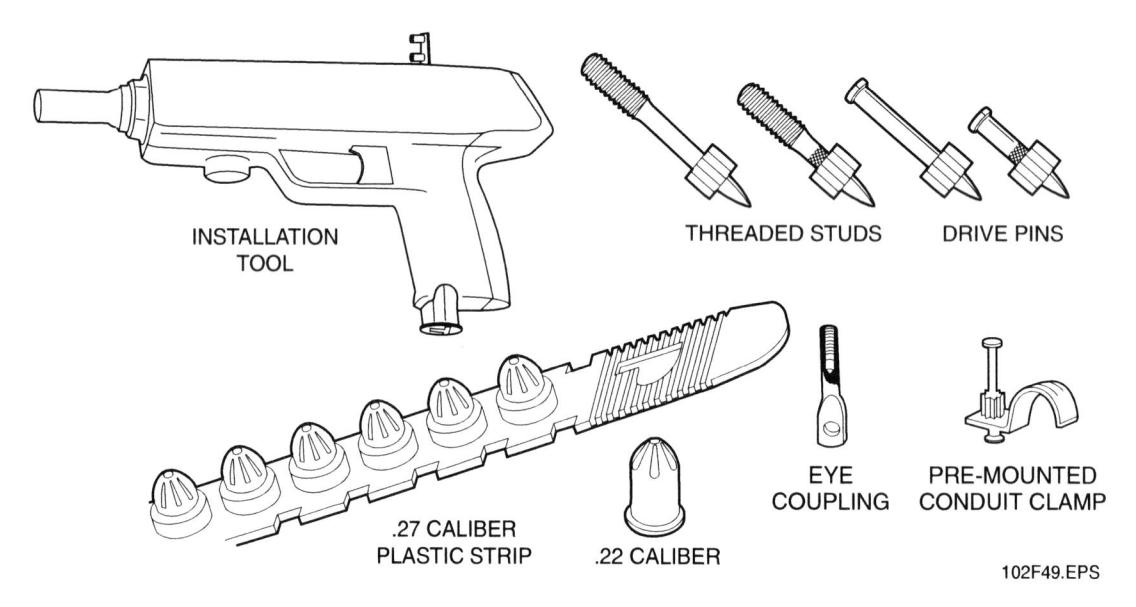

Figure 49 ♦ Components of powder-actuated driver.

6.1.0 Mortar Mixer

On most commercial jobs, mortar is mixed in a powered mortar mixer or pug mill. The mixer has an electric or gasoline engine and is usually on a set of wheels, as shown in *Figure 50*. The mixer portion consists of a drum with a turning horizontal shaft inside. Blades are attached to the shaft and revolve through the mix. The dump handle and drum release are used to empty the mortar onto a pan or board. Mixing mortar using a power mixer is explained later in the module entitled *Mortar*.

2.24   MASONRY LEVEL ONE — TRAINEE MODULE 28102-04

Figure 50 ◆ Mortar mixer.

Figure 51 ◆ Masonry pump.

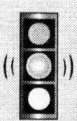

> **CAUTION**
>
> Check the oil, gas, and other fluid levels every time you use a mixer with a gasoline engine. Running out of gas with a batch of mortar in progress will spoil the batch and could damage the machine.

> **WARNING!**
>
> Always wear eye protection and other appropriate personal protective equipment when using a power mixer. Never place any part of your body in the mixer.

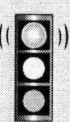

Figure 52 ◆ Handheld vibrator.

Mixers have capacities ranging from 1 to 12 cubic feet; the typical mixer holds about 4 cubic feet. The mixer drum needs to be washed out immediately after each use to keep mortar from hardening inside it.

6.2.0 Masonry Pump and Vibrator

The masonry pump (*Figure 51*) is used to deliver mortar or grout to a high location. Grout is usually pumped when it is used to fill the cores in a block wall. Grout is pumped from the mixer to the intake hopper of the grout pump to the delivery hose.

At the deposit site, the mason guides the grout into the cores. After the grout is delivered, it is sometimes vibrated to eliminate air holes. A typical handheld vibrator is shown in *Figure 52*.

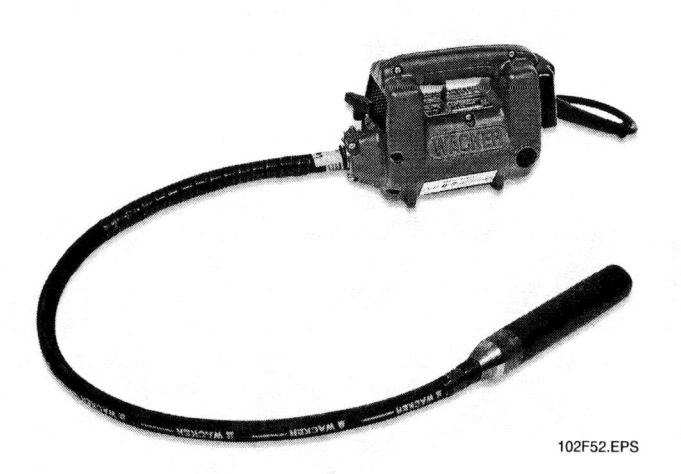

The snake-like end of the vibrator is inserted into the core. The mason inserts the vibrator into each core to make sure that air pockets are removed and the grout is consolidated. Steel reinforcement may also be placed in the cores before the grout is added.

6.3.0 Pressurized Cleaning Equipment

Pressurized cleaning equipment has effectively replaced bucket-and-brush cleaning for masonry structures. Pressurized cleaning uses abrasive material under pressure to scour the face of the masonry. The generic equipment for pressurized cleaning includes an air compressor, a tank or reservoir for pressurizing, a delivery hose, and a nozzle or tip. There are three types of pressurized cleaning systems, differing mostly in the abrading material they deliver. The following sections describe these pressurized systems: pressure washing, steam cleaning, and sandblasting.

MASONRY TOOLS AND EQUIPMENT

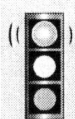

> **WARNING!**
> Pressurized cleaning equipment creates dangerous conditions. Read the manufacturer's operating manual before using. Make sure that you know how to use all pressure-release valves and safety switches. Wear safety glasses and other appropriate personal protective equipment.

6.3.1 Pressure Washing

Pressure washing can be the gentlest method for cleaning masonry structures. Also called high-pressure water cleaning, this is a newer cleaning technique. The pressure washer uses a compressed air pump to pressurize water and to deliver it in a focused, tightly controlled area. Sometimes pressure washing is done after manual cleaning.

Pressure washing has the best results when the operator uses a fan-type tip, dispersing the water through 25 to 50 degrees of arc. The amount or volume of water has more effect than the amount of pressure. The minimum flow should be 4 to 6 gallons per minute. Usually, the compressor should develop from 400 to 800 pounds per square inch (psi) water pressure for the most effective washing. It is important to keep the water stream moving to avoid damaging the wall.

Pressure washing can be used in combination with various cleaning compounds. Training and practice are necessary to properly control the mix of chemicals, pressure, and spray pattern.

6.3.2 Steam Cleaning

Steam cleaning (*Figure 53*) is another method of cleaning masonry. The steam-cleaning equipment is similar to the pressure washer, with the addition of a boiler to heat the water. Steam cleaning requires less water and is used extensively on interiors and ornately carved stonework. Steam can be used in combination with chemicals to remove applied coatings, such as paint, from masonry.

Figure 53 ◆ Typical steam-cleaning operation.

Steam cleaning is dangerous because the steam is not only scalding but under pressure. The steam also obscures the vision of the equipment operator. Steam cleaning is a highly specialized field. This work is usually done by a trained operator.

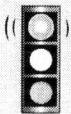

> **WARNING!**
> Steam can cause severe burns. Read the manufacturer's operating manual before using steam-cleaning equipment. Make sure that you know how to use all pressure-release valves and safety switches. Wear safety glasses and other appropriate personal protective equipment.

6.3.3 Sandblasting

Sandblasting is the oldest method of pressurized cleaning. It has the most capability of damaging or scarring the brick face. Sandblasting can deface brick, so it is best done by a trained operator.

Sandblasting employs abrasives, such as wet or dry grit, round or sharp-grained sand, crushed nut shells, rice hulls, egg shells, silica flour, ground corncobs, and other softer abrasives.

7.0.0 ◆ LIFTING EQUIPMENT

Once the mason's work reaches higher than 4', it is not efficient to work standing on the ground. Platforms or scaffolding lifts the mason and masonry units. An above-grade masonry work station usually has a mortar pan or board, the mason's tools, and a stack of masonry units. A typical raised work station is shown in *Figure 54*.

Typically, masonry units arrive at the job site bound or bundled and palletized. Each bundled cube may contain 500 standard bricks, or 90 standard blocks, depending on the manufacturer.

Figure 54 ◆ Typical raised work station.

Depending on the job, different types of equipment will move the cubes to the work station. The following sections describe lifting equipment for material handling.

> **WARNING!**
> Material handling procedures are some of the most hazardous activities on the job site. Stay clear of moving equipment and material in transport.

The following are some general rules for moving materials safely:

- Establish clear pathways for materials movement.
- Use a consistent set of signals to alert workers to materials movement.
- Do not ride on materials as they are moved.
- Stay out of the area between the moving materials and any wall or heavy equipment.

7.1.0 Mounted Hoists

A pulley system, or block and tackle, is the oldest aid to moving materials. However, simple pulley systems do not have safety features to control heavy loads. Adding a motor and supports to a simple pulley system allows the mason to safely move materials. This system is commonly known as a hoist. The hoist can be mounted on a scaffold as shown in *Figure 55*.

A motorized hoist can also be attached to a ladder to lift materials (*Figure 56*). This small hoist has a gasoline or electric engine, a pulley system, a take-up reel for lifting cable, a ladder, a lifting trolley, and hand controls. The combination is lightweight enough for one person to set up.

The typical ladder hoist can lift 400 pounds to a height of 16' to 40'. A plywood board can be put over the steel trolley for raising non-palletized materials. Some hoists come with gravel or mortar hoppers that fit on the trolley. The trolley is raised by the pulley and uses the sides of the ladder as rails.

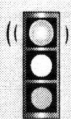

> **WARNING!**
> Every hoist and scaffold must be marked with a capacity rating. A hoist can fail due to overloading. This will cause serious injury and damage. Always check the rated capacity of a hoist before use.

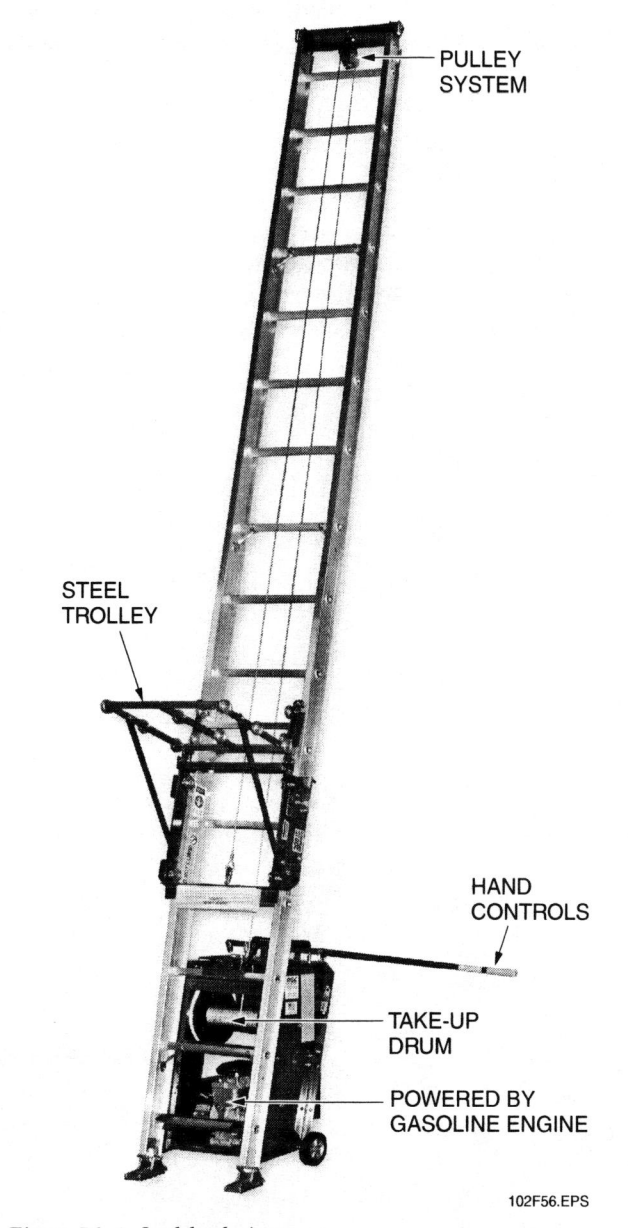

Figure 55 ◆ Hoist.

Figure 56 ◆ Ladder hoist.

MASONRY TOOLS AND EQUIPMENT 2.27

7.2.0 Portable Materials Hoists

A portable materials hoist is used to lift materials up to a mason on a scaffold. A materials hoist is also called a brick hoist (*Figure 57*). The materials hoist includes a lift platform, lift cabling, and a gasoline, diesel, or electric motor. There is usually a pulley system as well. The lift platform may have a cage around it. Materials hoists can lift from 1,000 to 5,000 pounds over a vertical distance of up to 300'.

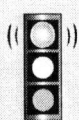

> **WARNING!**
> Do not overload a hoist. Hoist failure can cause serious injury and property damage.

Hoists can be mounted on wheels and towed to the job site. Portable hoists can also be mounted on a truck bed. The larger hoists are usually not portable but are attached to the side of the structure being built.

Materials hoists are not for lifting people. Personnel hoists have guardrails, doors, safety brakes, and hand controls in addition to the features of the materials hoist. Personnel hoists can be used for materials, but material hoists cannot be used for personnel. A personnel hoist is also known as a man lift.

> **WARNING!**
> Never ride on a materials hoist. Do not use the materials hoist as a work platform. A materials hoist has no safety features or brakes.

7.3.0 Hydraulic Lift Materials Truck

The hydraulic lift materials truck has a hydraulic boom arm for unloading masonry and other materials. The truck operator lowers stabilizing arms to prevent the truck from overturning. The hydraulic lift arm attaches to the load and swings out to unload each cube of masonry. If the masonry is not palletized, the cubes have openings in the bottom to accommodate the lift arm attachment prongs.

The hydraulic lift materials truck generally is not used for lifting materials above ground level. It is used to stage materials close to the work site or into a stockpile.

7.4.0 Cranes and Derricks

Larger and more versatile versions of the boom arm mechanism can be hydraulic or operate by cabling. This class of equipment includes cranes and derricks. Some of these use gasoline-powered motorized cables with hooks for raising and lowering heavy weights. They can also move over horizontal distances. The cabling runs through pulleys and a boom arm that also moves. There are several different types of cranes used to move masonry units on large construction projects:

- Tower cranes stand alongside, or in the middle of, the building under construction.
- Mobile cranes are mounted on crawler tracks or truck beds and move around the job site.
- Conventional cranes and derricks are placed away from the building depending on the length of the boom arm.

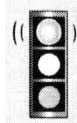

> **WARNING!**
> Cranes are a necessary part of large-scale construction projects and generate hazards. In high-rise construction, materials movement by crane poses the greatest safety hazard to workers on the job.

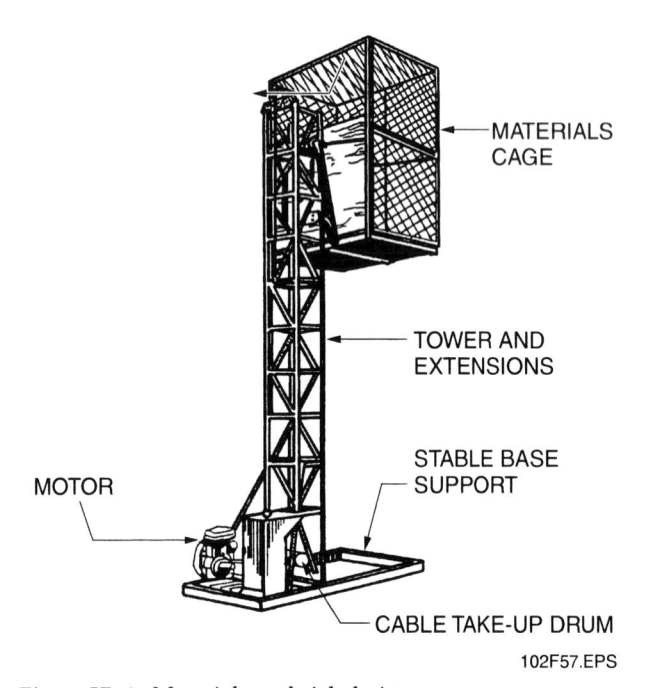

Figure 57 ◆ Materials, or brick, hoist.

Work safety calls for the following:

- Establishing clear pathways for materials movement
- Using a consistent set of signals to alert workers to movement
- Assigning individual responsibilities to each worker during materials movement

7.4.1 Tower Cranes

Tower cranes, as shown in *Figure 58*, have a vertical tower and horizontal swinging boom or jib. The boom can move in a full-circle horizontal swing unobstructed by the tower. The crane has a counterweight on the end of the boom to balance the load, and a cab on the tower for the operator.

Tower cranes are the most popular because of their range of movement. They are self-balancing, so they do not need to be tied to the structure. These cranes can be mounted in the middle of a high-rise building under construction and moved from one level to the next as the structure rises.

7.4.2 Mobile Cranes

Mobile cranes lift with swinging booms that are raised and lowered by a boom hoist cable, as shown in *Figure 59*. This particular mobile crane has a hydraulic boom arm that extends its reach. The boom and operator's cab can be mounted on a truck bed or crawler tracks. The counterweight is usually at the bottom end of the boom or under the operator's cab. The main advantage of this crane is that it can move around the site.

7.5.0 Forklifts, Pallet Jacks, and Buggies

The most common method for lifting heavy materials on a job site is a forklift (*Figure 60*). Forklift tractors, or forklifts, have hydraulic lifting arms that move up and down. They are called forklifts because of the fork shape of the prongs on the lifting arm. The prongs fit into the openings in a pallet, a large mortar pan, or the bottom of a cube of masonry. These mobile lifters are used to move masonry from a stockpile to a work station. If less

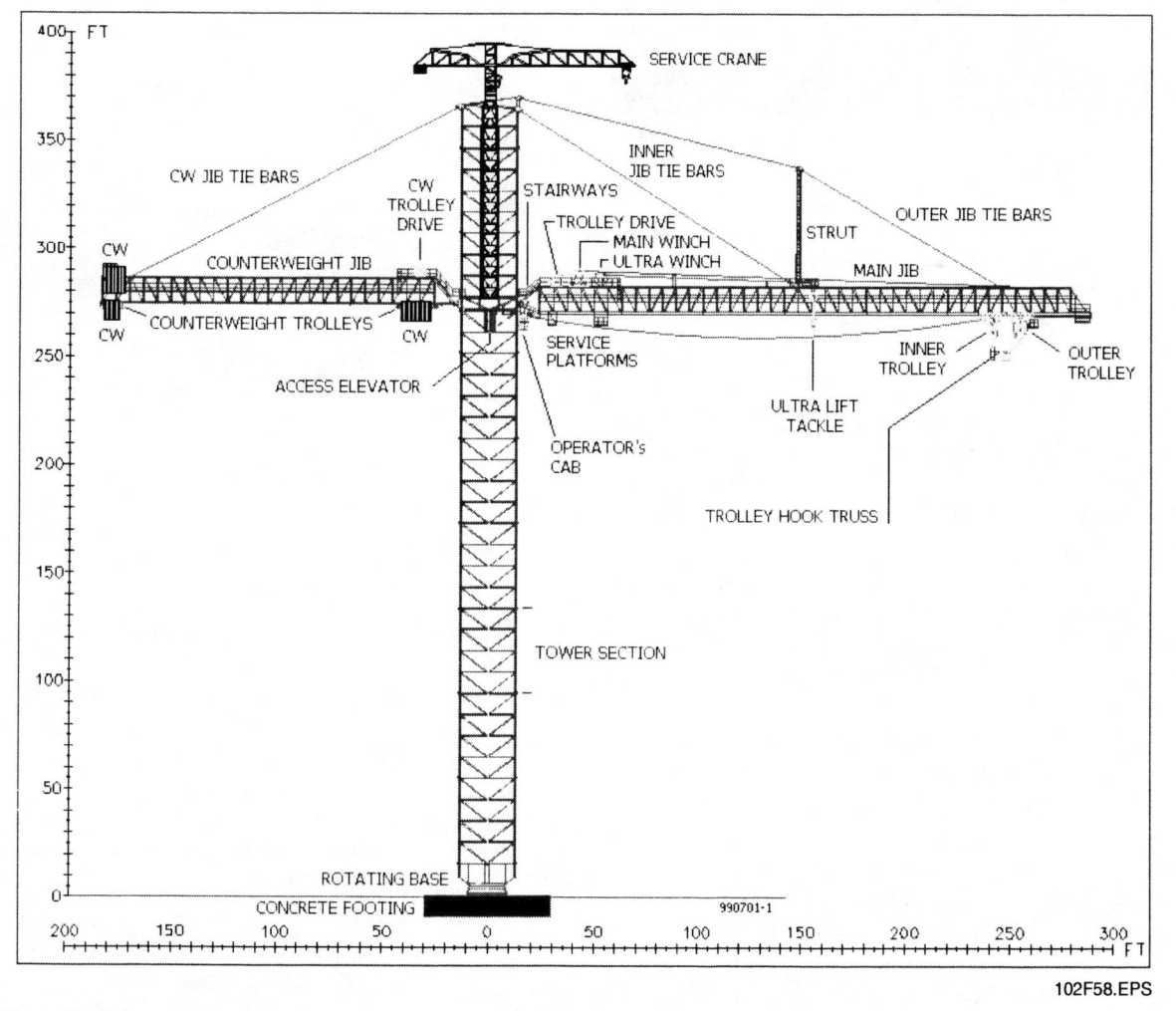

Figure 58 ♦ Tower crane.

MASONRY TOOLS AND EQUIPMENT   **2.29**

Figure 59 ◆ Mobile crane performing a lift.

Figure 60 ◆ Mason's reach-type forklift.

Figure 61 ◆ Motorized buggy.

than a full cube of masonry is needed, the materials must be stacked on a pallet. Make sure that the materials are secure before they are raised.

The forklift consists of a gasoline, propane-powered, or electric engine with a seat for the operator and a hydraulic powered lift. Because of their size, they are not used for raising cubes above the first floor. Because most forklifts can only lift materials straight up, the truck must be moved to maneuver the load.

Mason's forklifts or pallet jacks are specialized for masonry handling. They carry material from a stockpile to a work station and can lift as high as one scaffold height. The pallet jack has a gasoline, diesel, or electric engine, with a hydraulic forklift. There is no seat for the operator, who stands. The load capacity of a typical pallet jack varies from one-half to one cube of masonry.

Motorized buggies carry mortar and masonry units around the job site. As shown in *Figure 61*, they are large, motorized wheelbarrows with two or four wheels in front and one wheel in back. The bin dumps its load mechanically. There is a shelf in the back for the operator to stand on as the buggy moves.

7.6.0 Conveyors

Conveyors (*Figure 62*) move materials from ground level to medium heights. They can be linked by swivels to carry materials over long distances. At an angle of 60 degrees, a typical conveyor can carry a load of 500 pounds at speeds of about 72 feet per minute; this material is moving fast.

> **WARNING!**
> Keep clear of the discharge end of a conveyor belt. Check that the materials discharge area is clear and secure before the conveyor starts.

2.30   MASONRY LEVEL ONE — TRAINEE MODULE 28102-04

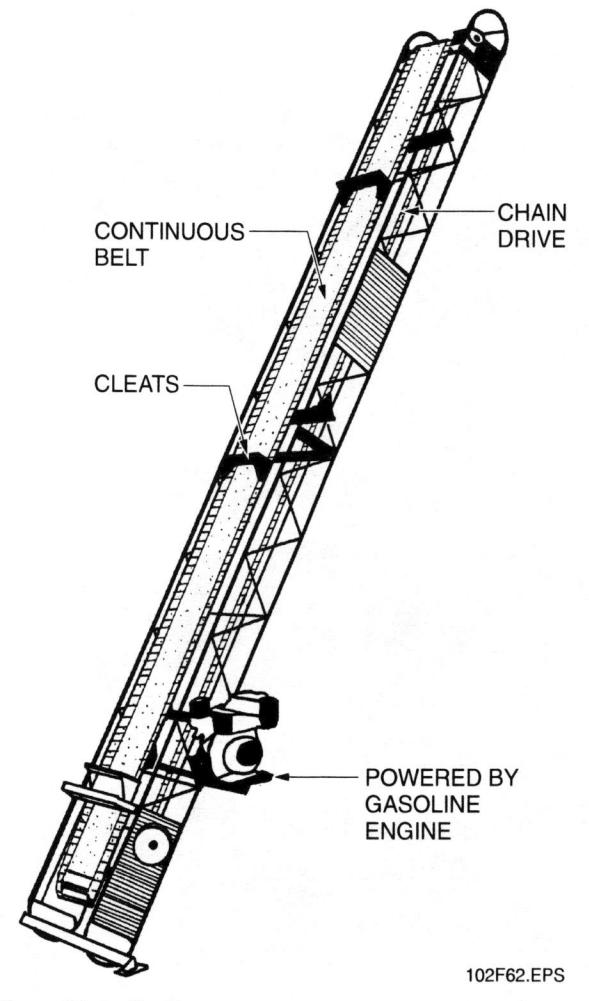

Figure 62 ♦ Conveyor.

Each span of a conveyor is called a flight. Mechanically, a conveyor flight has a continuous belt, moved by a chain or hydraulic drive, and powered by a gasoline or electric engine. The belt is typically about 16" wide. Specialized masonry conveyor belts have cleats, as shown in *Figure 62*, to keep the masonry from slipping down the belt. Other types of belts have troughs or bins to carry mortar or mortar ingredients.

8.0.0 ♦ SCAFFOLDING

Scaffolding is any elevated, temporary work platform. Most masonry jobs require some type of scaffolding. As the working level rises above 4', masons cannot work as efficiently. It is cost- and time-effective to raise the mason and the materials close to the work. Masons sometimes erect the scaffolding they work on, to make sure it is safe and stable.

All scaffolding must be erected, moved, and disassembled under the supervision of a competent person. All scaffolding must be assembled according to federal safety regulations.

There are three primary types of scaffolding in common use:

- Supported scaffolding, which has one or more platforms supported by rigid loadbearing members, such as frames, built on the ground
- Suspended scaffolding, which is composed of one or more platforms suspended by ropes anchored to the roof
- Man-lifts, personnel hoists, and other machinery

8.1.0 Tubular Steel Sectional Scaffolding

Tubular steel sectional scaffolding is the most common type of scaffolding. It is strong, lightweight, durable, and easy to erect. The steel frame sections come in several heights, with a typical 4' or 5' width. As shown in *Figure 63*, two commonly used scaffold frame types are the ladder-braced and the walk-through frames. The walk-through allows the mason an easy passage along the scaffold planks.

As tubular steel sectional scaffolding is set up, each frame is connected to the next frame at ground level with a horizontal diagonal brace.

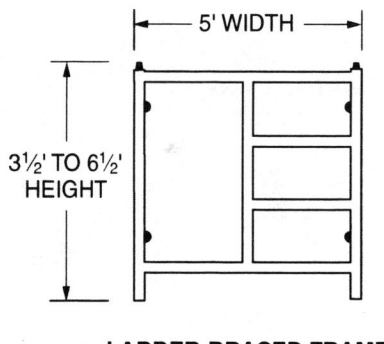

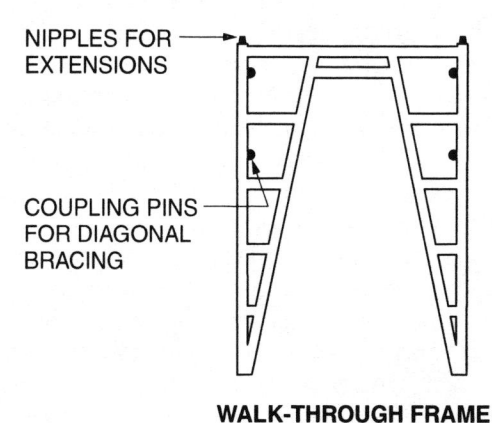

Figure 63 ♦ Scaffold frame types.

MASONRY TOOLS AND EQUIPMENT   2.31

Building Blocks

Fall-Protection Equipment

Fall-protection equipment is required on many types of scaffolding. Ask your supervisor if fall-protection equipment is needed. A body harness and line, as shown here, will protect you if the scaffolding fails.

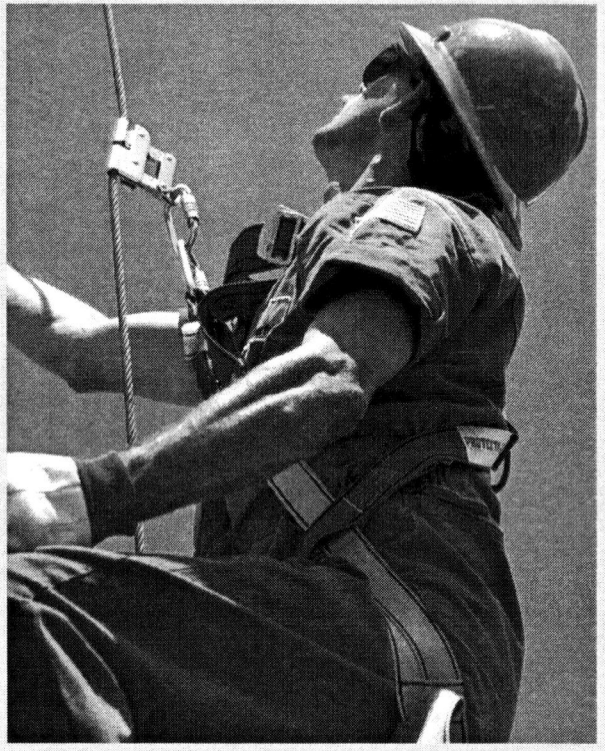

In addition, frames are connected at upper levels by overlapping diagonal braces. These overlapping braces must be secured to the frames by a wing nut or bolt or by a lock device that fits over the frame coupling pin.

To connect each section of scaffold framing vertically, a steel pin, or nipple, is put in the hollow tubing at the top of the lower section. The bottom of the upper section fits over this pin and is secured with a slip bolt, as shown in *Figure 64*. Each level must be braced as it is installed. All connections must be secured as they are made.

After bracing, the platform flooring or decking is put in place. Regular dimensional lumber planking can be used for the platforms on steel **trestles**, or hook-mount scaffold planks can be used (*Figure 65*).

Toeboards keep materials from sliding off the platform. Guardrails, to protect masons, must be placed and secured before work from the scaffold can start. Under some conditions, roofboards can also be added. The finished product looks like *Figure 66*.

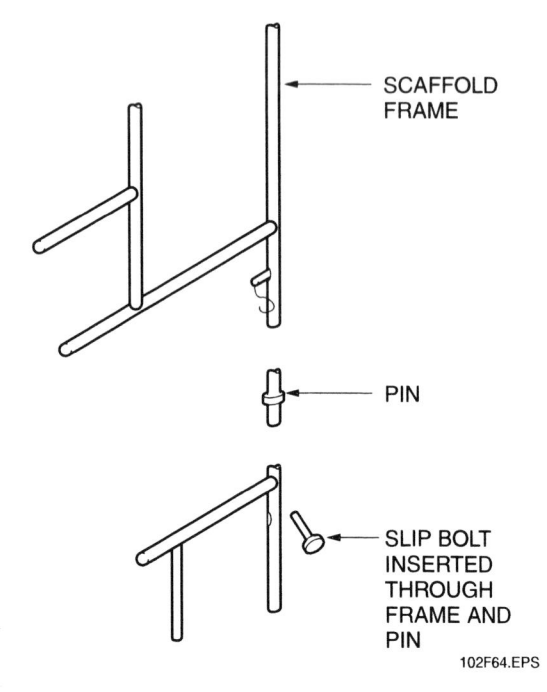

Figure 64 ♦ Connecting scaffold sections.

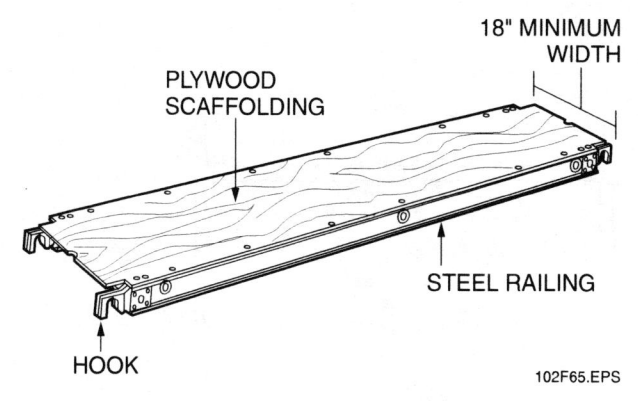

Figure 65 ◆ Hook-mounted scaffold plank.

Note that screw jacks have been put under the base plates. This is a good practice to level the framework after it is assembled. A sill board can be added for ground that is soft, damp, or liable to shift.

Tubular steel framing is very versatile and can be configured many ways with different accessories. Mason's side brackets, or extenders, extend the working area. Wheels are available and can be inserted into the bottom of framing members so that the structure becomes a rolling scaffold. Rolling scaffolds should only be used for pointing and striking joints, not for laying units. Even with wheel locks, they tend to move when heavily loaded with masonry units.

8.1.1 Putlog

Sometimes called a bridge, a putlog is a wooden beam supporting scaffolding. Putlogs are used where there is no solid base on which to set the scaffold frame. Because of grade conditions at the job site, one or more putlogs may be needed to stabilize scaffolding. Putlogs are set with their greater thickness vertical and must extend at least 3" past the edge of the scaffold frame.

If there is not enough space to rest the scaffold, holes are left in the first few courses of a wall. The putlog is inserted in the hole with the other end resting on solid ground (*Figure 67*) or on a constructed base.

The scaffold frame is erected on the putlogs, which provide a solid footing. The scaffolding is braced as usual, and the base plates may be fastened to the putlogs. After the scaffolding is disassembled, the putlogs are withdrawn from the holes, and the holes are filled with masonry units.

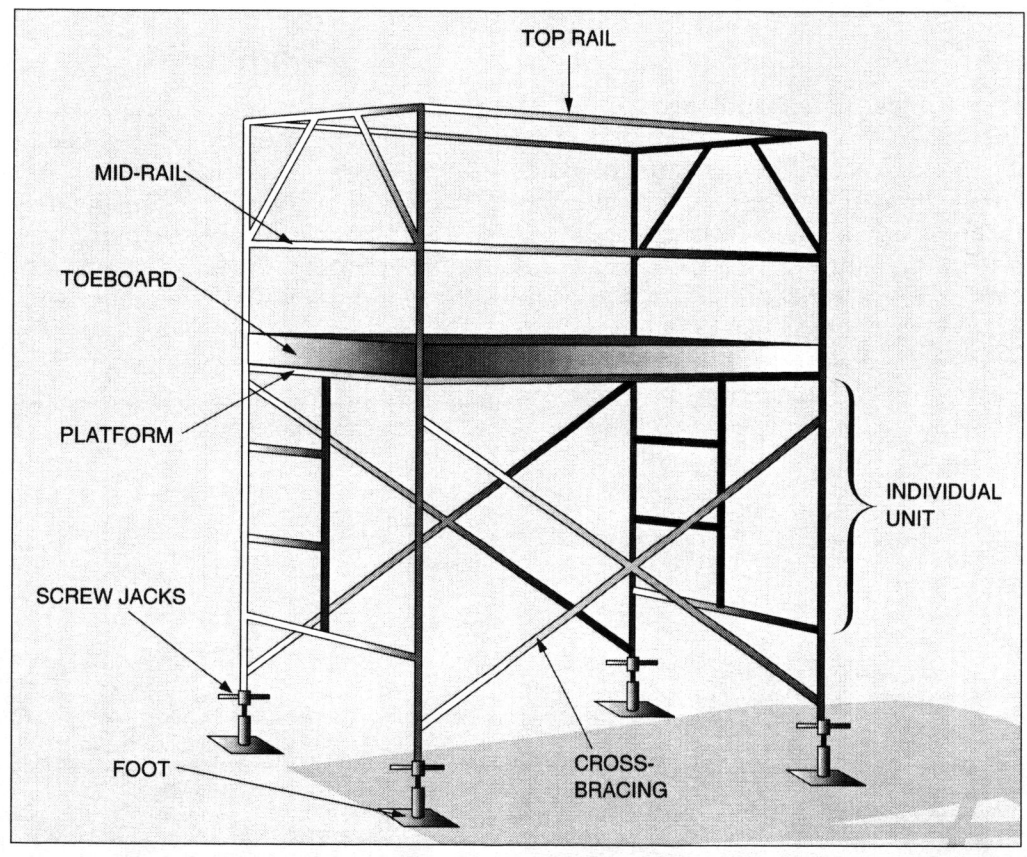

Figure 66 ◆ Completed tubular steel scaffolding.

MASONRY TOOLS AND EQUIPMENT

2.33

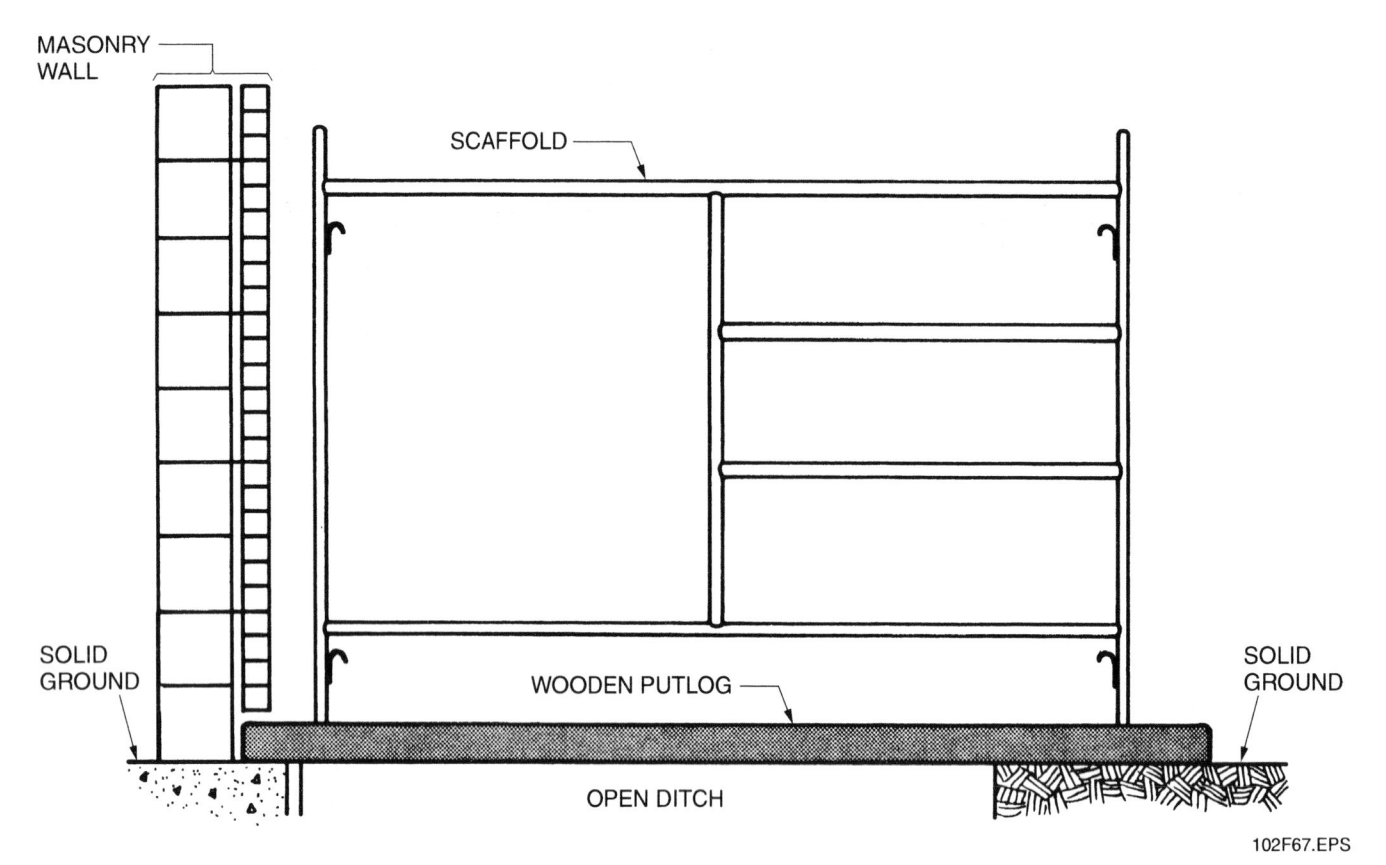

Figure 67 ◆ Putlog at base of scaffold.

8.1.2 Steel Tower

Adjustable steel tower, or self-climbing, scaffolding (*Figure 68*) is built of vertical members braced or tied to the structure itself. The ties must be positioned 30' apart or less. The scaffolding framing is attached to the vertical members. Metal planking is used for a working platform, which has guardrail and toeboard safety features, and a winch mechanism. The working platform can be raised as the work progresses. Because it requires attachment to the building, tower scaffolding must be put up by specialized workers.

8.2.0 Swing Stage

Swing stage scaffolding (*Figure 69*) is used on multi-story buildings. Steel beams are fastened to the roof, and steel cables are dropped to the ground. A steel cage is suspended from the cables with hangers, and a planking floor is added over the frame. Guardrails, toeboards, and an overhead canopy of plywood or metal mesh complete the cage. A winch moves the cage up and down, so it is always at the level of the work.

Unlike the other types of scaffolding, the swing stage is usually erected by specialists. This type of

Figure 68 ◆ Adjustable steel tower scaffolding.

Figure 69 ◆ Swing stage scaffolding.

scaffolding is also the safest because the mason is completely enclosed by steel, and there are redundant backup systems on the cabling and brakes.

8.3.0 Hydraulic Personnel Lift

On some jobs, a hydraulic personnel lift is used as scaffolding. As discussed earlier, a hydraulic materials lift does not have safety features needed for lifting people. Hydraulic personnel lifts *(Figure 70)* have guardrails, toeboards, hand controls, and stabilizer pads. They have backup safety systems, free-fall brakes, and safety ladders.

> **WARNING!**
> Do not use a materials lift as a scaffold.

8.4.0 Scaffold Safety

Masons spend a great deal of time working from scaffolding. The chance of an accident occurring because of poor work practices is greater on scaffolding than at any other time or place. Tubular steel scaffolding is less likely to give way than the old-fashioned wood scaffolding but requires more care in its assembly and use.

The federal safety rules focusing on worker safety are summarized here. These rules are only reminders for the safe assembly and use of scaffolding.

- Platforms on all working levels must be fully decked between the front uprights and the guardrail supports.
- The space between planks and the platform and uprights can be no more than 1" wide.
- Platforms and walkways must be at least 18" wide; the ladder jack, top plate bracket, and pump jack scaffolds must be at least 12" wide.
- For every 4' a scaffold is high, it must be at least 1' wide. If it is not, it must be protected from tipping by tying, bracing, or guying per the OSHA rules.
- Supported scaffolding must sit on base plates and mud sills or other steady foundations.
- The inboard ends of suspension scaffolding outriggers must be stabilized by bolts or other direct connections to the floor or roof deck, or they must be stabilized by counterweights.
- Access to and between scaffold platforms more than 2' above or below the point of access must be made by the following:
  - Portable ladders, hook-on ladders, attachable ladders, scaffold stairways, stairway-type ladders (such as ladder stands), ramps, walkways, integral prefabricated scaffold access, or equivalent means; or
  - Direct access from another scaffold, structure, personnel hoist, or similar surface
- Never use cross-braces to gain access to a scaffold platform.

The OSHA rules for scaffolding systems require that all workers who use or work around scaffolding must be trained to recognize the hazards of these systems, including proper erection and use. All employees who work on scaffolding must be trained by a person qualified in scaffolding hazards and uses, so they understand the procedures to control or minimize any hazards. All personnel who erect, take down, move, operate, maintain, repair, or inspect scaffolding must be trained by a competent person.

Figure 70 ♦ Hydraulic personnel lift.

MASONRY TOOLS AND EQUIPMENT

Summary

Masons use a variety of hand tools on the job. It is up to you to learn to use and maintain these tools properly. Masons handle course layout, alignment, and measuring tools constantly during a workday to check that a masonry structure is plumb. Measures and measuring tools include levels, squares, mason's line and fasteners, corner poles, chalkline, and two types of mason's spacing rules. Mortar mixing equipment includes water buckets and barrels, hoes, shovels, and mortar boards, boxes, and pans. Mortar boxes, boards, and pans must be cleaned after use, as should hand tools and power equipment.

Power tools and equipment make the mason's job easier. Power mixers quickly and thoroughly mix mortar. Power saws and hydraulic splitters are used to cut masonry units. Material needs to be lifted to the work location by hoists, cranes, forklifts, pallet jacks, and conveyors. Masons need to be lifted to the work location by scaffolding. OSHA has strict rules for scaffold safety as it is a leading cause of injury. A competent person must be available to supervise scaffold safety.

The mason's job is one that takes great skill and practice. Learning to properly use and maintain your tools and equipment is the first step toward becoming a successful mason.

Review Questions

1. The wide end of a trowel closet to the handle is known as the _____.
   a. toe
   b. heel
   c. shank
   d. ferrule

2. The most commonly used, everyday mason's hammer is known as a _____.
   a. mason's mash
   b. maul
   c. brick hammer
   d. tile hammer

3. The brick set and the bolster chisels have blades _____ wide.
   a. 2" to 3"
   b. 4" to 5"
   c. 6" to 7"
   d. 7" to 8"

4. Slickers, rakers, and sled runners are used to _____.
   a. finish or point the surface of mortar joints
   b. chip mortar from between masonry units
   c. cut masonry units
   d. prepare the ground before masonry units are placed

5. Acid brushes are made of stiff plastic and have _____.
   a. scrapers on the end
   b. removable handles
   c. long handles
   d. skatewheels

6. A plumb line is _____ with relation to the Earth's surface.
   a. horizontal
   b. vertical
   c. diagonal
   d. parallel

7. Masons prefer a folding rule because it _____.
   a. is more accurate
   b. can bend around corners
   c. can be used by one person
   d. is lightweight

8. To quickly mark a 45-degree angle you should use a _____.
   a. folding rule
   b. course spacing rule
   c. combination square
   d. torpedo level

9. Trigs and blocks _____.
   a. hold bevels taut and in place
   b. keep mortar in place
   c. are specialized masonry units
   d. keep mason's line taut and in place

10. A line pin holds the line secure but _____.
    a. is easily knocked off
    b. leaves a hole in the mortar
    c. must have a corner to be snugged against
    d. is held in place by line tension

11. A hawk or hod is used by the mason to _____.
    a. hold mortar
    b. finish joints
    c. face masonry units
    d. measure courses

12. Hydraulic splitters _____ as a handheld power saw.
    a. do not have as much cutting power
    b. do not create as much dust
    c. cannot cut as large of a block
    d. are not as accurate

13. The gentlest way to clean masonry surfaces is to use a _____.
    a. tuckpointer
    b. steam cleaner
    c. pressure washer
    d. sandblaster

14. A materials hoist has all of the following *except* _____.
    a. cages and cabling
    b. safety brakes and guardrails
    c. platforms and lift engines
    d. truck bed mounts

15. Cubes are quickly unloaded by a _____.
    a. hydraulic splitter
    b. materials hoist
    c. hydraulic lift materials truck
    d. barrow

16. On a tower crane, the boom can move _____.
    a. in a full circle horizontal swing
    b. as long as it is tied to the structure
    c. on crawler tracks
    d. on wheels around the job site

17. A pallet jack is a type of _____.
    a. forklift
    b. buggy
    c. crane
    d. conveyor

18. Putlogs are used to _____.
    a. keep materials from sliding off a scaffold
    b. create a solid base for a scaffold
    c. hold the mason's line taut
    d. transfer mortar to a hod

19. The safest type of scaffolding is the _____, because the mason is completely enclosed by steel, and there are redundant backup systems on the cabling and brakes.
    a. steel tower
    b. swing stage
    c. tubular steel sectional
    d. personnel lift

20. For every _____ a scaffold is high, it must be at least 1' wide.
    a. 2'
    b. 4'
    c. 5'
    d. 6'

MASONRY TOOLS AND EQUIPMENT

PROFILE IN SUCCESS

Garrett Hood, Masonry Foreman
McGee Brothers Masonry
Charlotte, NC

Before he reached the age of 21, Garrett Hood had already realized greater success than many achieve in their lifetimes. He won a national bricklaying competition while still in high school. He went on to become the foreman of a masonry crew at the age of 20, earning an income that many mid-career workers in the U.S would envy. Garrett Hood is an excellent example of what can be achieved simply by determination and a desire to succeed.

How did you get started in the masonry trade?
My high school had a masonry program and I decided to try it. I really liked it. While I was in the program, a friend recommended me to McGee Brothers, which is the biggest masonry contractor around. I started working summers for McGee Brothers while I was still in high school. I worked briefly as a laborer but quickly moved up to laying brick. After I graduated, I went to work for McGee Brothers full time.

Did you think about going to college?
Yes. In fact, I was accepted in the mechanical engineering program at North Carolina State, but I decided that I liked bricklaying so much that I wanted to stick with it.

Tell us about your experience in national competitions.
I won the SkillsUSA-VICA competition at the state level in 2001 and 2002, which qualified me to compete at the national competition in Kansas City. I won the national championship in 2001 and came in second in the nationals in 2002. I received about $2,000 in cash and tools combined for winning the championship.

What do you do in your current job?
I'm foreman of a seven-man masonry crew. I lay out the job, supervise the crew, and make sure materials and equipment are available when we need them. I'm also responsible for making sure that the job is done on schedule and that the quality of the job is up to company standards.

What would you advise people just entering the trade?
First of all, get some training. Masonry isn't easy, and you really need to develop your skills and knowledge of the trade. Also, don't do it if you don't really like it. Masonry work is hard work and it's often dirty work. You have to really care about it to do it well, and you have to be determined to stick with it. Masons have high standards. You won't succeed if you don't go to work every day committed to doing the best job you can.

GLOSSARY

Trade Terms Introduced in This Module

Bed joint: A horizontal joint between two masonry units.

Corner pole: Any type of post braced into a plumb position so that a line can be fastened to it. Also called a deadman.

Deadman: See *corner pole.*

Lead: The two corners of a structural unit or wall, built first and used as a position marker and measuring guide for the entire wall.

Parge: A thin coat of mortar or grout on the outside surface of a wall. Parging prepares a masonry surface for attaching veneer or tile, or parging can waterproof the back of a masonry wall.

Pointing: Troweling mortar or a mortar-repairing material, such as epoxy, into a joint after masonry is laid.

Retempering: Adding water to mortar to replace evaporated moisture and restore proper consistency. Any retempering must be done within the first two hours after mixing, as mortar begins to harden after 2½ hours.

Skewback: A sloping surface against which the end of an arch rests; may be brick cut on an angle.

Temper: To remix mortar by adding water to make it more workable.

Trestle: A system of scaffolding with diagonal legs; a split-leg support for a system of scaffolding.

REFERENCES

Additional Resources

This module is intended to be a thorough resource for task training. The following reference works are suggested for further study. These are optional materials for continued education rather than for task training.

Bricklaying: Brick and Block Masonry. Reston, VA: Brick Industry Association.

Concrete Masonry Handbook. Skokie, IL: Portland Cement Association.

Masonry Construction. David L. Hunter, Sr., Upper Saddle River, NJ: Prentice-Hall.

ACKNOWLEDGMENTS

Figure Credits

Topaz Publications, Inc.	102SA01, 102F37A, 102F38, 102F48, 102F54, 102F68
Associated General Contractors	102F01, 102F57, 102F62–102F65
Bon Tool Company	102F02–102F22, 102SA02, 102F24, 102F26, 102F27, 102SA04, 102F31, 102F33, 102F36, 102F37B, 102F39, 102SA05, 102F41, 102F42, 102F46, 102F47, 102F56, 102F61
The Stanley Works	102SA03, 102F28, 102F34
Portland Cement Association	102F29, 102F30, 102F32
Granite City Tool Co.	102F43, 102F45, 102F50, 102F52
Makita USA, Inc.	102F44
EZ Grout	102F51
Hotsy	102F53
Beta Max Hoists, Inc.	102F55
Towercrane.com	102F58
National Crane	102F59
CareLift Equipment Limited	102F60
Stone Mountain Access	102F69
Protecta International	102SA07
Unidex, Inc.	102F70

CONTREN® LEARNING SERIES — USER UPDATES

The NCCER makes every effort to keep these textbooks up-to-date and free of technical errors. We appreciate your help in this process. If you have an idea for improving this textbook, or if you find an error, a typographical mistake, or an inaccuracy in NCCER's Contren® textbooks, please write us, using this form or a photocopy. Be sure to include the exact module number, page number, a detailed description, and the correction, if applicable. Your input will be brought to the attention of the Technical Review Committee. Thank you for your assistance.

Instructors – If you found that additional materials were necessary in order to teach this module effectively, please let us know so that we may include them in the Equipment/Materials list in the Instructor's Guide.

Write: Product Development and Revision
National Center for Construction Education and Research
P.O. Box 141104, Gainesville, FL 32614-1104

Fax: 352-334-0932

E-mail: curriculum@nccer.org

Craft _____ Module Name _____

Copyright Date _____ Module Number _____ Page Number(s) _____

Description _____

_____

_____

_____

_____

(Optional) Correction _____

_____

_____

(Optional) Your Name and Address _____

_____

_____

_____

BUILDING BLOCKS

Brick MASONRY

Brick can be used to create decorative barriers such as the entry to this gated community.

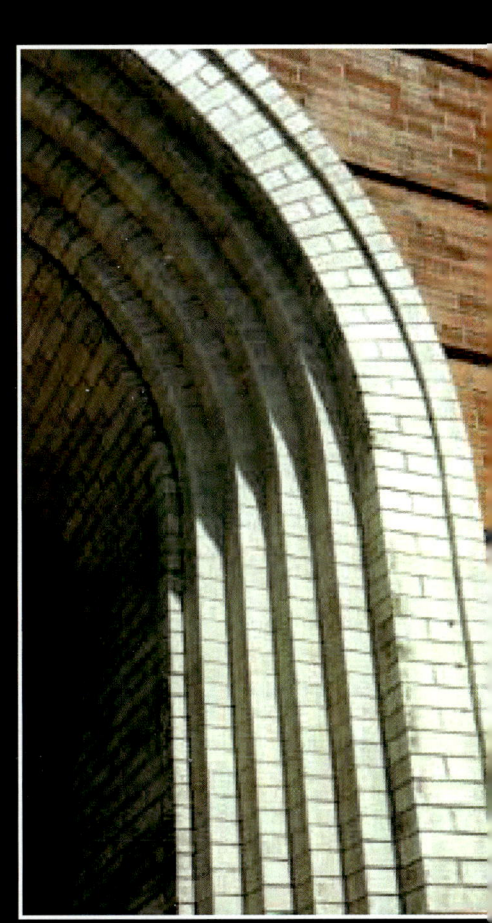

Since early Roman times, decorative and structural arches have been a hallmark of the mason's craft.

In many parts of the country, brick is the building material of choice for churches, schools, and commercial structures.

Brick gives this home a sense of strength and permanance.

1

Basic Bricklaying Technique

1. Once a straight line is established, the mortar bed is laid out and furrowed. Furrowing is not always permitted.

2. The end of the brick must be carefully buttered.

3. The bricks are tapped with the trowel handle to level the course.

4. The mason is careful to never touch the line when placing a brick.

5. The wall must be plumbed as each course is laid.

6. A closure brick is buttered on each end and carefully slid into place.

7. The head joints are struck with a short jointer.

8. A sled jointer is used to strike bed joints after the head joints are tooled. Timing is important.

Laying Concrete Block

1. A wall is placed by laying corner leads, starting with a good mortar bed.

2. Mortar is spread on the face shell to provide a bed for the next course.

3. Block masons usually butter the ears of the three blocks at a time and lay the mortar bed for three blocks.

4. Alignment is critical, just as it is with a brick wall.

5. Special blocks can be used to form corners.

6. Special blocks can also be used when a control joint is needed.

7. Ties are an important part of a cavity wall.

8. Mortar burrs can be removed with a trowel when they are mostly dry.

BUILDING BLOCKS

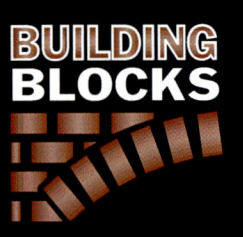

Concrete MASONRY

Freeway soundbarrier wall near San Francisco. The design reflects the San Francisco skyline.

Wall made with split-face and polished-face masonry units and a few bright red glazed units. Math & Sciences building at Oakwood school in North Hollywood, California.

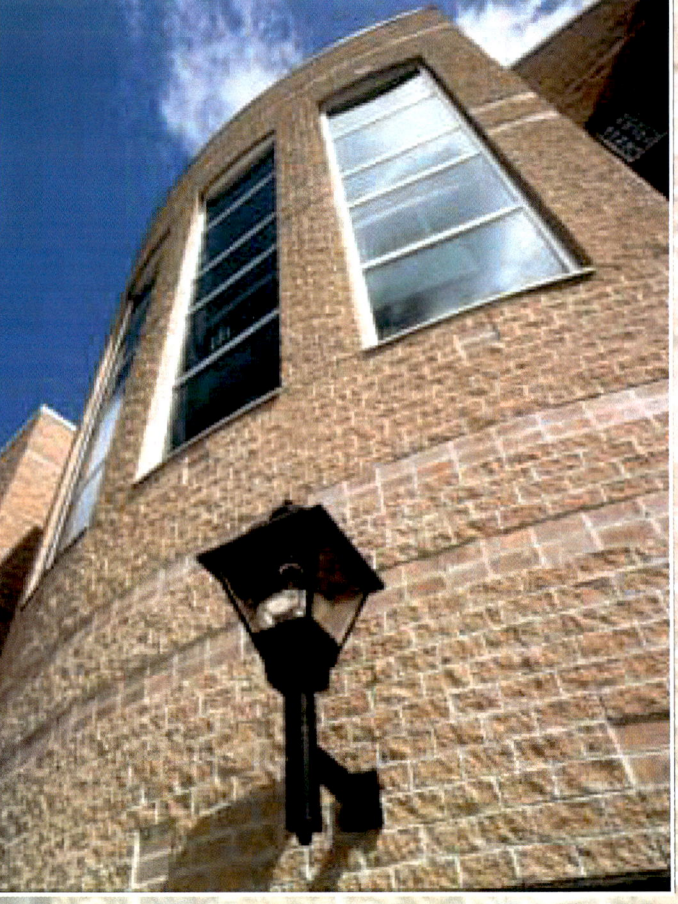

Curved wall with lamps at Limoilou College, Charlesbourg Campus Montreál, Quebec.

This dining room incorporates colored block as a design feature. Residence near San Diego.

4

Module 28103-04

Measurements, Drawings, and Specifications

COURSE MAP

This course map shows all of the modules in the first level of the *Masonry* curriculum. The suggested training order begins at the bottom and proceeds up. Skill levels increase as you advance on the course map. The local Training Program Sponsor may adjust the training order.

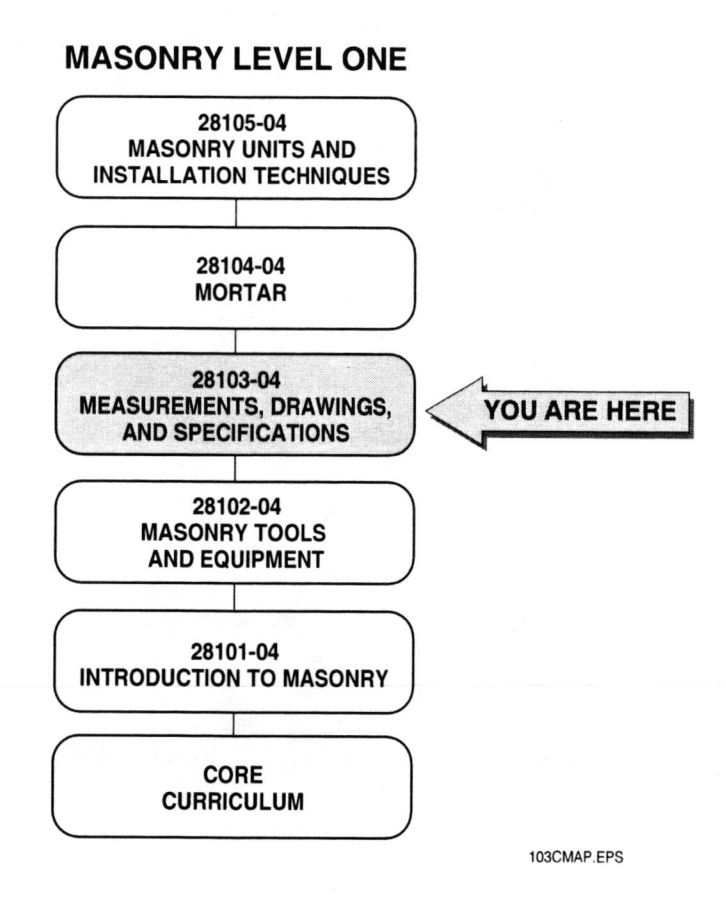

MASONRY LEVEL ONE

- 28105-04 MASONRY UNITS AND INSTALLATION TECHNIQUES
- 28104-04 MORTAR
- 28103-04 MEASUREMENTS, DRAWINGS, AND SPECIFICATIONS ← YOU ARE HERE
- 28102-04 MASONRY TOOLS AND EQUIPMENT
- 28101-04 INTRODUCTION TO MASONRY
- CORE CURRICULUM

Copyright © 2004 National Center for Construction Education and Research, Gainesville, FL 32614-1104. All rights reserved. No part of this work may be reproduced in any form or by any means, including photocopying, without written permission of the publisher.

MODULE 28103-04 CONTENTS

1.0.0 **INTRODUCTION** 3.1
    1.1.0 Masonry Math 3.1
    1.2.0 Reading Drawings 3.1
    1.3.0 Applying Specifications 3.2

2.0.0 **MASONRY MATH** 3.2
    2.1.0 Denominate Numbers 3.2
    2.1.1 Denominate Addition 3.2
    2.1.2 Denominate Subtraction 3.4
    2.1.3 Other Denominate Measures 3.5
    2.2.0 Fractions 3.6
    2.2.1 Finding the Lowest Common Denominator 3.6
    2.2.2 Adding Fractions 3.6
    2.2.3 Subtracting Fractions 3.7
    2.2.4 Multiplying Fractions 3.7
    2.2.5 Dividing Fractions 3.7
    2.3.0 Mason's Denominate Measures 3.7
    2.3.1 The Course System 3.7
    2.3.2 The Modular System 3.8
    2.4.0 Denominate Metric Measurements 3.11
    2.4.1 SI Units of Measure 3.11
    2.4.2 Converting SI Metric 3.12
    2.4.3 Conversion Examples 3.12
    2.5.0 Plane Figures and Area Measures 3.14
    2.5.1 Four-Sided Figures 3.14
    2.5.2 Three-Sided Figures 3.14
    2.5.3 One-Sided Figures 3.15
    2.5.4 Many-Sided Figures 3.16
    2.6.0 Solid Figures and Volumes 3.17
    2.7.0 Working with Right Triangles 3.18

3.0.0 **DRAWINGS** 3.19
    3.1.0 Understanding Drawings 3.19
    3.1.1 Lines as Symbols 3.19
    3.1.2 Symbols and Abbreviations 3.21
    3.1.3 Scales and Dimensions 3.21
    3.2.0 Residential Drawings 3.26
    3.2.1 Plot or Site Plans 3.27
    3.2.2 Floor Plans 3.27
    3.2.3 Foundation Plans 3.27
    3.2.4 Elevation Drawings 3.27
    3.2.5 Section Drawings 3.28
    3.2.6 Schedules 3.28

MODULE 28103-04 CONTENTS (CONTINUED)

   3.2.7 Specialty Plans 3.28
   3.2.8 Metric Drawings 3.28
 4.0.0 SPECIFICATIONS, STANDARDS, AND CODES 3.28
   4.1.0 Technical Specifications 3.29
   4.2.0 Standards 3.31
   4.3.0 Codes 3.32
   4.4.0 Inspection and Testing 3.32
SUMMARY 3.34
REVIEW QUESTIONS 3.34
PROFILE IN SUCCESS 3.36
GLOSSARY 3.37
APPENDIX A, Answers to Practice Exercises 3.38
APPENDIX B, Figures and Mathematical Formulas 3.40
APPENDIX C, ASTM Standards for Masonry Construction 3.41
REFERENCES & ACKNOWLEDGMENTS 3.43

Figures

Figure 1	Converting units	3.2
Figure 2	Borrowing units	3.4
Figure 3	Find missing dimensions	3.5
Figure 4	Brick spacing rule and oversize brick spacing rule	3.8
Figure 5	Reading the brick spacing rule	3.9
Figure 6	Reading the modular spacing rule	3.10
Figure 7	Common metric equivalents	3.12
Figure 8	Four-sided figures	3.14
Figure 9	Triangles named by sides	3.15
Figure 10	Triangles named by angles	3.15
Figure 11	Divide figures to find the area	3.15
Figure 12	Parts of a circle	3.15
Figure 13	Regular polygons	3.16
Figure 14	Wall with circular window	3.17
Figure 15	Volumes of solid figures	3.17
Figure 16	Conical, triangular, and spherical solids	3.17
Figure 17	Right triangle	3.18
Figure 18	Telephone pole with supporting cable	3.18
Figure 19	Lines used for construction drawings	3.19
Figure 20	Dimension line styles	3.20
Figure 21	Geometric drawing lines	3.21
Figure 22	Architectural material drawings	3.22
Figure 23	Electrical symbols	3.23
Figure 24	Plumbing symbols	3.24
Figure 25	Dimension lines for different wall types	3.26
Figure 26	Metric drawing	3.29
Figure 27	Standard FmHA/VA specification form	3.30
Figure 28	Specification contents	3.31
Figure 29	Building code requirements	3.33

Tables

Table 1	Common Measures	3.5
Table 2	Sizes of Bricks	3.9
Table 3	Common SI Prefixes	3.11
Table 4	Common Metric Measures	3.11
Table 5	U.S. Customary to SI Metric Conversions	3.13
Table 6	SI Metric to U.S. Customary Conversions	3.13
Table 7	Architectural Abbreviations	3.25

MODULE 28103-04

Measurements, Drawings, and Specifications

Objectives

When you have completed this module, you will be able to do the following:

1. Work with denominate numbers.
2. Read a mason's measure.
3. Convert measurements in the U.S. Customary (English) system into their metric equivalents.
4. Recognize, identify, and calculate areas, circumferences, and volumes of basic geometric shapes.
5. Identify the basic parts of a set of drawings.
6. Discuss the different types of specifications used in the building industry and the sections that pertain to masonry.

Recommended Prerequisites

Core Curriculum; Masonry Level One, Modules 28101-04 and 28102-04

Required Trainee Materials

1. Pencil and paper
2. Tape measure
3. Calculator

1.0.0 ♦ INTRODUCTION

This module covers the math tools you will need in your work as a mason. It reviews three basic skills that are essential to your success:

- Calculations for masons
- Reading plans and drawings
- Reading and meeting specifications

1.1.0 Masonry Math

Math skills are needed at every step of your work. From understanding a set of drawings to measuring and mixing mortar, math skills enable you to work quickly and efficiently. You use math to find out how much sand to put into the mortar mix or how many bricks will be needed. Calculation replaces trial and error to save you time, energy, materials, and money.

Denominate numbers are those that have a unit of measure associated with them; for example, 17 feet. Units of measure in the **U.S. Customary system** include inches, feet, and yards. Units of the **International System (SI)** include millimeters, centimeters, and meters. The units are an important part of the number. Five tons is very different than five inches. You must pay attention to the units to find the correct answer. This module gives you the tools you need to work with denominate numbers.

Masons often measure with a course rule or modular rule. These also form denominate numbers. This module will teach you to be comfortable with these rules.

You should know basic addition, subtraction, multiplication, and division of whole numbers, decimals, and fractions. You should also be able to convert fractions. This module briefly reviews working with fractions. If you are not comfortable with fractions or decimals, review *Introduction to Construction Math* in the *Core Curriculum*. Complete your review before beginning this module.

1.2.0 Reading Drawings

Working drawings are an on-the-job guide to what must be done. They detail the finished product and, in some cases, the intermediate steps.

This module introduces basic working drawings and how to read them. A set of residential plans is discussed in detail. This module will provide you with a working knowledge of how to read drawings and plans.

1.3.0 Applying Specifications

Specifications form part of the contract between the builder and the client. They contain detailed descriptions of the finished product. If there is a conflict between the plans and the specifications, the specifications are followed. This module describes specifications and how they are developed. You will use them to flesh out plans and drawings.

2.0.0 ◆ MASONRY MATH

Everyone uses mathematics. It is essential for a skilled mason or tradesman to be able to accurately solve math problems. This module reviews basic math concepts masons need every day. In addition, it covers special calculations used in masonry work.

2.1.0 Denominate Numbers

Denominate numbers include a unit of measure, such as feet, pounds, meters, kilos, and so on. Most of the world uses the metric system, meters, kilograms, and liters. In the United States, measurements are made with feet, pounds, and gallons. Masonry units are sold by the cubic yard or ton.

The unit name denominates, or identifies, what is to be added, subtracted, multiplied, or divided.

You can only add and subtract denominate numbers with the same units. If more are needed, you must borrow from larger units and convert them. When you are finished doing the arithmetic, you may need to convert smaller units into larger ones.

Converting is the process used to change feet to inches or yards to feet. You may also need to change feet back to inches. This is similar to carrying in addition as shown in *Figure 1*. The next sections gives examples.

2.1.1 Denominate Addition

The following section shows the steps for solving an addition problem with denominate numbers.

Add these denominate numbers:

```
  1 yard    2 feet   9 inches
 +2 yards   2 feet   5 inches
```

Step 1   Add the inches.

```
  1 yard    2 feet   9 inches
 +2 yards   2 feet   5 inches
                    14 inches
```

Step 2   There are 12 inches in a foot. Subtract 12 from 14 inches. Place the remaining inches in the inches column, and carry the foot to the foot column. The rule is to convert inches to feet when the number of inches is greater than 12.

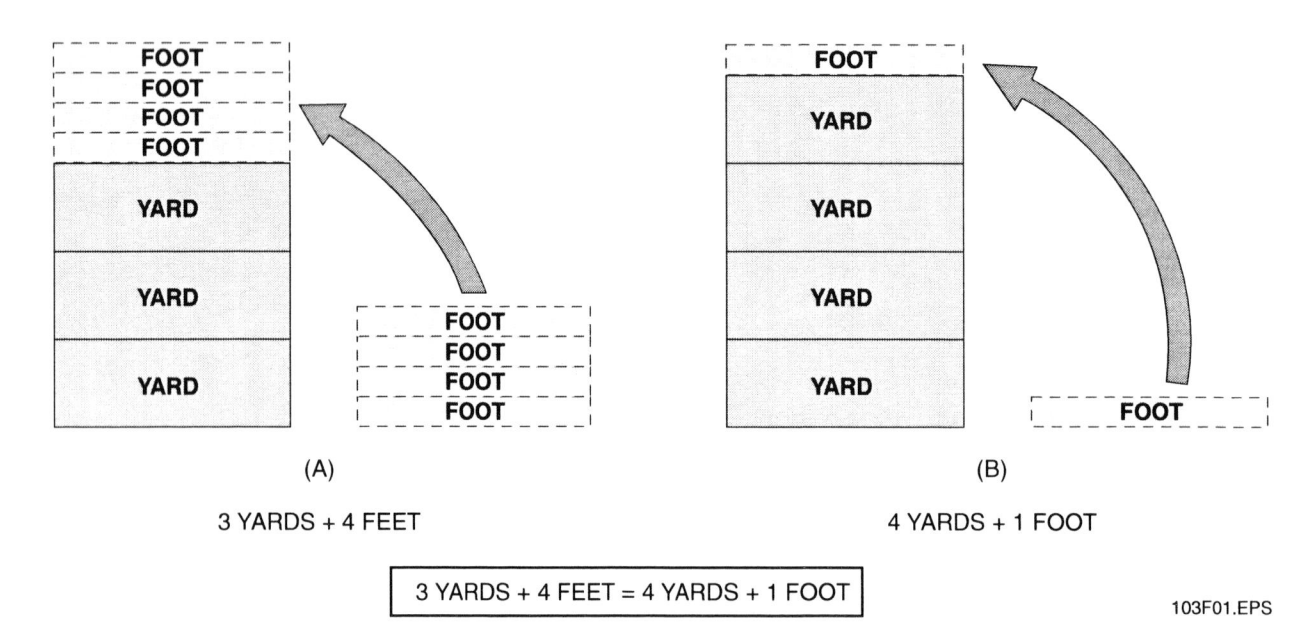

(A) 3 YARDS + 4 FEET

(B) 4 YARDS + 1 FOOT

3 YARDS + 4 FEET = 4 YARDS + 1 FOOT

103F01.EPS

Figure 1 ◆ Converting units.

Building Blocks

Apples and Oranges

People often say that "you can't add apples and oranges." In the same way, you can't add yards and inches. You need to have numbers with the same units before you can add or subtract. Use the following conversion table to convert inches, feet, and yards.

Inches	Feet	Yards
1	1/12	1/36
12	1	1/3
36	3	1

Step 3   Add the foot column.

```
              1 foot
    1 yard    2 feet    9 inches
   +2 yards   2 feet    5 inches
              5 feet   14 inches
                      -12 inches
                        2 inches
```

Step 4   Convert the feet to yards. Subtract 3 feet from 5 feet. Place the remaining feet in the feet column, and carry the yard to the yard column. The rule is to convert feet to yards when the number of feet is greater than 3.

```
    1 yard    1 foot
    1 yard    2 feet    9 inches
   +2 yards   2 feet    5 inches
              5 feet   14 inches
             -3 feet   12 inches
              2 feet    2 inches
```

Step 5   Add the yard column.

```
    1 yard    1 foot
    1 yard    2 feet    9 inches
   +2 yards   2 feet    5 inches
              5 feet   14 inches
             -3 feet   12 inches
    4 yards   2 feet    2 inches
```

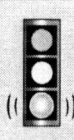

NOTE
Inches must be converted to feet when they exceed 12, and feet must be converted to yards when they exceed 3.

Practice Exercise (2.1.1)

This exercise will give you practice in adding denominate numbers. You can check your work by looking up the answers in *Appendix A*.

1.   2 yards   1 foot   10 inches
        +3 yards   2 feet   5 inches

2.   4 yards   2 feet   9 inches
        +6 yards   2 feet   4 inches

3.   7 yards   2 feet   7 inches
        +4 yards   1 foot   6 inches

4.   8 yards   1 foot   9 inches
        +4 yards   1 foot   7 inches

MEASUREMENTS, DRAWINGS, AND SPECIFICATIONS

> **BUILDING BLOCKS**
>
> ## Rule of Thumb
>
> Before tape measures, people used their bodies as a gauge. A small length was measured using the thumb. Larger distances were measured in feet or paces. You can still use your body to estimate distances.
>
> - 1 inch = the width of your thumb
> - 10 inches = the length of your foot
> - 5 feet = a pace (two steps)
> - 1 yard = the distance from your nose to your fingertips
>
> You might wish to see how closely these rules of thumb apply to you.

2.1.2 Denominate Subtraction

Subtraction problems call for similar conversions. In subtraction, remember that borrowing does not give the 10 units that you get working with ordinary numbers. The borrowed amount depends on the denominate measure you are converting (*Figure 2*).

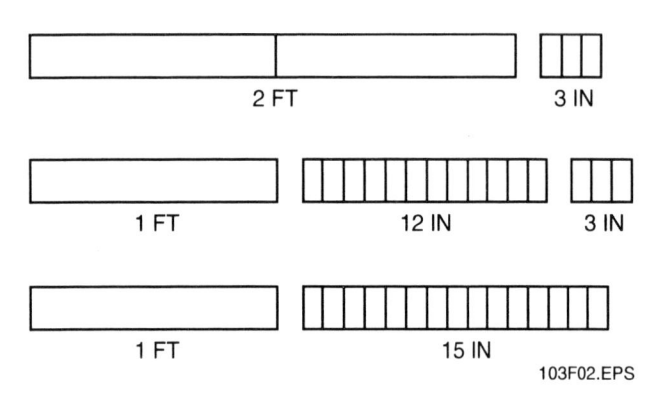

Figure 2 ◆ Borrowing units.

The following section shows the steps for solving a subtraction problem with denominate numbers.

Subtract these denominate numbers:

```
  5 yards   2 feet   7 inches
− 2 yards   2 feet   8 inches
```

Step 1 Subtract the inches. Since 7 is less than 8, borrow 12 inches from the foot column, leaving 1 foot. Add 12 inches to 7 inches to get 19 inches. Now subtract 8 inches from 19 inches.

```
                       (12 + 7) =
             1 foot    19 inches
  5 yards    2 feet    7 inches
− 2 yards    2 feet    8 inches
                       11 inches
```

Step 2 Subtract the foot column. Since 1 is less than 2, borrow 3 feet from the yard column, leaving 4 yards. Add 3 feet to 1 foot to get 4 feet. Now subtract 2 feet from 4 feet.

```
             (3 + 1) =
  4 yards    4 feet
                       (12 + 7) =
             1 foot    19 inches
  5 yards    2 feet    7 inches
− 2 yards    2 feet    8 inches
             2 feet    11 inches
```

Step 3 Since you borrowed 1 yard from 5 yards, you have 4 yards left. Now subtract the yard column.

```
             (3 + 1) =
  4 yards    4 feet
                       (12 + 7) =
             1 foot    19 inches
  5 yards    2 feet    7 inches
− 2 yards    2 feet    8 inches
  2 yards    2 feet    11 inches
```

Practice Exercise (2.1.2)

This exercise will give you practice in subtracting denominate numbers. You can check your work by looking up the answers in *Appendix A*.

1.
```
    6 yards   2 feet   6 inches
  − 2 yards   1 foot   8 inches
```

2.
```
    5 yards   2 feet   5 inches
  − 3 yards   2 feet   7 inches
```

Sometimes, plans will not give all of the dimensions used. Find the missing dimensions in *Figure 3* using Questions 3 and 4.

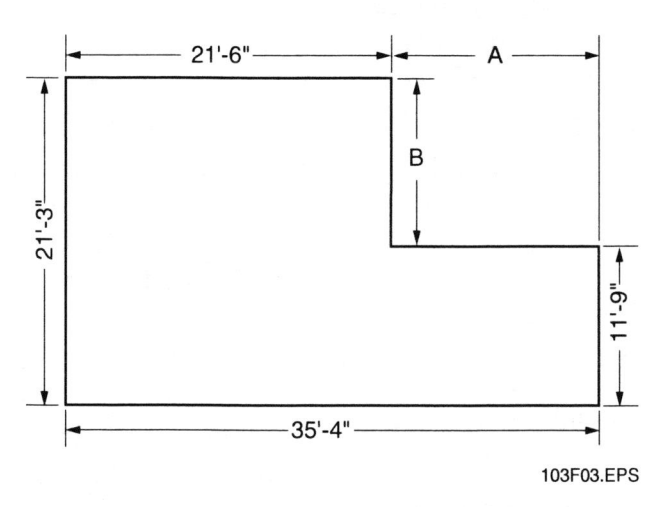

Figure 3 ◆ Find missing dimensions.

3. Subtract the known width from the total width to find A.

    ```
     35 feet    4 inches
    – 21 feet    6 inches
    ```

4. Subtract the known length from the total length to find B.

    ```
     21 feet    3 inches
    – 11 feet    9 inches
    ```

2.1.3 Other Denominate Measures

Denominate numbers commonly used in the United States are listed in *Table 1*. You should note that there are two types of ounces: the dry ounce and the fluid ounce. A fluid ounce is a measure of volume. It measures the size of a container. A dry ounce is a measure of weight. It measures the weight of the items in the container.

Table 1 Common Measures

WEIGHT UNITS
1 ton = 2,000 pounds
1 pound = 16 dry ounces |//
LENGTH UNITS
1 yard = 3 feet
1 foot = 12 inches
VOLUME UNITS
1 cubic yard = 27 cubic feet
1 cubic foot = 1,728 cubic inches
1 gallon = 4 quarts
1 quart = 2 pints
1 pint = 2 cups
1 cup = 8 fluid ounces
AREA UNITS
1 square yard = 9 square feet
1 square foot = 144 square inches

BUILDING BLOCKS

One Inch Equals Three Barleycorns

In early England, a small length was measured in several ways, including using barleycorn. Three barleycorns was called a ynce. However, after the Norman invasion, King Henry I standardized these measurements.

King Henry made the 12-inch foot official. A 12-inch foot was inscribed on the base of a column of St. Paul's Church in London. Measurements were said to be "by the foot of St. Paul's." Henry I established 3-foot standards called yards. William of Malmsebury wrote that the yard was the measure of the king's arm. This started the rumor that the yard was defined as the distance from the king's nose to his fingertips. In fact, both the foot and the yard were based on the ynce. A foot is 36 barleycorns, and the yard is 108 barleycorns.

MEASUREMENTS, DRAWINGS, AND SPECIFICATIONS

> **DID YOU KNOW?**
>
> ## *How Many Gallons*
>
> The traditional volume units are the names of standard containers. Until the eighteenth century, it was very difficult to measure the capacity of a container. Standard containers were defined by the weight of a particular item, such as wheat or beer, that they could carry. This custom led to several standard units. These included the barrel, the hogshead, and the peck. The gallon was originally the volume of eight pounds of wheat.
>
> The situation was still confused during the American colonial period. The Americans chose two of the many gallons. These two were the most common. For dry commodities, the Americans were familiar with the Winchester bushel. The corresponding gallon is one-eighth of this bushel.
>
> For liquids, Americans used the traditional British wine gallon. As a result, the U.S. volume system includes both dry and fluid units. The dry units are about one-sixth larger than the corresponding liquid units.
>
> In 1824, the British established a new system based on the Imperial gallon. The Imperial gallon was designed to hold exactly ten pounds of water. Unfortunately, Americans did not adopt this new, larger gallon. So the traditional English system actually includes three different gallons: U.S. liquid, U.S. dry, and British Imperial.

2.2.0 Fractions

Masons often use fractions in measuring and mixing. A fraction divides whole units into parts. They are usually written as two numbers, such as $\frac{1}{4}$, $\frac{1}{2}$, or $\frac{5}{8}$. The lower number is the denominator. The upper number is the numerator. This section will review how to add, subtract, divide, and multiply fractions.

2.2.1 Finding the Lowest Common Denominator

Fractions are similar to denominate numbers: you must have the same units in order to perform mathematical operations. You cannot simply add $\frac{5}{8}$ and $\frac{3}{4}$. The denominators (the bottom numbers) must be the same. The fractions must be converted before they are added together. The conversion process is known as finding a common denominator. The lowest common denominator is the smallest number that the denominators can be evenly divided into.

To find the lowest common denominator, follow these steps:

Step 1   Reduce each fraction to its lowest terms.

Step 2   Find the lowest common multiple of the denominators. Sometimes it is simple; one number is a multiple of the other. That means you can multiply by a whole number to get the larger number. If this is the case, all you have to do is find the equivalent fraction for the fraction with the smaller denominator.

Step 3   If neither of the denominators is a multiple of the other, you must multiply the two together to get a common denominator.

The following example will walk you through the steps to find the lowest common denominator between $\frac{3}{4}$ and $\frac{5}{8}$:

Step 1   In this example, $\frac{3}{4}$ and $\frac{5}{8}$ are already in their lowest terms.

Step 2   Looking at the denominators, you see that 8 is a multiple of 4. You need to find the equivalent fraction for $\frac{3}{4}$ that has a denominator of 8. Remember that you can multiply a fraction by 1 without changing the value. In this example, $\frac{2}{2} = 1$.

$$\frac{3}{4} \times \frac{2}{2} = \frac{6}{8}$$

Step 3   Alternatively, you can multiply the two denominators together to get a common denominator of 32. Then you convert each of the two fractions to fractions having a denominator of 32.

$$\frac{3}{4} \times \frac{8}{8} = \frac{24}{32}$$

$$\frac{5}{8} \times \frac{4}{4} = \frac{20}{32}$$

In this example, 8 is less than 32. So the lowest common denominator is 8. If one denominator is a multiple of the other, it will be the lowest common denominator. In that case, you do not need to perform Step 3.

2.2.2 Adding Fractions

How many inches will you have if you add $\frac{3}{4}$ of an inch and $\frac{7}{16}$ of an inch? To answer this question, you will have to add the fractions using the following steps:

Step 1   Find the lowest common denominator for the two fractions. Since 4 is a multiple of 16, the lowest common denominator is 16.

Step 2 Convert the fractions to equivalent fractions with the same denominator.

¾ × ⁴⁄₄ = ¹²⁄₁₆

Step 3 Add the numerators of the fractions.

¹²⁄₁₆ + ⁷⁄₁₆ = ¹⁹⁄₁₆

Step 4 Reduce the fraction to its lowest terms. If the numerator is larger than the denominator, the answer is greater than one.

¹⁹⁄₁₆ = ¹⁶⁄₁₆ + ³⁄₁₆ = 1³⁄₁₆

2.2.3 Subtracting Fractions

You follow the same steps to subtract fractions. Say you have ¾ of a bag of mortar mix. You need ½ a bag for a small batch of mortar. How much mortar mix will you have left?

Step 1 Find the common denominator. In this case, it is 4.

¾ − ½

Step 2 Multiply to convert the fractions to equivalent fractions.

½ × ²⁄₂ = ²⁄₄

Step 3 Subtract the numerators.

¾ − ²⁄₄ = ¼

2.2.4 Multiplying Fractions

Multiplying and dividing fractions is very different from adding and subtracting fractions. You do not need to find a common denominator. Say you have ¾ of a bag of mortar mix. You need to make three even batches. How much mix is in each batch? You want to know how much is ⅓ of ¾. The word *of* lets you know to multiply.

Step 1 Multiply the numerators together to get a new numerator. Multiply the denominators together to get a new denominator.

¾ × ⅓ = ³⁄₁₂

Step 2 Reduce the fraction to its lowest terms.

³⁄₁₂ = ¼

2.2.5 Dividing Fractions

Dividing fractions is similar to multiplying fractions with one added step. You must invert or flip the fraction you are dividing by. Use ½ ÷ ¾.

Step 1 Invert the fraction you are dividing by.

¾ becomes ⁴⁄₃

Step 2 Change the division sign to a multiplication sign. Multiply as instructed earlier.

½ ÷ ¾ becomes

½ × ⁴⁄₃ = ⁴⁄₆

Step 3 Reduce the fraction to its lowest terms.

⁴⁄₆ = ⅔

2.3.0 Mason's Denominate Measures

Masons have two denominate numbering systems of their own: the course system and the modular system. As noted in the *Masonry Tools and Equipment* module, masons have two kinds of rules for these measures. When working with these rules (or any measuring tools), it is important to do the following:

- Familiarize yourself with the scale.
- Take readings carefully, and take them again to avoid making costly mistakes.

The old carpentry rule "measure twice and cut once" applies here.

2.3.1 The Course System

The course system predates the modular system of measurement. The course system is also called the brick spacing rule or system. The course counter rule numbers the courses of different sizes of brick that will fill a vertical space.

This rule is used to lay out and space standard brick courses to nonmodular dimensions. The rule has inches on the other side, marked in sixteenths of an inch. Nonmodular course spacing is measured and estimated with the brick spacing rule.

Figure 4 shows the standard brick spacing rule and an oversize brick spacing rule. The rule has a gauge at the beginning that measures the size of one brick. It is used to identify the size of the brick so you will know which scale to read. On the stan-

THINK ABOUT IT

Dividing Fractions

You want to divide two-thirds of a bag of cement mix in half. What is the mathematical equation and correct answer?

dard brick spacing rule, all the reference measures fall between 2⅜ and 3 inches. For larger bricks, you need to use the oversize brick spacing rule.

The large figures on the rule are references for the nominal sizes of standard brick and mortar. The small figures, at right angles to the black figures, count the number of courses for that size brick. The number of courses are marked for reference numbers in *Figure 5*. The large figures are in black, and the small ones are in red on the actual rule.

The standard spacing rule is useful for marking a corner pole or deadman. The first step in using the standard spacing rule is to check the specifications for the size of brick and/or the course height. Then determine the height to be filled by the courses. The third step is to transfer the correct markings from the spacing rule to the corner pole.

2.3.2 The Modular System

Today, brick is made for use on the modular grid system. The dimensions are based on a 4-inch unit called a module. The grid system makes it easier to combine different materials in a construction job. It creates a standard measurement, so different materials can be easily measured or calculated.

In modular design, the nominal dimension of a masonry unit is the manufactured dimension plus the thickness of the mortar joint. The nominal dimension is a multiple of 4 inches. For example, bricks are designed so that the size of the brick and the mortar joint equals a multiple of 4 inches.

A modular brick with a nominal length of eight inches will have a manufactured dimension of 7½ inches if it is designed to be laid with a ½-inch mortar joint. It will have a manufactured dimension of 7⅝ inches if it is designed to be laid with a ⅜-inch mortar joint. The brick and mortar will fit in two 4-inch modules.

Table 2 shows nominal and actual manufactured dimensions for nominal brick sizes and actual dimensions for nonmodular brick sizes. It includes

BUILDING BLOCKS

Reading Rules Accurately

Rules are used to make accurate measurements. The markings on a standard rule are in inches and feet. The markings on a metric rule are in millimeters and centimeters. Some have standard markings on one side and metric markings on the other.

A standard ruler is divided into whole inches and then halves, fourths, eighths, and sixteenths. Some rules also include thirty-seconds and sixty-fourths. You need to pay close attention when measuring. Your projects are only as accurate as your measurements.

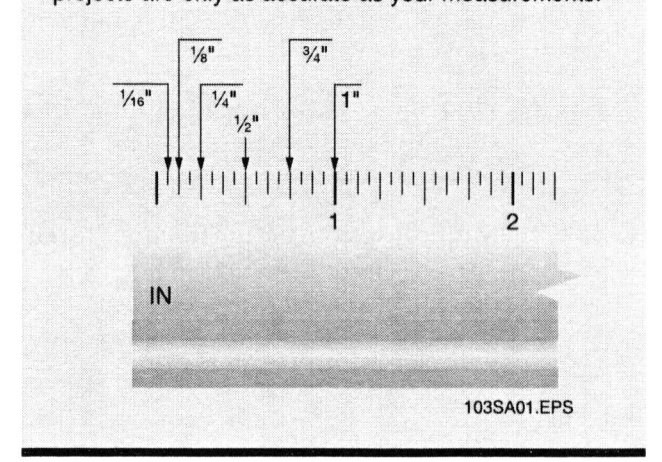

the planned joint thickness of ⅜ of an inch or ½ of an inch. The last column shows the number of courses required for each size of brick to equal a 4-inch modular unit or a multiple of a 4-inch unit.

Most masonry materials will tie and level off at a height of 16 inches vertically. Two courses of blocks with mortar will be 16 inches high. Six courses of standard bricks with mortar will also be 16 inches vertically. As a result, the wythes can be tied together at 16-inch intervals or at multiples of 16 inches.

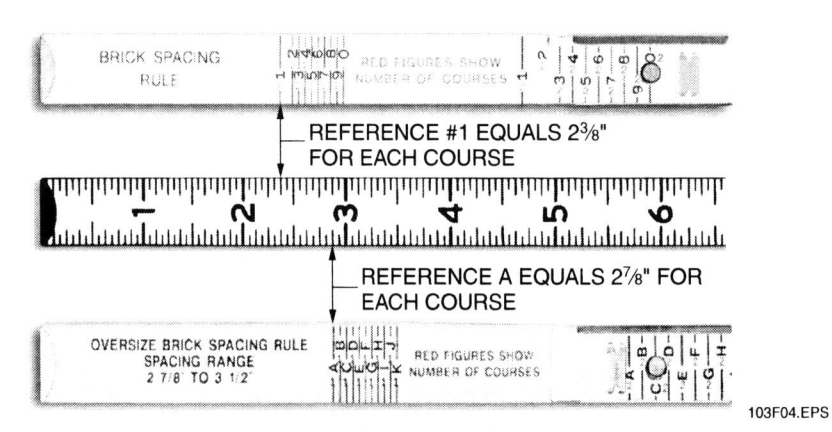

Figure 4 ◆ Brick spacing rule and oversize brick spacing rule.

3.8   MASONRY LEVEL ONE — TRAINEE MODULE 28103-04

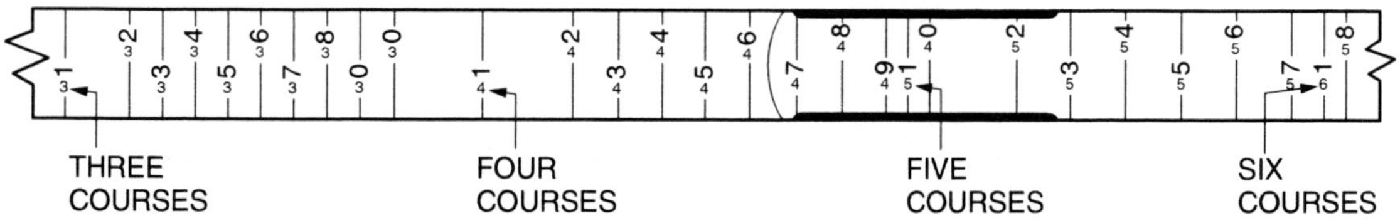

103F05.EPS

Figure 5 ◆ Reading the brick spacing rule.

The following example will show you how to determine how many courses of brick are needed to build a wall 8 feet high using standard brick.

Step 1 Determine the number of 16-inch sections.

8 ft × 12 in/ft = 96 in

96 in ÷ 16 in = 6 (16-inch sections)

Step 2 Use *Table 2* to find the number of courses in a 16-inch section for standard brick.

Step 3 Multiply the number of courses per 16-inch section by the number of sections to find the total number of courses.

6 × 6 = 36 courses

Practice Exercise (2.3.2)

You are building a wall 6 feet high. How many courses will you need for each type of brick? You can check your work by looking up the answers in *Appendix A*.

1. Modular brick ________
2. Roman brick ________
3. Utility brick ________
4. Norman brick ________

Table 2 Sizes of Bricks

Unit Designation	Nominal Dimenions Inches w	h	l	Joint Thickness Inches	Specified Dimenions Inches w	h	l	# of Courses in 16"
MODULAR BRICK SIZES								
Modular	4	2⅔	8	⅜ ½	3⅝ 3½	2¼ 2¼	7⅝ 7½	6
Engineer Modular	4	3⅕	8	⅜ ½	3⅝ 3½	2¾ 2 13/16	7⅝ 7½	5
Closure Modular	4	4	8	⅜ ½	3⅝ 3½	3⅝ 3½	7⅝ 7½	4
Roman	4	2	12	⅜ ½	3⅝ 3½	1⅝ 1½	11⅝ 11½	8
Norman	4	2⅔	12	⅜ ½	3⅝ 3½	2¼ 2¼	11⅝ 11½	6
Engineer Norman	4	3⅕	12	⅜ ½	3⅝ 3½	2¾ 2 13/16	11⅝ 11½	5
Utility	4	4	12	⅜ ½	3⅝ 3½	3⅝ 3½	11⅝ 11½	4
NONMODULAR BRICK SIZES								
Standard				⅜ ½	3⅝ 3½	2¼ 2¼	8 8	6
Engineer Standard				⅜ ½	3⅝ 3½	2¾ 2 13/16	8 8	5
Closure Standard				⅜ ½	3⅝ 3½	3⅝ 3½	8 8	4
King				⅜	3 3	2¾ 2⅝	9⅝ 9⅝	5
Queen				⅜	3	2¾	8	5

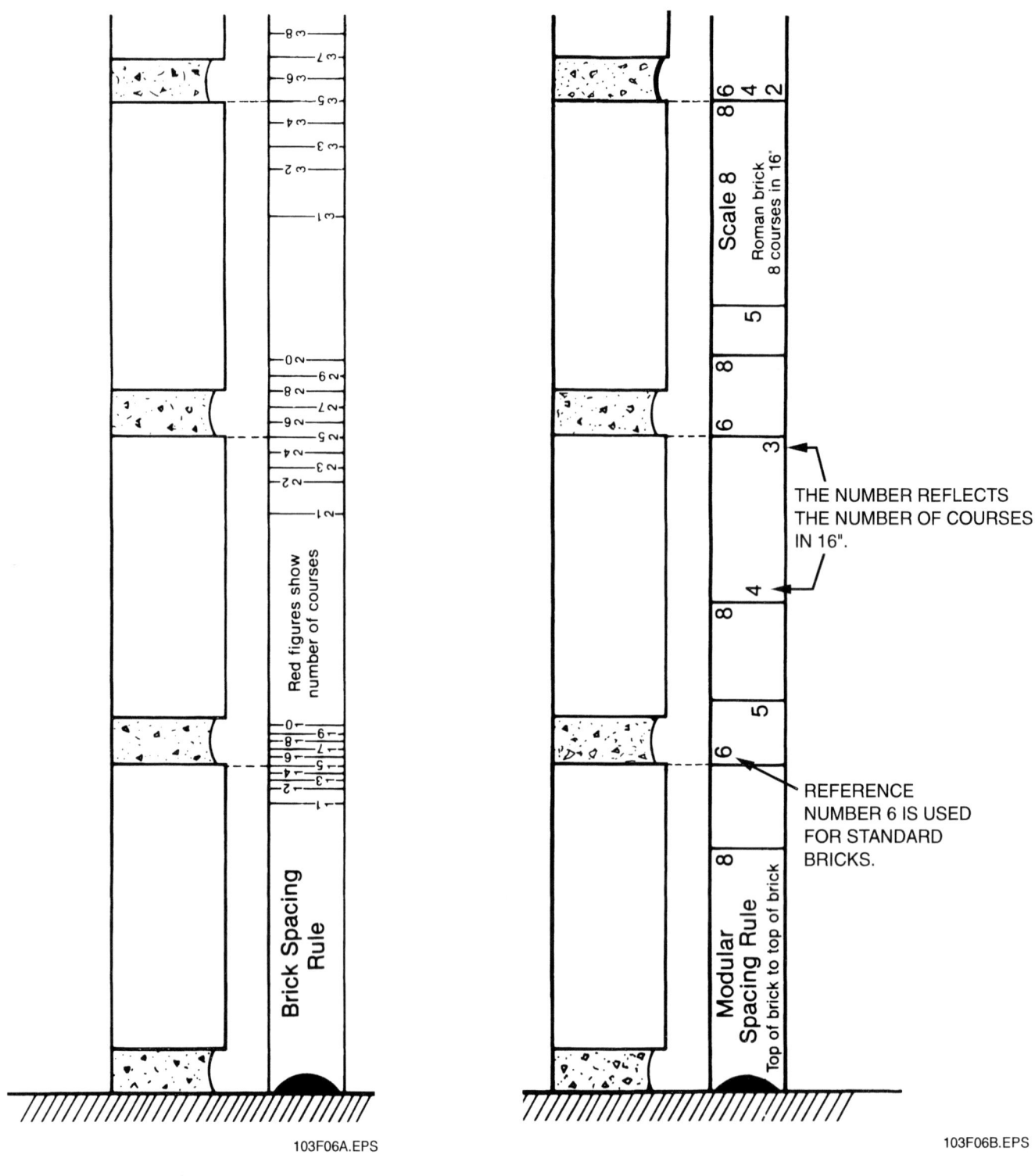

Figure 6 ◆ Reading the modular spacing rule.

    The modular spacing rule (*Figure 6*) has a modular scale on one side and inches marked into sixteenths on the other side.

    The black figures are the references for the nominal sizes of modular brick and block. The modular markings give course numbers for different sizes. Scale 2 is for regular block or any brick with 2 courses equal to 16 inches in height. Scale 3 measures 3 courses in 16 inches and so on. Again, the specifications are the place to find the brick size or planned course height.

> **DID YOU KNOW?**
> *The Metric System*
>
> The metric system is a revolutionary idea. It was first adopted by the French revolutionary assembly in 1795. There was considerable resistance to the system at first. The first countries to actually require use of the metric system were Belgium, the Netherlands, and Luxembourg in 1820. It was adopted by all industrialized countries in 1875.

2.4.0 Denominate Metric Measurements

The metric system is another measuring system. It has been used worldwide since 1875. It is being used more often in the United States, but private industry has been slow to change. Efforts to convert to the metric system have been growing. The General Services Administration, which oversees all federal building projects, began requiring bids with metric specifications. In September 1996, all federally assisted highway construction projects were required to use metric standards.

2.4.1 SI Units of Measure

The official name of the metric system is Système International d' Unités. It is abbreviated as SI. The SI, or metric system, is a very convenient and logical system of measurements and weights. It is based on the number 10. This is similar to our system of currency: 10 pennies are equal to a dime, and 10 dimes are equal to a dollar.

Terms such as kilometer, centimeter, and millimeter are denominate units of metric measure. SI metric units of measure include base units:

- Meter for length
- Gram for weight
- Liter for liquid volume

Smaller or larger units are noted by adding a prefix to the base unit. The prefix expresses a multiple of 10. *Table 3* lists the more common SI prefixes.

An SI measurement is a base unit plus a prefix. The same prefixes are used with all base units. A kilometer is 1,000 meters, while a millimeter is 1/1000 of a meter. A kilogram is 1,000 grams, while a milligram is 1/1000 of a gram. *Table 4* shows common metric measures. Compare this to *Table 1*. The metric system is very easy to use.

Table 3 Common SI Prefixes

PREFIX	SYMBOL	NUMBER	MULTIPLICATION FACTOR
giga	G	billion	$1,000,000,000 = 10^9$
mega	M	million	$1,000,000 = 10^6$
kilo	k	thousand	$1,000 = 10^3$
hecto	h	hundred	$100 = 10^2$
deka	da	ten	$10 = 10^1$
			BASE UNITS $1 = 10^0$
deci	d	tenth	$0.1 = 10^{-1}$
centi	c	hundredth	$0.01 = 10^{-2}$
milli	m	thousandth	$0.001 = 10^{-3}$
micro	μ	millionth	$0.000001 = 10^{-6}$
nano	n	billionth	$0.000000001 = 10^{-9}$

Table 4 Common Metric Measures

WEIGHT UNITS		
1 kilogram	=	1,000 grams
1 hectogram	=	100 grams
1 dekagram	=	10 grams
1 gram	=	1 gram
1 decigram	=	0.1 gram
1 centigram	=	0.01 gram
1 milligram	=	0.001 gram

LENGTH UNITS		
1 kilometer	=	1,000 meters
1 hectometer	=	100 meters
1 dekameter	=	10 meters
1 meter	=	1 meter
1 decimeter	=	0.1 meter
1 centimeter	=	0.01 meter
1 millimeter	=	0.001 meter

LIQUID VOLUME UNITS		
1 kiloliter	=	1,000 liters
1 hectoliter	=	100 liters
1 dekaliter	=	10 liters
1 liter	=	1 liter
1 deciliter	=	0.1 liter
1 centiliter	=	0.01 liter
1 milliliter	=	0.001 liter

2.4.2 Converting SI Metric

The United States uses both the metric units and U.S. Customary units. Sometimes measurements must be converted from one system to the other.

How do millimeters, centimeters, and meters compare with the units in the U.S. Customary system? A centimeter is about ⅗ of an inch; a meter is about 1.1 yards; a kilometer is just a little more than 0.6 of a mile. *Figure 7* gives some common equivalents.

Table 5 shows some common U.S.-to-metric conversions. To convert from U.S. Customary to SI metric, follow this procedure:

Step 1 Change the U.S. measurement from fractions to decimal form.

Step 2 Multiply the decimal by the factor in the right-hand column of *Table 5*.

Step 3 Round the answer to the accuracy required.

Table 6 shows metric-to-U.S. conversions. To convert from SI metric to U.S. Customary, follow this procedure:

Step 1 Multiply the quantity by the factor in the right-hand column of *Table 6*.

Step 2 Round the answer to the accuracy required.

Step 3 Convert the decimal part of the answer to the nearest common fraction.

2.4.3 Conversion Examples

A spread footing for a foundation measures 1 foot, 6 inches wide by 22 feet long. What are its metric dimensions?

Step 1 Convert the fractional units to a decimal.

    1 foot, 6 inches = 1⁶⁄₁₂ or 1.5 feet

Step 2 Multiply by the correct conversion factor using *Table 5*.

    1.5 feet × 30.48 centimeters = 45.7 centimeters

    22 feet × 30.48 centimeters = 670.6 centimeters

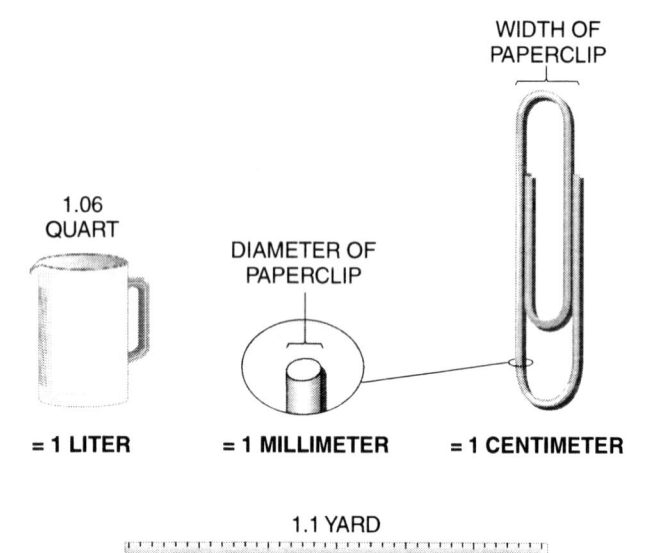

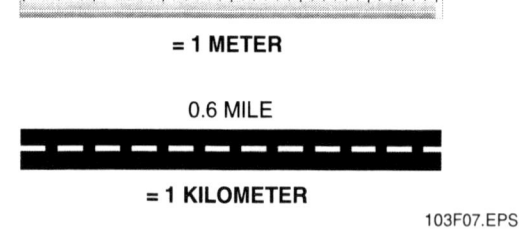

Figure 7 ♦ Common metric equivalents.

DID YOU KNOW?
The Yottameter

What happens when we want a number larger than the largest prefix? Every so often, new prefixes are added by the International Bureau of Weights and Measures in Paris. Added in 1991, the largest prefix is the yotta. It is one septillion or 10^{24}. That is, 1 with 24 zeros after it.

How far is a yottameter? It is about 10 billion light years. Only the Hubble telescope can see that far.

DID YOU KNOW?
Measuring the World

The metric units were designed in a scientific manner. The Earth itself was selected as the measuring stick. The meter was defined as one ten-millionth of the distance from the equator to the north pole. The liter was the volume of one cubic decimeter. The kilogram was the weight of a liter of pure water.

Scientific methods in 1795 were not accurate enough to measure the world. Modern measurements have shown that the world is slightly larger than was first estimated. However, the difference does not change the basic metric system.

Table 5 U.S. Customary to SI Metric Conversions

U.S. CUSTOMARY		SI METRIC
WEIGHTS		
1 ounce (oz)	=	28.35 grams
1 pound (lb)	=	435.6 grams or 0.4536 kilograms
1 (short) ton	=	907.2 kilograms
LENGTHS		
1 inch (in)	=	2.540 centimeters
1 foot (ft)	=	30.48 centimeters
1 yard (yd)	=	91.44 centimeters or 0.9144 meters
1 mile	=	1.609 kilometers
AREAS		
1 square inch (in^2)	=	6.452 square centimeters
1 square foot (ft^2)	=	929.0 square centimeters or 0.0929 square meters
1 square yard (yd^2)	=	0.8361 square meters
VOLUMES		
1 cubic inch (in^3)	=	16.39 cubic centimeters
1 cubic foot (ft^3)	=	0.02832 cubic meter
1 cubic yard (yd^3)	=	0.7646 cubic meter
LIQUID MEASUREMENTS		
1 (fluid) ounce (fl oz)	=	0.095 liter or 28.35 grams
1 pint (pt)	=	473.2 cubic centimeters
1 quart (qt)	=	0.9263 liter
1 (US) gallon (gal)	=	3,785 cubic centimeters or 3.785 liters
TEMPERATURE MEASUREMENTS		
To convert degrees Fahrenheit to degrees Celsius, use the following formula: C = 5/9 × (F − 32).		

Table 6 SI Metric to U.S. Customary Conversions

SI METRIC		U.S. CUSTOMARY
WEIGHTS		
1 gram (G)	=	0.03527 ounces
1 kilogram (kg)	=	2.205 pounds
1 metric ton	=	2,205 pounds
LENGTHS		
1 millimeter (mm)	=	0.03937 inches
1 centimeter (cm)	=	0.3937 inches
1 meter (m)	=	3.281 feet or 1.0937 yards
1 kilometer (km)	=	0.6214 miles
AREAS		
1 square millimeter	=	0.00155 square inches
1 square centimeter	=	0.155 square inches
1 square meter	=	10.76 square feet or 1.196 square yards
VOLUMES		
1 cubic centimeter	=	0.06102 cubic inches
1 cubic meter	=	35.31 cubic feet or 1.308 cubic yards
LIQUID MEASUREMENTS		
1 cubic centimeter (cm^3) =		0.06102 cubic inches
1 liter (1,000 cm^3)	=	1.057 quarts, 2.113 pints, or 61.02 cubic inches
TEMPERATURE MEASUREMENTS		
To convert degrees Celsius to degrees Fahrenheit, use the following formula: F = (9/5 × C) + 32.		

Step 3 Centimeters are too small for measuring on a large scale, so it may be necessary to convert them to n appropriate unit. If a number is less than one, add a leading zero. Round to the nearest hundredth.

    100 centimeters to a meter

    45.7 centimeters ÷ 100 = 0.457 meters

    670.6 centimeters ÷ 100 = 6.706 meters

    The footing is 0.46 × 6.71 meters.

A drawing shows the height of a wall to be 2.5 meters. What is its height in feet and inches?

Step 1 Multiply by the correct conversion factor using *Table 6*.

    2.5 meters × 3.281 feet = 8.2025 feet

Step 2 Convert the decimal portion of feet to inches.

    0.2025 feet × 12 inches = 2.43 inches = 2 inches + 0.43 inches

Step 3 Convert the 0.43 inches to a fraction using $^{16}/_{16}$ as the equivalent fraction.

    $^{0.43}/_1$ × $^{16}/_{16}$ = $^{6.88}/_{16}$, rounds to $^{7}/_{16}$ inches

    2.5 meters = 8 feet, 2$^{7}/_{16}$ inches

Practice Exercise (2.4.2 and 2.4.3)

Use the information in *Table 5* to complete the following conversion exercises.

1. ½ inch = _____ centimeters
2. ⅝ inch = _____ centimeters
3. ⁹⁄₁₆ inch = _____ centimeters
4. 5 gallons of water = _____ liters
5. 94 pound bag of cement = _____ kilograms
6. Convert your height to meters and centimeters.

MEASUREMENTS, DRAWINGS, AND SPECIFICATIONS

2.5.0 Plane Figures and Area Measures

Knowledge of geometry is useful on a construction site. For example, if a wall is to have two windows in it, it will not need bricks in these areas. Geometry allows the mason to calculate how many bricks will be needed. This knowledge saves time and money when ordering materials. It can also be used to save steps when carrying bricks to the workstation.

The next sections review calculating areas for common geometric shapes. The measurements of these shapes may be denominated in square meters or square feet, or square centimeters or square inches depending on the job.

Plane figures are figures drawn in only two dimensions. Rectangles, triangles, and circles are common plane figures. The area of a plane figure is expressed in square units of the appropriate denomination.

2.5.1 Four-Sided Figures

Squares, rectangles, and parallelograms are four-sided regular polygons (*Figure 8*). They are figures with opposite parallel sides of the same length.

A rectangle is a polygon that has four sides of two different lengths that meet at right angles. The formula for finding the area of a rectangle is length × width, or:

A = lw

A square has four sides of the same length that meet at right angles. The formula for finding the area of a square is also length × width, simply expressed as side times side, or:

A = s^2 *or* A = ss

A parallelogram has four sides that do not meet at right angles. The formula for finding the area of a parallelogram is base × height. The base (b) is the longest side, and the height (h) is the shortest distance between the upper and lower bases. This formula is expressed as:

A = bh

For example, a drawing shows a wall to be built that is 4 feet high and 10 feet long. What is the surface area of the wall?

Step 1 Find the formula for the calculation. The wall is a rectangle. The formula for the surface area of a rectangle is A = lw.

SQUARE

A = s^2
A = ss

RECTANGLE

A = lw

PARALLELOGRAM

A = bh

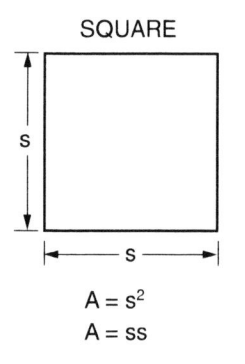

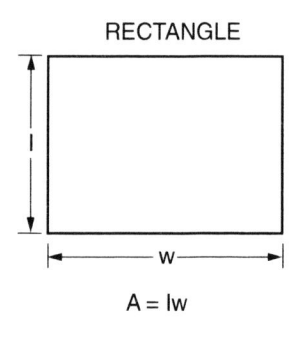

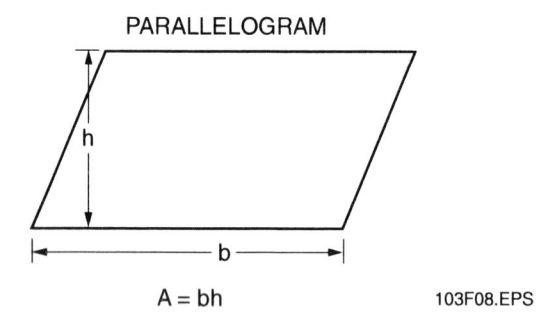

Figure 8 ◆ Four-sided figures.

Step 2 Calculate the answer using the data from the drawing.

A = lw

A = 4 feet × 10 feet

A = 40 square feet

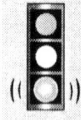

> **NOTE**
> Before performing a calculation, make sure that the two numbers have the same units. If the units are different, you must convert them to the same units.

2.5.2 Three-Sided Figures

Triangles are three-sided figures. They take many shapes. In all triangles, the three internal angles add up to 180 degrees. This is useful to know. If you know two of the angles, you can calculate the third. For example, if two angles of a triangle are 25 degrees and 75 degrees, the unknown angle is 80 degrees:

180 degrees − 25 degrees − 75 degrees = 80 degrees

Triangles can be identified by the relationships of the sides. Three examples are shown in *Figure 9*.

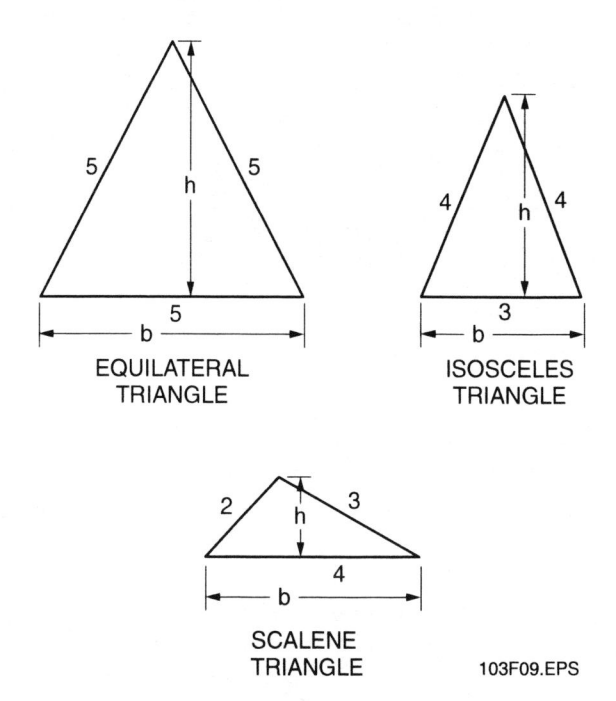

Figure 9 ◆ Triangles named by sides.

All three sides of an equilateral triangle are the same size. Only two sides of an isosceles triangle are the same size. None of the sides of a scalene triangle are the same size.

Triangles are also classified according to their interior angles (*Figure 10*). If one of the angles is 90 degrees, it is called a right triangle. If one of the angles is greater than 90 degrees, it is called an obtuse triangle. If each of the interior angles is less than 90 degrees, it is called an acute triangle.

Every triangle is really an exact half of a parallelogram. Remembering this makes it easy to calculate the area of a triangle. The height of a triangle is the length of a line drawn from one angle to the side opposite, or base, that meets the base at a right angle. The area of a triangle is half the area of a parallelogram with the same base and height. The area of a triangle is expressed as:

$A = \frac{1}{2} b \times h$ *or* $A = bh \div 2$

Some areas are a combination of shapes. The house in *Figure 11* contains several shapes. The base of the house is a rectangle. The top portion is a triangle. Two of the windows are rectangles and the vent is a circle. If you know the dimensions, you can calculate the area for the rectangle and the triangle to find the total surface area. You would then need to calculate the area of the two windows and vent and subtract that from the total surface area. Then you can find out how much brick you would need for the side of the house.

Figure 11 ◆ Divide figures to find the area.

2.5.3 One-Sided Figures

Circles are single, closed lines with all points the same distance from the center. *Figure 12* shows the parts of a circle. Note that the radius is half of the diameter. Either one of these dimensions may be given on a plan or drawing to indicate the size of the circle. The circumference is the outside line that defines the circle; the area is the space within it.

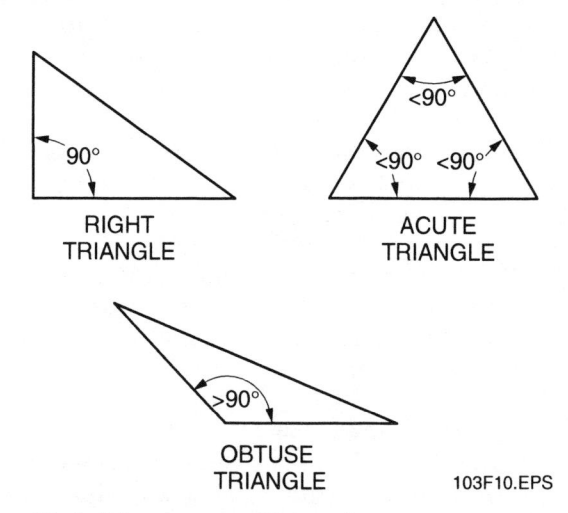

Figure 10 ◆ Triangles named by angles.

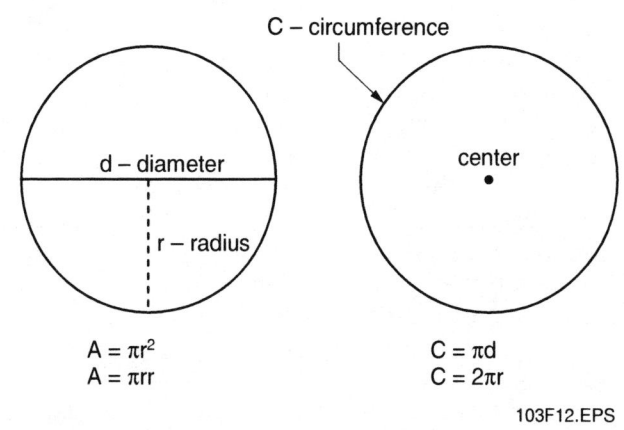

Figure 12 ◆ Parts of a circle.

MEASUREMENTS, DRAWINGS, AND SPECIFICATIONS

> **THINK ABOUT IT**
>
> ### Circles
>
> If a circle has a radius of 10 feet, is the circumference 62.8 feet or 62.8 square feet? What about the area?

If you are building a circular garden wall, you will need to know its circumference to figure out how much material will be needed. The circumference of a circle is its diameter (d) times the constant pi (π), or twice its radius (r) times the constant π. The rounded value of π is 3.14, or 22/7. The formula for the circumference of a circle is expressed as:

$C = \pi d$ *or* $C = 2\pi r$

> **NOTE**
>
> Remember to include the units, such as feet or inches, in your calculations.

You may need to know the area of a circular patio to be paved with bricks. The area of a circle is expressed in square units. The formula for the area of a circle is expressed as:

$A = \pi r^2$ *or* $A = \pi r r$

2.5.4 Many-Sided Figures

Many-sided figures with all sides the same size and the same distance from the center are called regular polygons. They are also named after their number of sides. A five-sided figure is a pentagon, a six-sided figure is a hexagon, and an eight-sided figure is an octagon. *Figure 13* shows some regular polygons.

Occasionally, a plan will incorporate a hexagonal window over a door or an octagon in a bathroom. Sometimes a structure will be in the shape of an octagon instead of a square. To calculate the circumference of any regular polygon, you need to know the number and the length of each side.

The formula for the area of a regular polygon is the sum of the lengths of the sides divided by 2 then multiplied by r. The r is the distance from the center to any one angle. You may have to use your ruler on the plan to approximate that distance. The formula for the area of a regular polygon is written as:

$A = (S_1 + S_2 + S_3 + S_n)\, r/2$

The small *n* indicates that the sides continue to the required number.

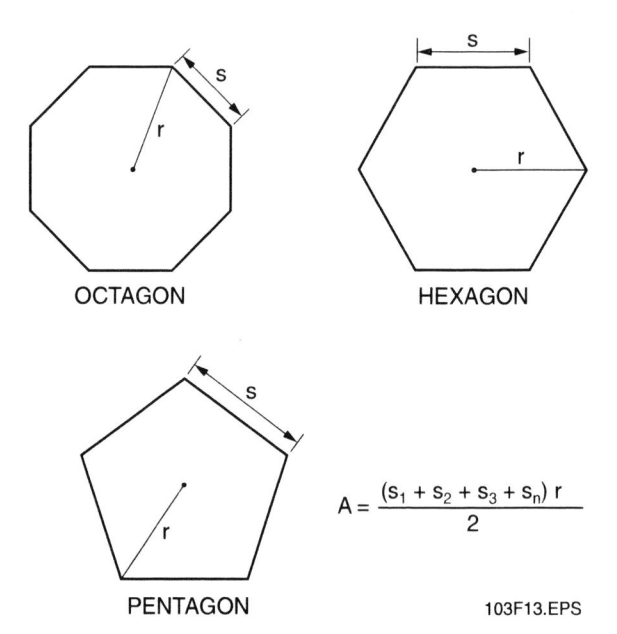

Figure 13 ◆ Regular polygons.

Practice Exercise (2.5.0)

Calculate the areas of the following plane figures. Then convert the answers to the other system of measure.

1. A square with a height of 1.56 meters

2. A rectangle twice as wide as it is high, with a height of 3 feet, 8 inches

3. A triangle with a base of 35 centimeters and a height of 0.5 meters

4. A circle with a diameter of 0.52 meters

5. A pentagon with a side of 24 inches and r of 28½ inches

6. You have a 3 feet wide by 6 feet high opening in a wall that was originally a window. The plans changed, and now you have to fill that space with standard block with an 8 × 16 inch face. How many blocks do you need to carry? If you add 5 percent for breakage, how many blocks does that make? If you use a Flemish bond pattern, you will need one-third more blocks. How many in all?

7. If you know that 1.125 standard blocks are needed to fill 1 square foot, how would you calculate Question 6 differently? What would the answer be?

8. A wall (*Figure 14*) that must be covered in brick veneer is 4 meters high and 5 meters long. There is a circular window with a diameter of 1 meter. What is the surface area of the wall?

Figure 14 ◆ Wall with circular window.

2.6.0 Solid Figures and Volumes

We calculate areas by measuring two-dimensional figures. The measures are usually of length and width. The calculations are written as square measurements, such as square feet or square meters. We calculate volumes by measuring three-dimensional figures. The measures are of height, width, and depth. The calculations are written as cubic measurements, such as cubic feet or cubic meters.

Figure 15 shows the most common solid figures in construction and the formulas for their volume calculations. Notice that these three figures—the cylinder, prism, and rectangular solid—have the same shape from top to bottom.

CAUTION

In most drawings, the depth of a solid is called the height and noted with an *H*. However, that can be confusing for prisms and pyramids, as the area of a triangle is usually calculated as base (b) times height (h). Unfortunately, *d* usually denotes the diameter of a circle and should not be used for depth.

Solid figures with the same shape from top to bottom are easy to measure. Calculate the volume by first getting the area of the top surface. Then, multiply that area by the depth or height to get the volume.

- Volume of a cylinder = π × the radius squared × depth
  $V = \pi r^2 \times H$
- Volume of a prism = (base × height ÷ 2) × depth
  $V = (bh \div 2) \times H$
- Volume of a rectangle = length × width × height
  $V = l \times w \times H$

The three figures shown in *Figure 16* are more complicated. The need to deal with the volume of these figures is rare. The formulas first calculate the area of one part of the figure, then make adjustments to the depth to get the volume.

Although the arithmetic is somewhat different, the underlying idea is the same as for the first three solid figures. These formulas are used to calculate cubic yards, cubic feet, cubic meters, or cubic centimeters for odd shapes.

$V = \pi r^2 \times H$

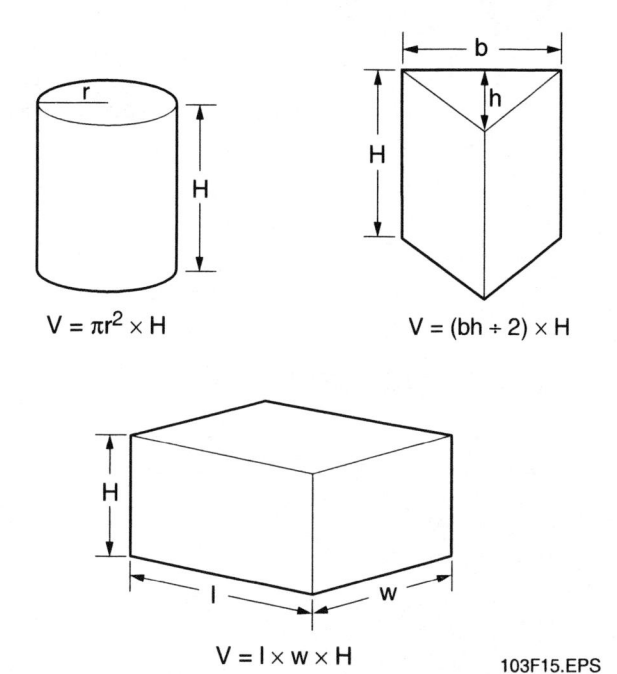

$V = (bh \div 2) \times H$

$V = l \times w \times H$

Figure 15 ◆ Volumes of solid figures.

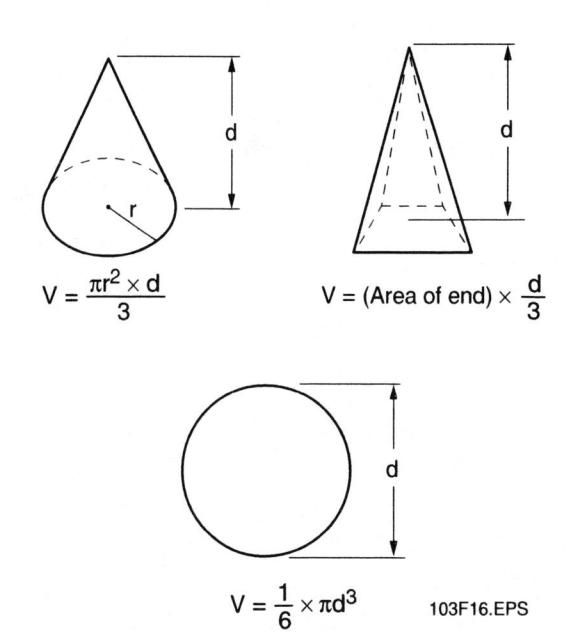

$V = \dfrac{\pi r^2 \times d}{3}$

$V = (\text{Area of end}) \times \dfrac{d}{3}$

$V = \dfrac{1}{6} \times \pi d^3$

Figure 16 ◆ Conical, triangular, and spherical solids.

MEASUREMENTS, DRAWINGS, AND SPECIFICATIONS

Practice Exercise (2.6.0)

Calculating cubic volumes is an important part of ordering supplies. The following exercises provide examples of how this is done.

1. You have just finished a cavity wall, 8 feet × 20 feet. You have a 4-inch cavity to fill with granular styrene insulation. How many cubic yards of insulation will you need?
2. You need to reinforce a single-wythe block wall 26 feet, 8 inches long. The block is 8 × 16 inches set 9 courses high. The blocks have two cores each, and each core measures 2 × 4 × 8 inches. How many cubic yards of grout will you need to fill every core in this wall?

2.7.0 Working with Right Triangles

Perhaps the most used shape in construction is the right triangle. Any vertical object or structure that forms a 90-degree angle with the ground is part of a right triangle. This can be a telephone pole or the wall of a building. If you draw an imaginary line from a point on the ground to the top of the structure, you have a right triangle. The line forms the hypotenuse of the right triangle. The base of the triangle extends from the bottom of the pole or structure to the starting point on the ground of your triangle.

Since the right triangle has one right angle, the other two angles are acute angles. They are also complementary angles. The sum of these two angles is 90 degrees. The right triangle has two sides perpendicular with each other, thus forming the right angle. To aid in writing equations, the sides and angles of a right triangle are labeled and shown in *Figure 17*. Normally, capital letters are used to label the angles, and lowercase letters are used to label the sides. The sides can be remembered as a for altitude and b for base. The third side (side c), opposite the right angle, is called the hypotenuse. It is always longer than the other two sides.

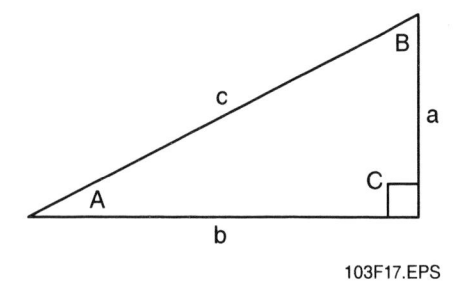

Figure 17 ◆ Right triangle.

If you know the length of any two sides of a right triangle, you can calculate the length of the third side. This equation is called the Pythagorean theorem. It states that the square of the hypotenuse c is equal to the sum of the squares of the remaining two sides (a and b). Expressed mathematically:

$$c^2 = a^2 + b^2$$

You may rearrange to solve for the unknown side as follows:

$$a = \sqrt{c^2 - b^2} \quad b = \sqrt{c^2 - a^2} \quad c = \sqrt{a^2 + b^2}$$

For example, you have a telephone pole with a support cable (*Figure 18*). You know the length of the support cable, but not the height of the pole. You can easily measure the distance on the ground between the end of the support cable and the pole.

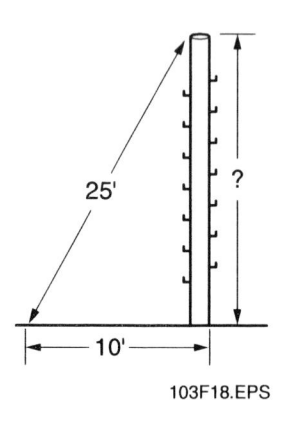

Figure 18 ◆ Telephone pole with supporting cable.

The support cable is the hypotenuse of the triangle. It is 25 feet. The distance on the ground, 10 feet, is the base.

$$a = \sqrt{c^2 - b^2}$$
$$a = \sqrt{25^2 - 10^2}$$
$$a = \sqrt{625 - 100}$$
$$a = \sqrt{525}$$
$$a = 22.9" \text{ or } 22'\text{-}10\tfrac{3}{16}"$$

Appendix B shows all of the figures and mathematical formulas listed in this section.

3.0.0 ◆ DRAWINGS

Architectural drawings are always part of project documentation. With the specifications, they form the written guidelines for the builder. In order to do well in your work, you must be able to read and understand the information on the project drawings.

This section is a review of blueprint material introduced in the *Core Curriculum*. It also presents some new information about drawings, their organization, and the symbols for construction materials.

3.1.0 Understanding Drawings

In order to read drawings, you have to recognize the various lines and symbols used in their preparation. The next sections review the keys to lines, symbols, and scales and the information they carry.

3.1.1 Lines as Symbols

Each line symbol used on drawings means something different. *Figure 19* shows the most common types of lines. However, these lines can vary. Always consult the legend or symbol list when referring to any drawing.

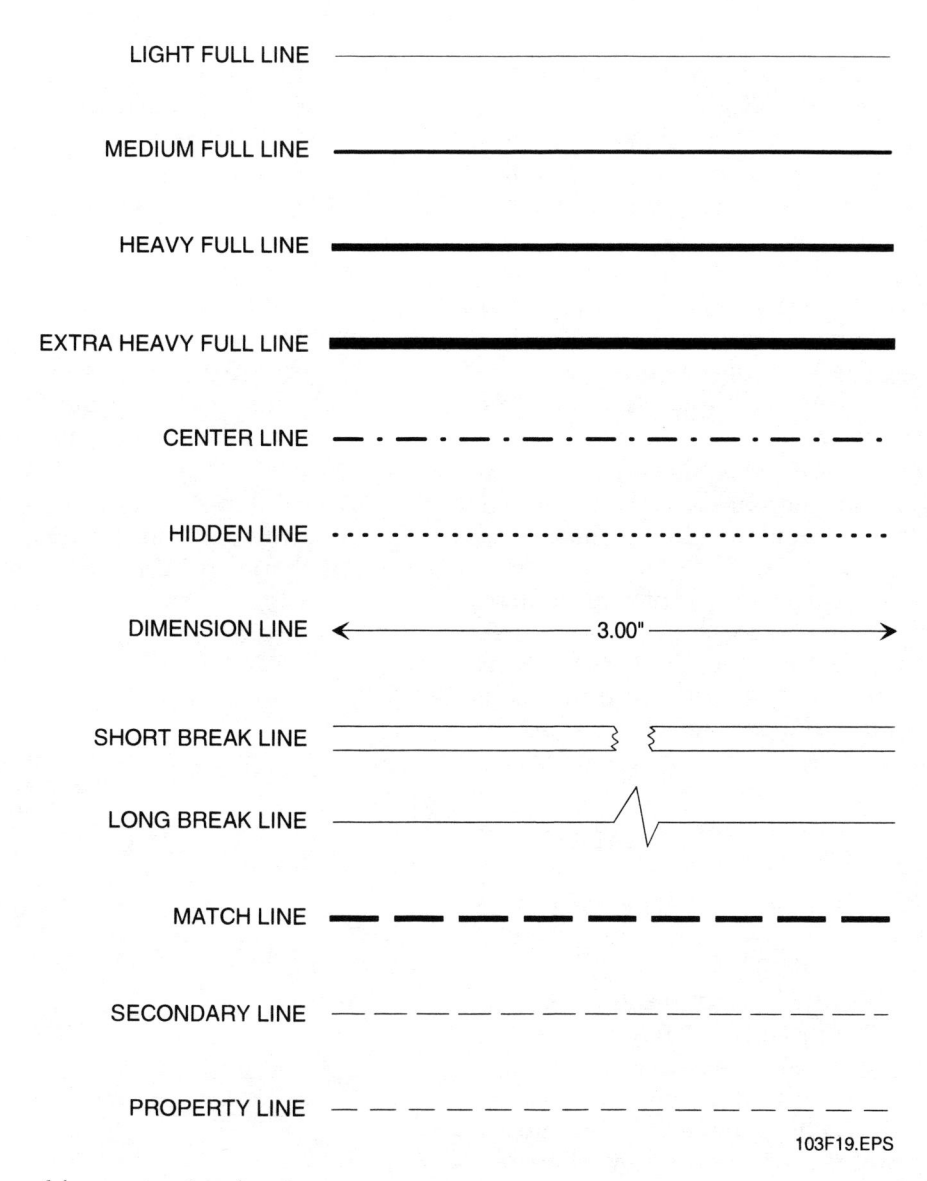

Figure 19 ◆ Lines used for construction drawings.

The drafting lines shown in *Figure 19* are used as follows:

- *Light full line* – This line is used for section lines, building background (outlines), and similar uses where the object to be drawn is secondary to the system being shown, such as HVAC or electrical.
- *Medium full line* – This type of line is frequently used for hand lettering on drawings. It is further used for some drawing symbols and circuit lines.
- *Heavy full line* – This line is used for borders around title blocks, schedules, and for hand lettering drawing titles. Some types of symbols are frequently drawn with a heavy full line.
- *Extra heavy full line* – This line is used for border lines on architectural/engineering drawings.
- *Center line* – A center line is a broken line made up of alternately spaced long and short dashes. It indicates the centers of objects, such as holes, pillars, or fixtures. Sometimes, the center line indicates the dimensions of a finished floor.
- *Hidden line* – A hidden line consists of a series of short dashes that are closely and evenly spaced. It shows the edges of objects that are not visible in a particular view. The object outlined by hidden lines in one drawing is often fully pictured in another drawing.
- *Dimension line* – These are thin lines used to show the extent and direction of dimensions. Dimension lines have three parts: a line, a dimension, and a termination symbol. The dimension is usually placed in a break inside the dimension lines. Normal practice is to place the dimension lines outside the object's outline. However, it may sometimes be necessary to draw the dimensions inside the outline depending on the available room. *Figure 20* shows some common dimension line styles.
- *Short break line* – This line is usually drawn freehand and is used for short breaks.
- *Long break line* – This line, which is drawn partly with a straightedge and partly with freehand zigzags, is used for long breaks.
- *Match line* – This line is used to show the position of the cutting plane. Therefore, it is also called the cutting plane line. A match or cutting plane line is a heavy line with long dashes alternating with two short dashes. It is used on drawings of large structures to show where one drawing stops and the next drawing starts.

- *Secondary line* – This line is frequently used to outline pieces of equipment or to indicate reference points of a drawing that are secondary to the drawing's purpose.
- *Property line* – This is a light line made up of one long and two short dashes that are alternately spaced. It indicates land boundaries on the site plan.

Other uses of the lines just mentioned include the following:

- *Extension lines* – Extension lines are lightweight lines that start about 1/16 inch away from the edge of an object and extend out. A common use of extension lines is to create a boundary for dimension lines. Dimension lines meet extension lines with arrowheads, slashes, or dots. Extension lines that point from a note or other reference to a particular feature on a drawing are called leaders. They usually end in either an arrowhead or a dot and may include an explanatory note at the end.
- *Section lines* – These are often referred to as cross-hatch lines. Drawn at a 45-degree angle, these lines show where an object has been cut away to reveal the inside.
- *Phantom lines* – Phantom lines are solid, light lines that show where an object will be installed. A future door opening or a future piece of equipment can be shown with phantom lines.
- *Geometric lines* – These are usually found in section drawings. They provide information about the shape and dimension of objects. *Figure 21* shows geometric lines found in drawings.

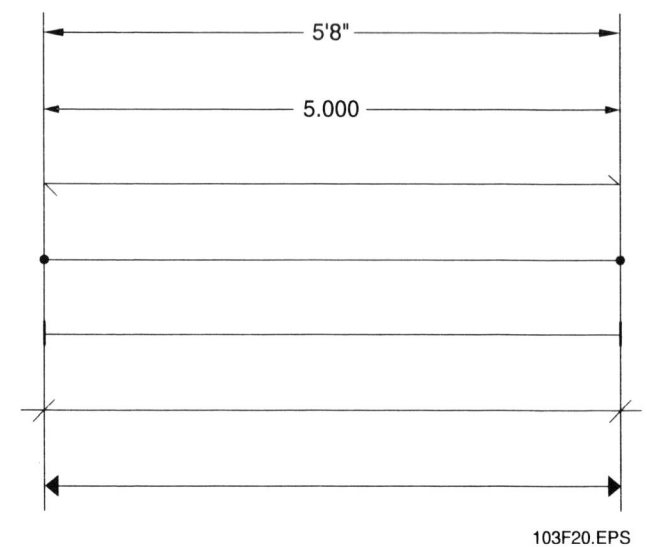

Figure 20 ❖ Dimension line styles.

3.1.2 Symbols and Abbreviations

Symbols and abbreviations are used throughout a set of drawings. They convey accurate, concise, and specific information about a certain object without using much space. Some symbols will look like the object they represent while others will not. You will tend to memorize those you use frequently.

Figure 22 shows architectural symbols for common building materials. Note that about one-half the symbols refer to masonry materials. Some symbols look alike, so it is important to check where they are and what is next to them.

Figures 23 and *24* show electrical and plumbing symbols. Sometimes, outlets or drains are located in or through masonry structures. It is important to check the drawing for any masonry structure to make sure there are no inserts for outlets or drains. If inserts are specified, they must be located where specified, so they will connect properly.

Table 7 is a list of common architectural abbreviations. These abbreviations are used to clarify details on drawings, especially section drawings. They also appear in specification lists.

3.1.3 Scales and Dimensions

Drawings are normally made using a certain scale. The inches or fractions of inches on the drawing represent real distances on the project. For residential drawings, the scale is usually ¼ of an inch equals 1 foot. This means that ¼ of an inch on the drawing equals 1 foot on the ground. For larger commercial projects, a smaller scale of ⅛ of an inch to 1 foot is normally used. Architectural renderings may use a scale of ⅒ of an inch to 1 foot. For detail or sectional drawings, a larger scale is used. Commonly, these detail scales are ½ inch to 1 foot, or 1 inch to 1 foot. These scales are referred to as the ¼-, ⅛-, ½-, and 1-inch scales. When reading a drawing, it is a good idea to first check the scale.

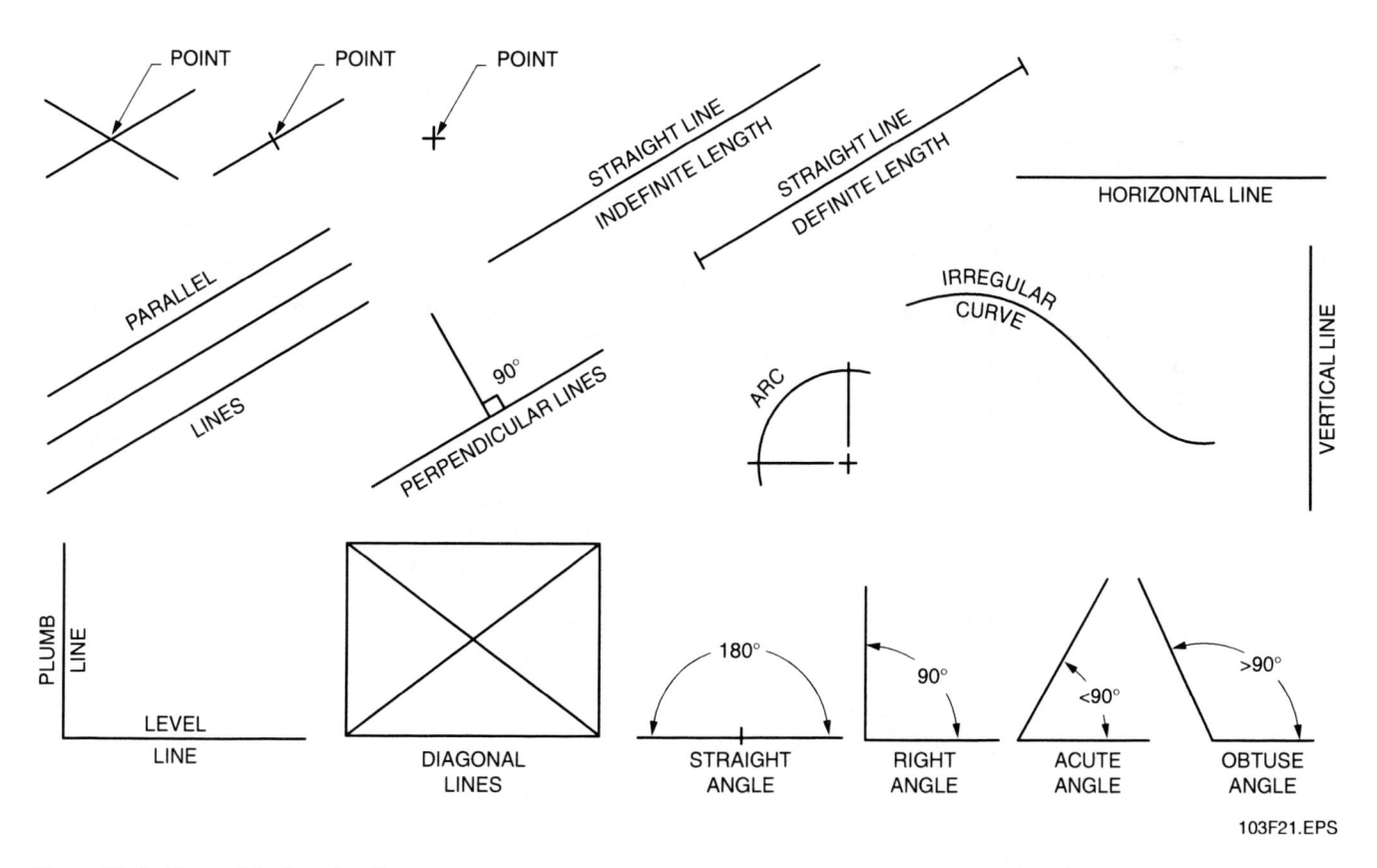

Figure 21 ♦ Geometric drawing lines.

MEASUREMENTS, DRAWINGS, AND SPECIFICATIONS

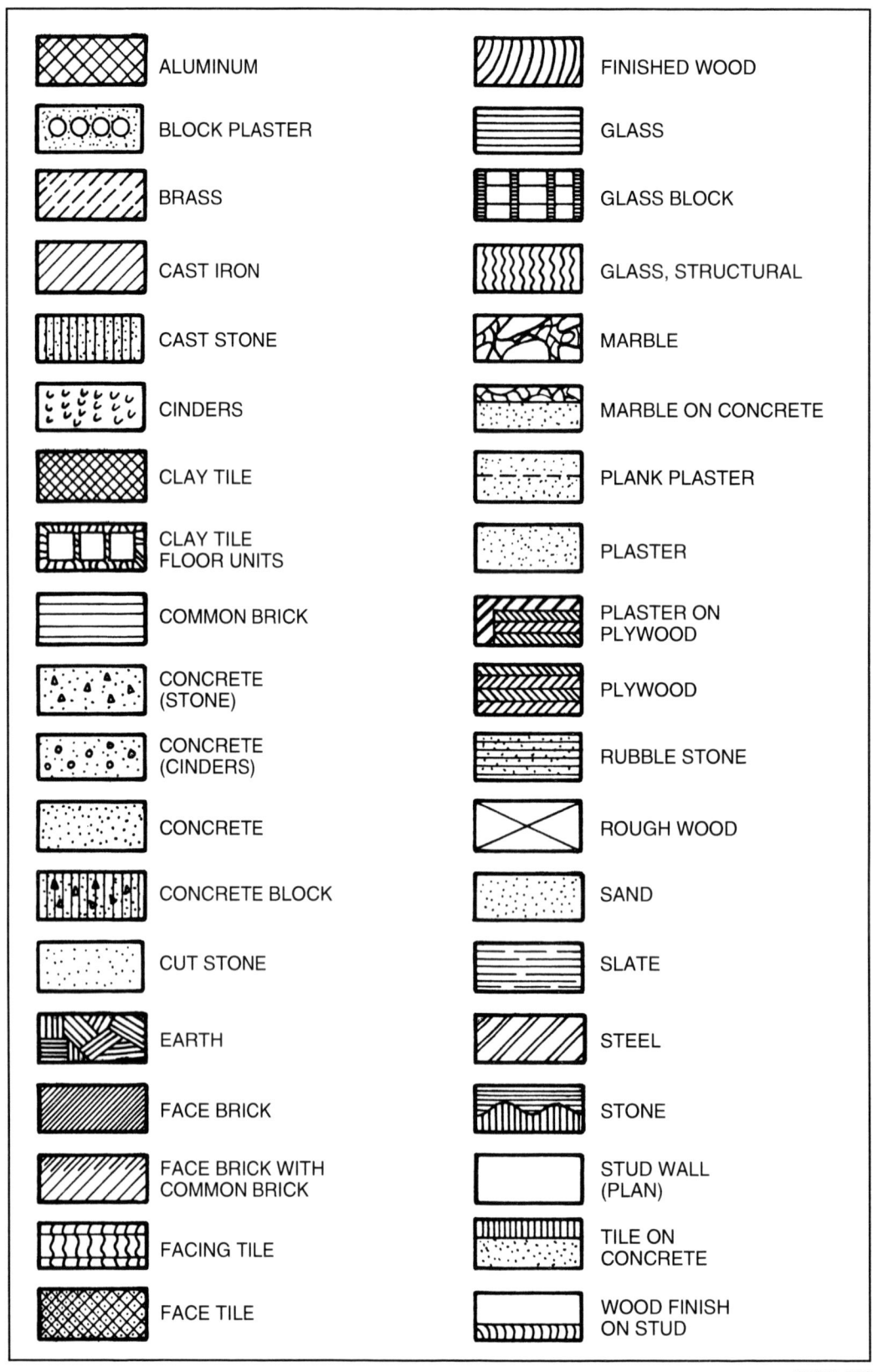

Symbol	Material	Symbol	Material
	ALUMINUM		FINISHED WOOD
	BLOCK PLASTER		GLASS
	BRASS		GLASS BLOCK
	CAST IRON		GLASS, STRUCTURAL
	CAST STONE		MARBLE
	CINDERS		MARBLE ON CONCRETE
	CLAY TILE		PLANK PLASTER
	CLAY TILE FLOOR UNITS		PLASTER
	COMMON BRICK		PLASTER ON PLYWOOD
	CONCRETE (STONE)		PLYWOOD
	CONCRETE (CINDERS)		RUBBLE STONE
	CONCRETE		ROUGH WOOD
	CONCRETE BLOCK		SAND
	CUT STONE		SLATE
	EARTH		STEEL
	FACE BRICK		STONE
	FACE BRICK WITH COMMON BRICK		STUD WALL (PLAN)
	FACING TILE		TILE ON CONCRETE
	FACE TILE		WOOD FINISH ON STUD

Figure 22 ◆ Architectural material symbols.

				CEILING	WALL
Single-Pole Switch	S		Surface Fixture	○	—○
Double-Pole Switch	S₂		Surface Fixt. w/Pull Switch	○ PS	—○ PS
Switch & Single Receptacle	⊖s		Recessed Fixture	Ⓡ	—Ⓡ
Switch & Duplex Receptacle	⊖s		Surface or Pendant Fluorescent Fixture	[○]	
Single Receptacle	⊖		Recessed Fluor. Fixture	[○ R]	
Duplex Receptacle	⊖		Surface Exit Light	Ⓧ	—Ⓧ
Single Special Purpose Recep.	⊖		Recessed Exit Light	(XR)	—(XR)
Range Receptacle	⊖R		Junction Box	Ⓙ	—Ⓙ
Special Purpose Connection or Provision for Connection. Subscript letters indicate Function (DW—Dishwasher; CD—Clothes Dryer, etc.)	⊖DW		Fixture Type (see schedule)	△A	
			Wiring Concealed in Ceiling or Wall	————	
			Wiring Concealed in Floor	– – – –	
Clock Receptacle w/Hanger	Ⓒ		Wiring Exposed	· · · · · ·	
Fan Receptacle w/Hanger	Ⓕ		Branch Circuit Homerun to Panelboard. Number of arrows indicates number of circuits in run.	◄◄———	
Single Floor Receptacle	⊟		Panel Lighting	▬▬▬▬	

RESIDENTIAL OCCUPANCIES

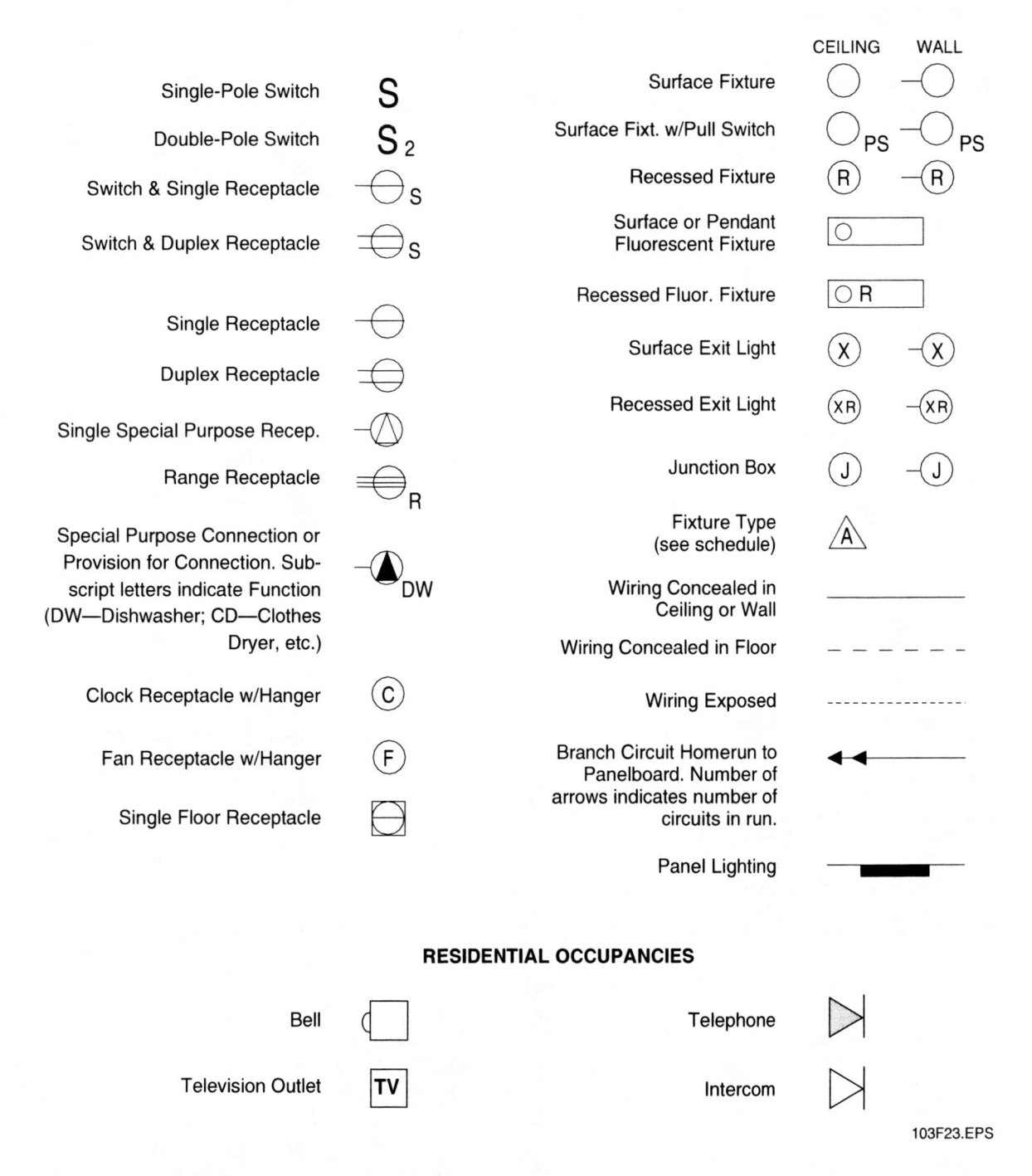

Bell	⊓		Telephone	▷
Television Outlet	TV		Intercom	▷

Figure 23 ◆ Electrical symbols.

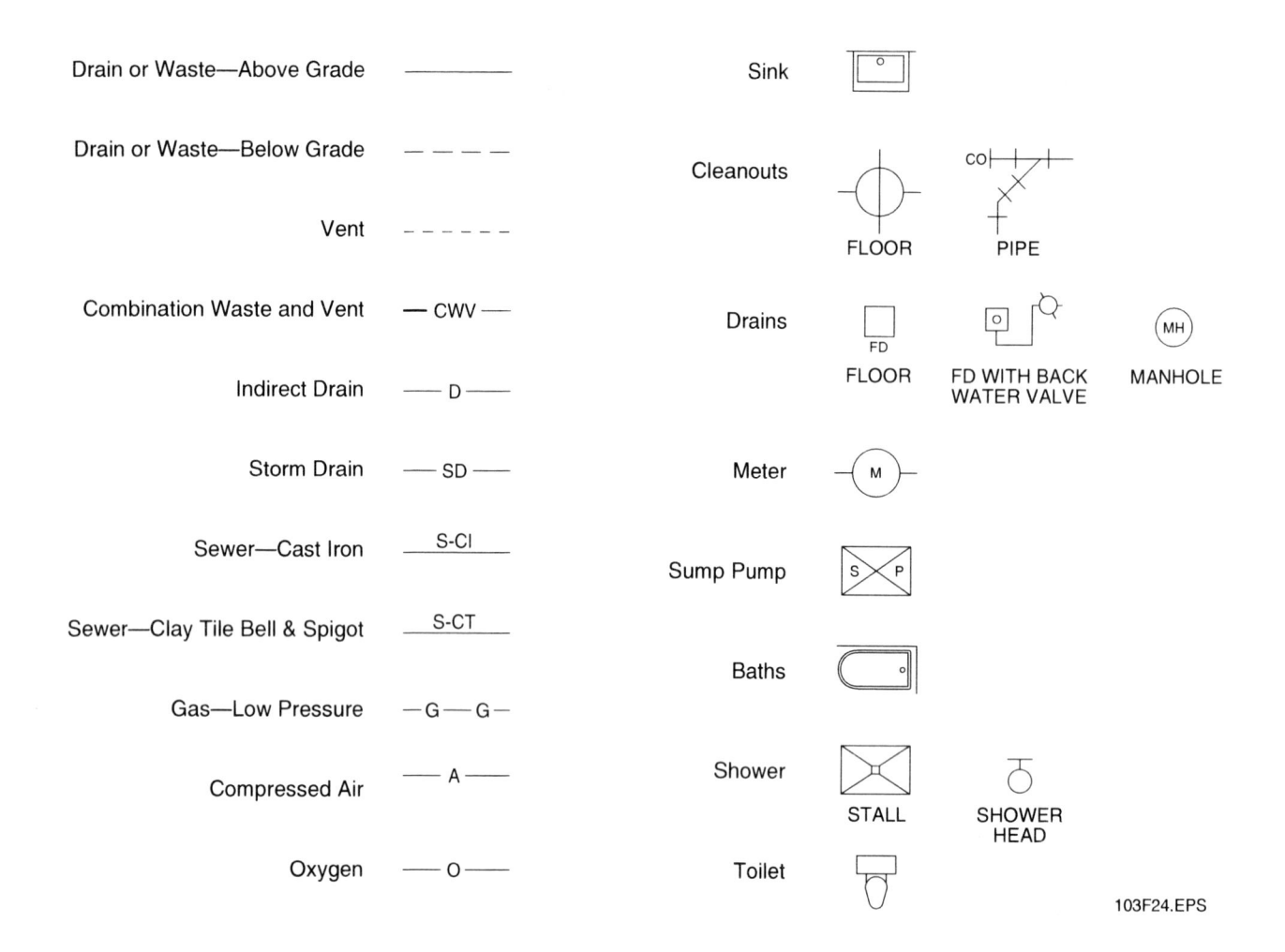

Symbol Name	Symbol
Drain or Waste—Above Grade	————
Drain or Waste—Below Grade	– – – –
Vent	- - - - -
Combination Waste and Vent	— CWV —
Indirect Drain	— D —
Storm Drain	— SD —
Sewer—Cast Iron	S-CI
Sewer—Clay Tile Bell & Spigot	S-CT
Gas—Low Pressure	— G — G —
Compressed Air	— A —
Oxygen	— O —

Symbol Name	Symbol
Sink	
Cleanouts	FLOOR, PIPE (CO)
Drains	FLOOR (FD), FD WITH BACK WATER VALVE, MANHOLE (MH)
Meter	(M)
Sump Pump	(S P)
Baths	
Shower	STALL, SHOWER HEAD
Toilet	

Figure 24 ◆ Plumbing symbols.

Another term you will hear is size. The actual size of an object is two times larger than what is shown on a half-size drawing, four times larger than on a quarter-size drawing, and so on. Size drawings usually show small details of assembly or finishing. Scale and size both need to be checked when reading a drawing.

Dimensions tell the builder how large an object is and its specific location in relation to other objects. Dimensions are usually shown in feet and inches (12'-6"), while dimensions less than one foot are shown in inches (6½). Dimensioning interior walls is generally done in one of three ways; they are dimensioned to the center of the stud, to the outside of the stud, or to the outside of the finished walls. Dimensioning to the center of the studs is the most common.

Masonry walls are dimensioned to the outside of the walls, never to the center. *Figure 25* shows some typical dimensioning practices. Note that windows and doors in wood frame walls are dimensioned to the outside edge of the exterior wall. Windows and doors in brick veneer and frame walls are dimensioned to the edge of the framing, not the edge of the brick.

When reading a set of drawings to determine a particular dimension, do not measure the drawing. It is best to calculate that dimension from the information on the drawing. Do not measure drawings because they may have shrunk or stretched from the reproduction process. If you must measure a drawing to determine a dimension, do it in several places if possible. If you are using a reduced set of drawings (they should be marked as reduced), never measure a dimension, always calculate it.

Table 7 Architectural Abbreviations

ABBREVIATIONS

A.B.	– ANCHOR BOLT	FDN.	– FOUNDATION	RM.	– ROOM
ADD'L	– ADDITIONAL	FIN.	– FINISH	SCHED.	– SCHEDULE
ADJ.	– ADJACENT	FLR.	– FLOOR	SECT.	– SECTION
AISC	– AMERICAN INSTITUTE OF STEEL CONSTRUCTION	F.O.B.	– FACE OF BRICK	SHT.	– SHEET
		F.O.CONC.	– FACE OF CONCRETE	SIM.	– SIMILAR
ALT.	– ALTERNATE	F.O.W.	– FACE OF WALL	S.L.V.	– SHORT LEG VERTICAL
ARCH.	– ARCHITECTURAL	FS	– FLAT SLAB	SPC.	– SPACE
ASTM	– AMERICAN SOCIETY FOR TESTING & MATERIALS	FT.	– FOOT	SPEC.	– SPECIFICATION
		FTG.	– FOOTING	SQ.	– SQUARE
BLDG.	– BUILDING	F.W.	– FILLET WELD	STD.	– STANDARD
BM.	– BEAM	GA.	– GAUGE	STIFF.	– STIFFENER
B.O.	– BOTTOM OF	GAL.	– GALVANIZED	STL.	– STEEL
BOT.	– BOTTOM	G.L.	– GLU-LAM BEAM	STOR.	– STORAGE
BSMT.	– BASEMENT	GR.	– GRADE	SYM.	– SYMMETRICAL
BTWN.	– BETWEEN	GR. BM.	– GRADE BEAM	T&B	– TOP AND BOTTOM
CANT.	– CANTILEVER	H.A.S.	– HEADED ANCHOR STUD	THK.	– THICKNESS
CB.	– CARDBOARD	HORIZ.	– HORIZONTAL	T.O.	– TOP OF
CH.	– CHAMFER	H.S.B.	– HIGH STRENGTH BOLT	TYP.	– TYPICAL
C.J.	– CONTROL/CONSTRUCTION JOINT	ID	– INSIDE DIAMETER	U.N.O.	– UNLESS NOTED OTHERWISE
CLR.	– CLEAR, CLEARANCE	IN.	– INCH	VAR.	– VARIES
CMU	– CONCRETE MASONRY UNIT	INT.	– INTERIOR	VERT.	– VERTICAL
COL.	– COLUMN	JNT.	– JOINT	V.I.F.	– VERIFY IN FIELD
CONC.	– CONCRETE	LB.	– POUND	WT.	– WEIGHT
CONN.	– CONNECTION	LIN. FT.	– LINEAL FEET		
CONST.	– CONSTRUCTION	L.L.V.	– LONG LEG VERTICAL	**SYMBOLS**	
CONT.	– CONTINUOUS	MAT'L	– MATERIAL		
CONTR.	– CONTRACTOR	MAX.	– MAXIMUM	℄	CENTER LINE
CTRD.	– CENTERED	MECH.	– MECHANICAL	⌀	DIAMETER
DET.	– DETAIL	MID.	– MIDDLE	⊕	ELEVATION
DIAG.	– DIAGONAL	MIN.	– MINIMUM		
DIAM.	– DIAMETER	MISC.	– MISCELLANEOUS	&	AND
DIM.	– DIMENSION	MTL.	– METAL	W/	WITH
DISCONT.	– DISCONTINUOUS	N.I.C.	– NOT IN CONTRACT	ℙ	PLATE
DWG.	– DRAWING	NO.	– NUMBER	X	BY
EA.	– EACH	NOM	– NOMINAL	#	NUMBER
E.F.	– EACH FACE	N.T.S.	– NOT TO SCALE	@	AT
EL.	– ELEVATION	O.C.	– ON CENTER	☐	SQUARE
ELECT.	– ELECTRICAL	OD	– OUTSIDE DIAMETER	∟	ANGLE
ELEV.	– ELEVATOR	O.H.	– OPPOSITE HAND		
EQ.	– EQUAL	OPNG.	– OPENING		
E.W.B.	– END WALL BARS	ℙ	– PLATE		
E.W.	– EACH WAY	PSF	– POUND PER SQUARE FOOT		
EXIST.	– EXISTING	PSI	– POUND PER SQUARE INCH		
EXP. JNT.	– EXPANSION JOINT	R	– RADIUS		
EXT.	– EXTERIOR	REINF.	– REINFORCEMENT		
F.D.	– FLOOR DRAIN	REQ'D	– REQUIRED		

MEASUREMENTS, DRAWINGS, AND SPECIFICATIONS

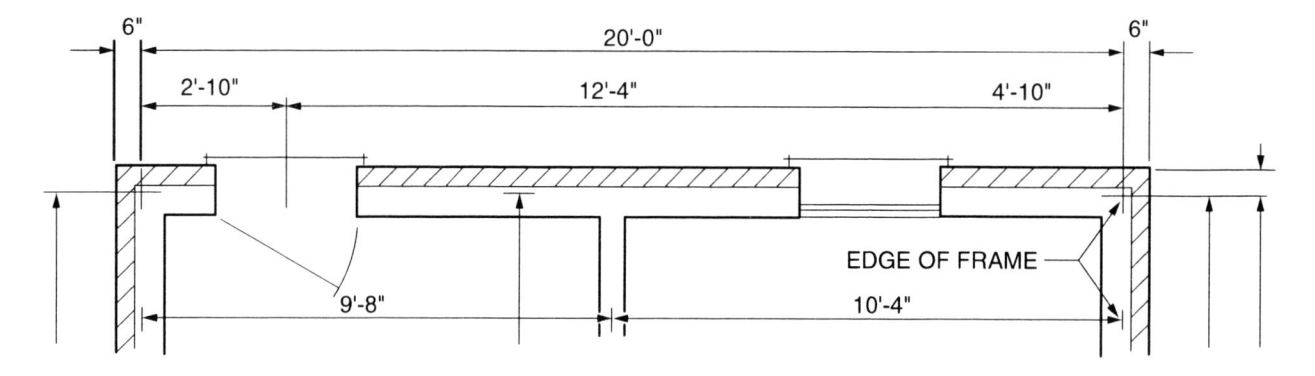

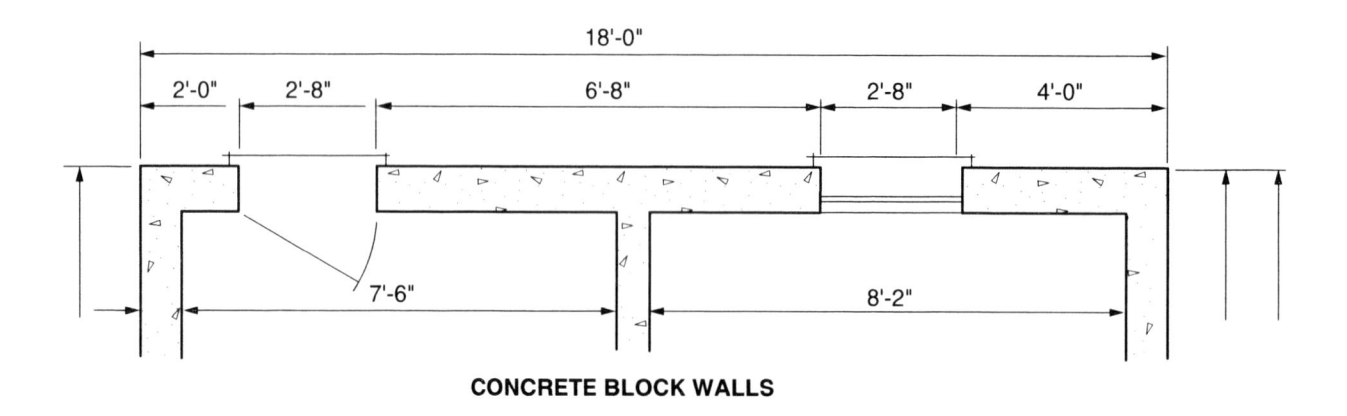

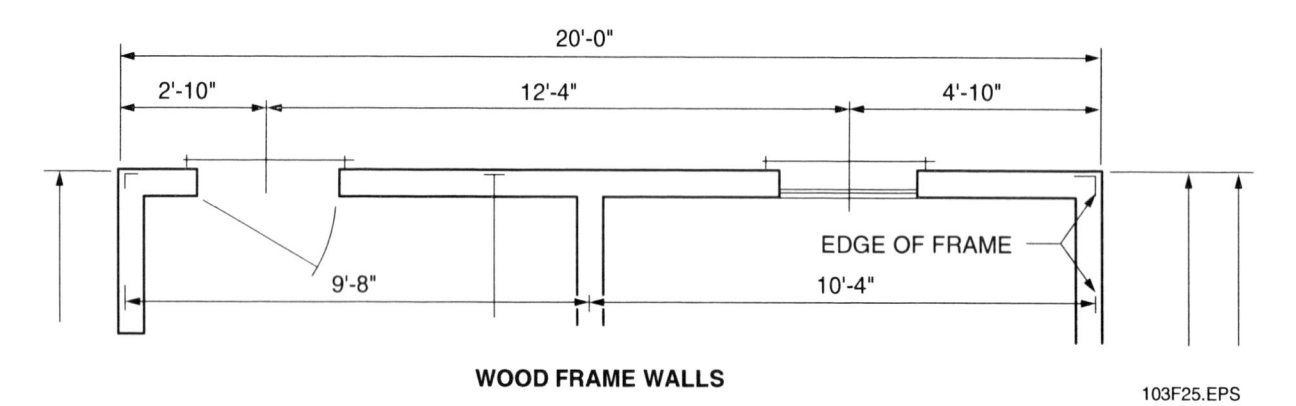

Figure 25 ♦ Dimension lines for different wall types.

3.2.0 Residential Drawings

This section discusses the set of construction drawings included with this module. These working drawings are typical of those used for residential jobs. They are numbered from 1–11. Detailed mechanical and plumbing drawings have been intentionally omitted.

Working drawings are a set of drawings that describe a construction project in sufficient detail to allow a contractor to bid the work and then build it. The drawings also show the craftworker exactly what is to be done. Just as a mason always uses a level to check for plumb, a mason always uses a drawing to check for detail, location, and measurements. These drawings are usually bound together in a package. The package contains a cover page, a table of contents page, and sheets showing the technical details of the project. Depending on the complexity of the project, the specifications may also be included as part of the working drawings. The title block of the drawings also contains useful information about revisions to the plans.

The number of drawings in a set of plans will vary with the complexity of the project. A private home will require fewer sheets than a commercial building, and a large industrial site will require even more drawings. A large commercial project may have site plans by the surveyor, architectural drawings by the architect, structural drawings by the structural engineer, mechanical drawings by the mechanical engineer, and electrical drawings by the electrical engineer, plus detailed specification sheets that are separate from the drawings.

Working drawings are frequently called construction drawings or blueprints. Working drawings are normally organized in the following order:

- Plot or site plans show the general layout of the land and information provided by the surveyor or site planner.
- Architectural/engineering plans include floor plans, foundation plans, elevations, section plans, roofing plans, and schedules for doors, windows, and other equipment.
- Specialty plans include mechanical, plumbing, electrical, and ductwork drawings. These may also include details for any custom features or unusual designs.

3.2.1 Plot or Site Plans

The plot or site plan shows the location of the building in relation to the property lines. It may show utilities, contour lines, site dimensions, other buildings on the property, walks, drives, and retaining walls. This plan also shows the finished floor elevation(s) and the North direction arrow.

Plan 1 is the plot plan for this project. The curved lines with numbers on the front and back lawns are elevation or contour lines. They show that the front of the lot slopes down on the right. At the left side of the house, the elevation lines show that the lot slopes very sharply down to the backyard.

3.2.2 Floor Plans

The floor plan provides the largest amount of information, perhaps making it the most important drawing of all. The floor plan shows all exterior and interior walls, doors, windows, patios, walks, decks, fireplaces, built-in cabinets, and appliances. The floor plan is actually a cross-sectional view taken horizontally between the floor and the ceiling. The height of the cross section is usually cut about 4 feet above the floor. Sometimes this is varied to show important details of the structure.

Plans 2 and 3 are the floor plans for this project. Each level gets its own plan. The doors are noted by diamonds, and the windows by circles. The letters inside the symbols refer to the drawing schedules, with sizes and other details listed in table form.

> **NOTE**
> The diagonal lines on the exterior walls are intended to represent face brick. Note that this architect did not use the standard symbol from *Figure 22*. This is a common occurrence. For this reason, architects will often include a legend in the drawing set to show the symbol they used.

Looking at the plans, how many windows and doors are scheduled for level 1? For level 2? Can you name some of the types of lines used on this plan? How many different lines are used on this plan?

3.2.3 Foundation Plans

The foundation plan is usually the first of the structural drawings. It shows the foundation size and materials. It shows details about excavation, waterproofing, and supporting structures, such as footings or piles. It can have sections of the footings and foundation walls. When a building has a basement, it is included here as well.

Plan 4 is the foundation plan. Notice that it has two section indicators, C/C and D/D. Section indicator D/D crosses a line for a hidden feature, a thickened area in the slab. The section drawings called out by those indicators are in the upper right corner of the plan.

3.2.4 Elevation Drawings

Elevation drawings are views of the exterior features of the building. They usually show all four sides of a building. Sometimes, for a building of unusual design, more than four elevations may be needed. The elevation drawings show outside features, such as placement and height of windows, doors, chimneys, and roof lines. Exterior materials are indicated, as well as important vertical dimensions.

Plans 5 and 6 are the exterior elevation drawings. Notice in these plans that the concrete block retaining wall is now a very obvious feature of the building. Can you find the retaining wall on the plot plan?

Interior elevation drawings show in greater detail the various cabinets, bookshelves, fireplaces, and other important interior features. These are sometimes called detail drawings.

Plan 7 is the interior elevation of the kitchen, vanities, and fireplaces. What are the hidden lines showing for the fireplace elevations?

3.2.5 Section Drawings

Section drawings give information about the construction of walls, stairs, or other items which cannot be easily given on the elevation or floor plan drawings. These drawings are usually drawn to a scale large enough to show the details without cluttering the drawing. A section taken through the narrow width of a building is known as a transverse section. A section taken through the entire length is known as a longitudinal section.

Go back to Plan 3. Locate the section indicators A/A and B/B. Then look at Plan 8. These are the section plans for the horizontal cuts along the section indicators. The section drawings show the building from the foundation footings to the roof.

Plan 9 shows sections through exterior and interior walls. This plan is at a different scale and a much finer level of detail. The working mason should be able to read this section plan and answer questions about the work. What is the foundation wall made of? What is the distance between the brick veneer and the backing walls? How frequently is the veneer tied to the backing? What is the distance between weepholes? Is there a specification here for the windowsills?

3.2.6 Schedules

Schedules are an important tool in a set of plans. Although they are tables, not drawings, they give specific information that would be impractical to include on the documents. They also make it convenient for ordering material by collecting this detailed information in one place.

Go back to Plan 7, which has the door and window schedules. What type of window will be in the kitchen? How many different sizes of door will be needed? What other kinds of items might be listed on schedules?

3.2.7 Specialty Plans

Specialty plans give details about plumbing, electrical, and duct work. It might appear that a mason does not have to know about these things, but this appearance is deceiving.

Plans 10 and 11 are the electrical plans for the first and second floors. Each floor has at least one electrical connector box required on the outside of that brick veneer wall. How many outside connector boxes are needed for both floors? What will be attached to those connector boxes? How will this influence the location of those boxes?

Masons are only one of the many kinds of craftworkers on the job. Understanding what other workers are doing and how they affect your work is important. This information is on the drawings. Use them to help you work smarter.

3.2.8 Metric Drawings

As discussed earlier, the metric system of measurement will eventually be the world standard. Even though no standard sizings have been set for the lumber and building industries in the U.S., some work is going on. New standard sizes for lumber and other materials must be approved. Clay and concrete masonry products have made a soft conversion, which means that the size of the products has not changed but metric sizes have been identified. Units sized to metric dimensions will be adopted in the future.

Figure 26 shows a construction plan labeled using the metric system. On this plan, units for linear measure are restricted to meters and millimeters. Thus, the figures to the left of the decimal indicate meters, while the figures to the right indicate millimeters. Note that the lines and symbols used in the remainder of the plan are not changed.

4.0.0 ◆ SPECIFICATIONS, STANDARDS, AND CODES

A mason must follow certain rules in addition to the building plans. These include specifications, standards, and codes. Construction plans or drawings usually include specifications. These are known as specs. They are written instructions or information needed to complete the work. Specs contain information not found on the drawings. The architect, the engineer, or the owner give direction on how the job must be done. In fact, such specifications are part of the building contract. The plans cannot show all the necessary information for the mason to complete the work. Specifications add the details. These include the following:

- Quality of materials
- Quality of workmanship (minimum tolerance)
- Procedures or techniques to be used during construction
- Various responsibilities of each subcontractor

Generally, specifications are divided into two major areas: general conditions and technical specifications. The general conditions cover such

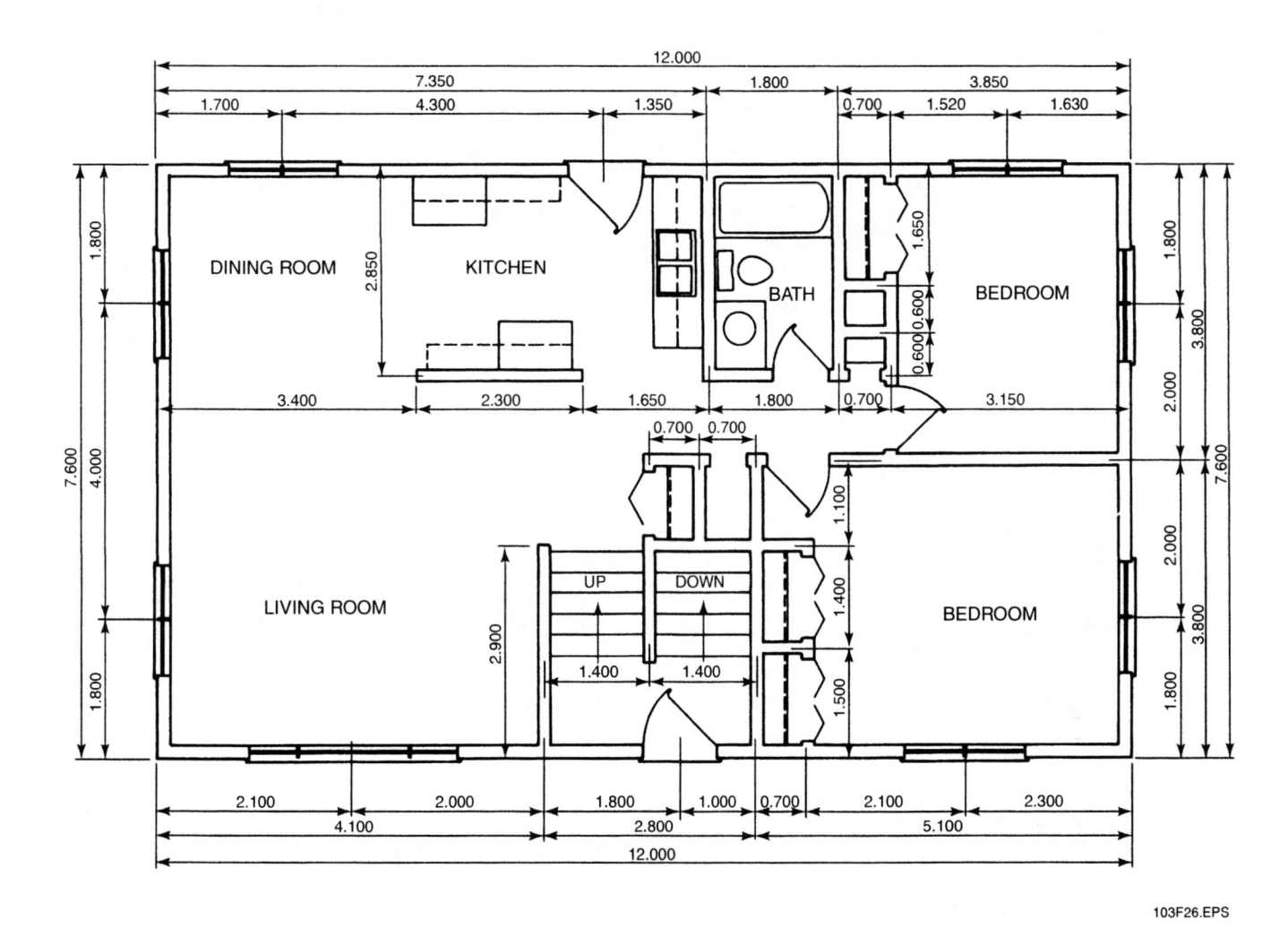

Figure 26 ♦ Metric drawing.

legal items as insurance, permit responsibilities, and payment schedules. The mason will have little contact with these general conditions. Masons do need to read the technical specifications, though, to do their work properly.

Specifications are prepared for specific projects. They reflect the special conditions of that project. Standards apply to all projects. They describe the best practices or minimum requirements for doing certain tasks. Standards are referenced by specifications and by codes.

Codes are legal requirements for building and construction. Model codes are set by a national organization. Local governments adopt model codes and may modify them. Local codes have the force of the law. Building inspectors enforce building codes.

4.1.0 Technical Specifications

Technical specifications directly affect the mason's work. They list how the job is to be completed. Each section includes information for the work of a single subcontractor. On every project, there will be a set of technical specifications for the masonry work.

Technical specifications are legal documents because they are part of a contract. The mason must use both the drawings and the specifications to construct the project. If there is a discrepancy between the two, the specifications have priority. When conflicts occur, consult your supervisor.

Figure 27 is the first page of a six-page form. It is a general technical specification form. It is used by the Farmers Home Administration (FmHA) and the Department of Veterans Affairs (VA) when they finance home building. Other agencies or private companies may use this form. It lists materials, lumber grades, framing dimensions, and so on. The masonry section is short. Only a few items are included. Projects using this form will supplement it with additional forms and specifications.

Regardless of the forms used, masonry trade specifications should include the information outlined in *Figure 28*. This information defines the mason's responsibility on the project.

Description of Materials

U.S. Department of Housing and Urban Development
Department of Veterans Affairs
Farmers Home Administrtation

HUD's OMB Approval No. 2502-0192 (exp. 04/30/2004)
and 2502-0313 (exp. 01/31/2005)

Public reporting burden for this collection of information is estimated to average 30 minutes per response, including the time for reviewing instructions, searching existing data sources, gathering and maintaining the data needed, and completing and reviewing the collection of information. This agency may not collect this information, and you are not required to complete this form, unless it displays a currently valid OMB control number.

The National Housing Act (12 USC 1703) authorizes insuring financial institutions against default losses on single family mortgages. HUD must evaluate the acceptability and value of properties to be insured. The information collected here will be used to determine if proposed construction meets regulatory requirements and if the property is suitable for mortgage insurance. Response to this information collection is mandatory. No assurance of confidentiality is provided.

☐ Proposed Construction ☐ Under Construction   No. _____ (To be inserted by HUD, VA or FmHA)

Property address (Include City and State)

Name and address of Mortgagor or Sponsor

Name and address of Contractor or Builder

Instructions

1. For additional information on how this form is to be submitted, number of copies, etc., see the instructions applicable to the HUD Application for Mortgage Insurance, VA Request for Determination of Reasonable Value, or FmHA Property Information and Appraisal Report, as the case may be.
2. Describe all materials and equipment to be used, whether or not shown on the drawings, by marking an X in each appropriate check-box and entering the information called for each space. If space is inadequate, enter ìSee misc.î and describe under item 27 or on an attached sheet. **The use of paint containing more than the percentage of lead by weight permitted by law is prohibiited.**
3. Work not specifically described or shown will not be considered unless required, then the minimum acceptable will be assumed. Work exceeding minimum requirements cannot be considered unless specifically described.
4. Include no alternates, ìor equalî phrases, or contradictory items. (Consideration of a request for acceptance of substitute materials or equipment is not thereby precluded.)
5. Include signatures required at the end of this form.
6. The construction shall be completed in compliance with the related drawings and specifications, as amended during processing. The specifications include this Description of Materials and the applicable Minimum Property Standards.

1. **Excavation**
   Bearing soil, type _____

2. **Foundations**
   Footings concrete mix _____ strength psi _____ Reinforcing _____
   Foundation wall material _____ Reinforcing _____
   Interior foundation wall material _____ Party foundation wall _____
   Columns material and sizes _____ Piers material and reinforcing _____
   Girders material and sizes _____ Sills material _____
   Basement entrance areaway _____ Window areaways _____
   Waterproofing _____ Footing drains _____
   Termite protection _____
   Basementless space ground cover _____ insulation _____ foundation vents _____
   Special foundations _____
   Additional information

3. **Chimneys**
   Material _____ Prefabricated (make and size) _____
   Flue lining material _____ Heater flue size _____ Fireplace flue size _____
   Vents (material and size) gas or oil heater _____ water heater _____
   Additional information

4. **Fireplaces**
   Type ☐ solid fuel ☐ gas-burning ☐ circulator (make and size) _____ Ash dump and clean-out _____
   Fireplace facing _____ lining _____ hearth _____ mantel _____
   Additional information

Retain this record for three years   Page 1 of 6   ref. HUD Handbook 4145.1 & 4950.1 form HUD-92005 (10/84)
VA Form 26-1852 and form FmHA 424-2

103F27.EPS

Figure 27 ◆ Standard FmHA/VA specification form.

> **Scope of the work**
> 	Type of work and materials to be used
> 	The mason's responsibilities (for instance, is the mason responsible for applying flashings? caulking? for job cleanup?)
>
> **Type of materials**
> 	Type of face and common brick, concrete block, and stone, and type of mortar mix specified for each
> 	Type of waterproofing for foundation wall and where it is to be applied to the wall
> 	Quality standards the materials should meet, such as compressive strength and color range
>
> **Masonry workmanship**
> 	Quality of workmanship expected
> 	Type of mortar joint finish and the type of bond desired
> 	Size of pieces of masonry units to be installed in the wall and smallest allowable piece (for example, no piece smaller than 2 inches in face brickwork)
> 	Cutting instrument to be used on masonry units, such as a chisel and hammer or masonry saw
> 	Care of the wall at the conclusion of the day's work, such as the stipulation that all masonry work be covered with tarpaulins or heavy plastic to protect it from the weather
> 	Minimum temperature that work may commence to prevent the freezing of mortar
> 	Specific instructions on the parging of the basement wall
>
> **Stone work**
> 	Type of stone and bond pattern desired
> 	Size and type of mortar joint required
> 	Method in which the stone is to be cut, chiseled, or sawed with the masonry saw
> 	Type of protection for the work
>
> **Concrete masonry unit**
> 	Specific shape, such as bullnose or corner block
> 	Type of mortar and mix
> 	Type of joint finish and allowable thickness
> 	Type of wall ties or joint reinforcement
> 	Type of lintel to be placed over openings, such as concrete masonry lintel, steel, or precast concrete
>
> **Building in and around mechanical work**
> 	Methods of installation of pipe chases, heating units, and electrical work
>
> **Caulking**
> 	Type of caulking and filler materials to be installed in back of the caulking if necessary
> 	Workmanship to be neat and excess caulking removed from frames
>
> **Cleaning and pointing masonry work**
> 	Type of cleaning agent to be used and instructions on how concrete block and stone are to be cleaned
> 	Directions indicating that all holes in mortar joints are to be pointed with fresh mortar as the work is cleaned

Figure 28 ♦ Specification contents.

Highly detailed technical specifications spell out not only the materials, but how they are to be assembled and finished. This ensures the long-term performance of the materials. It is based on engineering studies about the properties of masonry construction.

Specifications take advantage of these engineering studies. They do this by referencing existing standards and codes. By referencing them, the specifications incorporate the provisions of the standards and codes.

Almost every type of building material has a multitude of size and performance standards and codes for it. The standards and codes guide and regulate manufacturers, designers, and builders. There are voluntary standards, national standards, international standards, national codes, local building codes, and model codes. The next sections discuss standards and codes in greater detail.

4.2.0 Standards

When you look at technical specifications, you will find items like "work to meet *ASTM C144-96*" or "as defined by *CSA A165.2*." These phrases refer to studies by the American Society for Testing and Materials (ASTM) International or the Canadian Standards Association (CSA). These are known as standards. They establish minimum requirements for all aspects of masonry units and masonry work.

A standard usually has a single subject. It defines one aspect of the subject. The standards are based on studies, research, and advances in materials and construction techniques. Typical standards include the following:

- *ASTM C1072, Method for Measurement of Masonry Flexural Bond Strength*
- *ASTM C315, Specification for Clay Flue Linings*

> **DID YOU KNOW?**
> *Earliest Building Code*
>
> The earliest building code was the Code of Hammurabi. The king of the Babylonian empire adopted this code in 2200 B.C.E. It assessed severe penalties, including death, if a building was not constructed safely.

Appendix C lists ASTM standards for masonry construction. Read through the list of titles in *Appendix C*. You can see how much study has gone into modern masonry materials and practices.

The ASTM and CSA are only two of the organizations publishing standards for the masonry industry. Consensus standards and codes are developed by other independent organizations, including the following:

- The American Society of Civil Engineers (ASCE)
- The American Institute of Steel Construction (AISC)
- The National Concrete Masonry Association (NCMA)
- The American Concrete Institute (ACI)

Standards are updated regularly. ASTM and the other organizations gather data on new materials and techniques and include them in the standards. For ASTM publications, the year of publication follows a dash after the title number. Always make sure you are using the most current standard.

4.3.0 Codes

Building codes are enforceable standards. They include all aspects of building construction. Their primary purpose is to protect the public. Safety is included in all aspects of building construction. Some areas have special concerns, like earthquakes. The local codes in these areas include special conditions for local hazards. Local codes can also include local requirements, such as conformance with a zoning plan, or lot setbacks.

You may have heard of the *Uniform Building Code*. Throughout the years, organizations have worked together to establish model building codes. Model codes are technical documents written by an organization. They are based on or incorporate standards published by ASTM and other organizations. Local communities or states can adopt a model code. The model code then becomes a local code or law. This provides communities with sound laws without the expense of research and investigation.

Prior to 2000, there were three model codes used throughout the U.S.:

- *The Southern Standard Building Code*, published by the Southern Building Code Congress (SBCC), typically used throughout the southeast
- *The National Building Code*, published by the Building Officials and Code Administrators International (BOCA), adopted mostly in the northeast and central states
- *The Uniform Building Code*, published by the International Conference of Building Officials (ICBO), used throughout the west

In 2000, these three organizations merged into the International Code Council. They issued the *International Building Code (IBC)*. This model code has been adopted and is now law throughout most of the U.S. The National Fire Protection Association also issued a model code called *NFPA 5000*. It is used in a few communities and has been adopted by the state of California.

Building Code Requirements for Masonry Structures, ACI 530-92/ASCE 5-92/TMS 402-92, consolidates several masonry codes. It is published by the ACI, ASCE, TMS, and NCMA. *Figure 29* is an outline of the code. Each of these topics refers to many standards. It reflects many years of testing and research.

Specifications for commercial and industrial projects are based on local codes. Unlike residential projects, commercial and industrial projects are always designed and managed by architects and engineers registered by the state. They must make sure that the project complies with all local codes. Codes for commercial projects are usually more detailed. They are more stringent about fire protection, loadbearing, and parking spaces than codes for residential buildings.

4.4.0 Inspection and Testing

Every construction contract has specifications to cover inspection and acceptance of the finished work. ASTM has published many specifications about testing. Acceptance of the finished product depends on the inspection and test results.

For residential and commercial projects, inspection and testing are usually done by the county building inspector. This inspector will be looking for compliance with the local building code. The inspector works closely with the contractor to coordinate inspections with the completion of critical phases of the work. Work cannot progress on the next phase until this inspection has been made and the inspector signs off on the work. For this reason, you may see an inspector's check-off sheet

Building Code Requirements for Masonry Structures

PART 1—GENERAL

Chapter 1—General Requirements
- 1.1 Scope
- 1.2 Permits and drawings
- 1.3 Approval of special systems of design or construction
- 1.4 Standards cited in this code

Chapter 2—Notation and Definitions
- 2.1 Notation
- 2.2 Definitions

PART 2—QUALITY ASSURANCE AND CONSTRUCTION REQUIREMENTS

Chapter 3—General
- 3.1 Materials, labor, and construction
- 3.2 Acceptance relative to strength requirements

Chapter 4—Embedded Items—Attachments to framing and existing construction
- 4.1 Embedded pipes and conduits
- 4.2 Attachment of masonry to structural frames and other construction
- 4.3 Connectors

Chapter 5—General Analysis and Design Requirements
- 5.1 Scope
- 5.2 Loading
- 5.3 Load combinations
- 5.4 Design strength
- 5.5 Material properties
- 5.6 Deflection of beams and lintels
- 5.7 Lateral load distribution
- 5.8 Multiwythe walls
- 5.9 Columns
- 5.10 Pilasters
- 5.11 Load Transfer
- 5.12 Concentrated loads
- 5.13 Section properties
- 5.14 Anchor bolts solidly grouted in masonry
- 5.15 Framed construction
- 5.16 Stack bond masonry

Chapter 6—Unreinforced Masonry
- 6.1 Scope
- 6.2 Stresses in reinforcement
- 6.3 Axial compression and flexure
- 6.4 Axial tension
- 6.5 Shear

Chapter 7—Reinforced Masonry
- 7.1 Scope
- 7.2 Steel reinforcement
- 7.3 Axial compression and flexure
- 7.4 Axial tension
- 7.5 Shear

Chapter 8—Details of Reinforcement
- 8.1 Scope
- 8.2 Size of reinforcement
- 8.3 Placement limits for reinforcement
- 8.4 Protection for reinforcement
- 8.5 Development of reinforcement embedded in grout

Chapter 9—Empirical Design of Masonry
- 9.1 Scope
- 9.2 Height
- 9.3 Lateral stability
- 9.4 Compressive stress requirements
- 9.5 Lateral support
- 9.6 Thickness of masonry
- 9.7 Bond
- 9.8 Anchorage
- 9.9 Miscellaneous requirements

Appendix A—Special Provisions for Seismic Design
- A.1 General requirements
- A.2 Special provisions for Seismic Zones 0 and 1
- A.3 Special provisions for Seismic Zone 2
- A.4 Special provisions for Seismic Zones 3 and 4

Figure 29 ◆ Building code requirements.

posted on the job site with the inspector's comments and signature for each phase. This may be in the job superintendent's office, or it may be posted at the next place the inspector is due to visit.

For residential projects, testing is usually limited to looking at the material tags and manufacturer's documents that come with the materials. For this reason, you should not throw these documents away until the inspector visits the site and signs off on the work.

On some projects, the architect will also inspect the work as it progresses. The architect will be looking for compliance with the drawings and specifications and will not worry about the building code. On some projects, a representative of the owner may check the craftsmanship of the work accomplished every day. Work that does not meet standards, drawings, specifications, or local codes must be done again. This can be very costly, so it is important to do it right the first time.

Inspection and testing on commercial and industrial projects is much more involved and, depending on the project size, will require a full-time staff hired by the owner. The representatives of the owner and/or architect will look for quality and compliance with the plan and specifications. The building inspector will look for compliance with the code. The contractor must pay to take out and replace work that is not up to specifications or code, just as with residential projects.

Summary

This module covered three topics: math, drawings, and specifications.

Math skills are useful for calculating materials and supplies. They are also needed to interpret project drawings. Masons need to know how to convert denominate fractions and metric measures. They must be able to calculate areas and volumes of common geometric figures. They also need to read mason's rules, both modular and course.

Project drawings include several categories. A project drawing package will include plot plans, floor plans, elevations, sections, materials schedules, structural drawings, and mechanical drawings. They provide details by using lines, symbols, and measurement notations.

Project drawings also include specifications that provide detail not available on the plans. Technical specifications refer to and include standards and codes. Standards are single-topic, nationally published, detailed measures. Model codes include references to many standards. Local codes are usually based on model codes and include local matters, such as zoning.

Review Questions

1. Add the following:

    3 yards 1 foot 11 inches
    +1 yard 2 feet   8 inches

    a. 4 yards 3 feet 7 inches
    b. 1 yard 2 feet 3 inches
    c. 5 yards 1 foot 7 inches
    d. 4 yards 3 feet 19 inches

2. Add 2 3/16 and 5/8.
    a. 2 8/16
    b. 2 3/4
    c. 2 13/16
    d. 2 1/2

3. How many courses are shown on the brick spacing rule at arrow A in *Figure 1*?
    a. One
    b. Two
    c. Three
    d. Four

4. A modular spacing rule is based on a module size of _____ inches.
    a. 2
    b. 4
    c. 8
    d. 16

5. If 1 foot equals 30.48 centimeters, convert 250 feet into meters.
    a. 76.2 meters
    b. 7,620 meters
    c. 762 meters
    d. 7.62 meters

6. If one square meter equals 10.76 square feet, convert 5 square meters into the appropriate U.S. Customary units.
    a. 583 square inches
    b. 53 square feet
    c. 53 square feet, 9 1/16 inches
    d. 53.8 square feet

7. A wall measures 4 feet high and 40 feet long. You are going to face the wall using a brick with a nominal size of 4 inches × 2 2/3 inches × 8 inches. From a chart, you determine you will need 675 bricks per 100 square feet. How many bricks will be needed assuming no wastage or breakage?
    a. 972
    b. 1,080
    c. 1,180
    d. 2,160

8. A foundation is to be circular and 20 centimeters thick. Its diameter is 4 meters. How many cubic meters of concrete will it take?
    a. 0.8 cubic meters
    b. 2.5 cubic meters
    c. 8 cubic meters
    d. 12.5 cubic meters

9. If the length of a window is 2 meters and the width is 1.6 meters, what is the total surface area of the window?
    a. 3.2 square meters
    b. 4.2 square meters
    c. 4.6 square meters
    d. 5.4 square meters

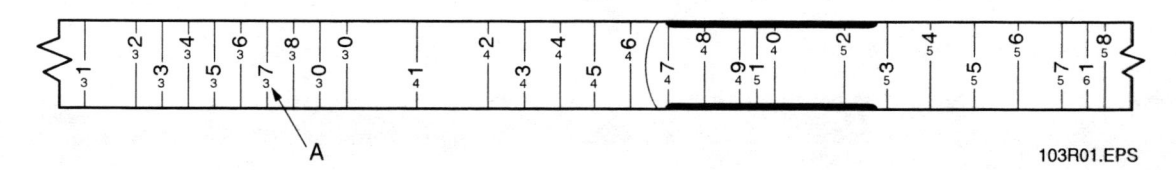

Figure 1

10. A wall is 20 feet high, and it must be secured by a cable 25 feet from the wall. How long is the cable? (Round your answer to the nearest foot.)
    a. 30 feet
    b. 32 feet
    c. 35 feet
    d. 40 feet

11. Heavy full lines are used for _____.
    a. extension lines
    b. edges of objects
    c. border lines
    d. section lines

12. After looking at a ¼ scale floor plan, you decide there is no way to calculate a certain dimension that you need. Using a ruler, you measure the dimension as 4⅞ inches. What dimension will you use when laying out the actual work?
    a. 4 feet, 10½ inches
    b. 9 feet, 9 inches
    c. 19 feet, 3 inches
    d. 19 feet, 6 inches

13. Which of the following plans would show the location of utilities, walks, and drives?
    a. Section drawing
    b. Floor plan
    c. Elevation drawing
    d. Plot or site plan

14. Ceiling heights are shown on which drawings?
    a. Section drawings
    b. Elevation drawings
    c. Foundation plans
    d. Schedule drawings

15. Once a building is completed, the county building inspector will look for compliance with the _____.
    a. ASTM Standards
    b. International Building Codes
    c. technical specifications
    d. local building codes

MEASUREMENTS, DRAWINGS, AND SPECIFICATIONS

PROFILE IN SUCCESS

Kenneth Cook, Vice President

Pyramid Masonry
Atlanta, GA

Ken Cook went to his first masonry project site when he was six years old and has never looked back. Today he is vice president of operations for a major masonry contracting company.

How did you get started in the Masonry trade?
My father was a mason, and he began teaching me the trade when I was just a child. I learned on the job, but I had a long time to learn and a patient teacher to bring me along. Not everyone can count on having the same opportunity, so it's best to learn the proper techniques through a training program before you go on a job.

Describe your job.
As vice president of operations, I handle the day to day running of the business. I do sales and estimating work as well. I keep my hand in by spending as much time as I can on job sites. I make sure the work is being done to the client's specifications, and I do whatever is necessary to make sure the job stays on schedule and within budget.

What do you think it takes to become a success?
In a word, teamwork. There are many tasks to do on the site: mixing mortar, laying brick or block, erecting and moving scaffolds, and keeping masons supplied with materials. Everyone has to do their part, and everyone has to pitch in and do what is needed to keep the job flowing smoothly. If there's one prima donna on a crew, it will drag down the entire crew.

What do you like most about your job?
I enjoy working directly with the crews on the job, especially teaching them proper technique and helping them upgrade their skills. A company is only as good as its employees, so it's important to make sure every employee is as effective and productive as possible.

What would you say to someone just entering the trade?
Get training. Do it through an apprentice program if you can, but get training in any way possible. Also, listen to experienced people. They can pass on knowledge and tricks of the trade they learned from experience. After you've had a couple of years of experience, you may think you know everything there is to know about the trade. In reality, just about everyone you encounter on the job will have a different perspective. No matter how long you've been in the trade, you will meet people who have ideas and techniques you've never thought of.

GLOSSARY

Trade Terms Introduced in This Module

Denominate numbers: Those numbers indicating a unit of measure, such as feet or tons.

Nominal dimension: The size of the masonry unit plus the thickness of one standard (½ inch or ⅜ inch) mortar joint; the nominal dimensions are used in laying out courses.

International System (SI): The metric-based units of measure used in most countries.

U.S. Customary system: The units of measure commonly used in the United States, such as inches, feet, yards, miles, quarts, and gallons. Also known as the English System.

APPENDIX A

Answers to Practice Exercises

Practice Exercise (2.1.1)

1. | | | |
    |---|---|---|
    | 1 yard | 1 foot | |
    | 2 yards | 1 foot | 10 inches |
    | +3 yards | 2 feet | 5 inches |
    | | 4 feet | 15 inches |
    | | −3 feet | 12 inches |
    | 6 yards | 1 foot | 3 inches |

2. | | | |
    |---|---|---|
    | 1 yard | 1 foot | |
    | 4 yards | 2 feet | 9 inches |
    | +6 yards | 2 feet | 4 inches |
    | | 5 feet | 13 inches |
    | | −3 feet | 12 inches |
    | 11 yards | 2 feet | 1 inch |

3. | | | |
    |---|---|---|
    | 1 yard | 1 foot | |
    | 7 yards | 2 feet | 7 inches |
    | +4 yards | 1 foot | 6 inches |
    | | 4 feet | 13 inches |
    | | −3 feet | 12 inches |
    | 12 yards | 1 foot | 1 inch |

4. | | | |
    |---|---|---|
    | 1 yard | 1 foot | |
    | 8 yards | 1 foot | 9 inches |
    | +4 yards | 1 foot | 7 inches |
    | | 3 feet | 16 inches |
    | | −3 feet | 12 inches |
    | 13 yards | 0 feet | 4 inches |

Practice Exercise (2.1.2)

1. | | | |
    |---|---|---|
    | 1 foot (12 + 7) | | = 18 inches |
    | 6 yards | 2 feet | 6 inches |
    | −2 yards | 1 foot | 8 inches |
    | 4 yards | 0 feet | 10 inches |

2. | | | |
    |---|---|---|
    | 4 yards (3 + 1) | | = 4 feet |
    | | 1 foot (12 + 5) | = 17 inches |
    | 5 yards | 2 feet | 5 inches |
    | −3 yards | 2 feet | 7 inches |
    | 1 yard | 2 feet | 10 inches |

3. | | | |
    |---|---|---|
    | | 34 feet (12 + 4) | = 16 inches |
    | | 35 feet | 4 inches |
    | | − 21 feet | 6 inches |
    | | 13 feet | 10 inches |
    | 4 yards | 1 foot | 10 inches |

4. | | | |
    |---|---|---|
    | | 20 feet (12 + 3) | = 15 inches |
    | | 21 feet | 3 inches |
    | | − 11 feet | 9 inches |
    | | 9 feet | 6 inches |
    | 3 yards | 0 feet | 6 inches |

Practice Exercise (2.3.2)

1. *Modular brick* – 6 × 4.5 = 27 courses
2. *Roman brick* – 8 × 4.5 = 36 courses
3. *Utility brick* – 4 × 4.5 = 18 courses
4. *Norman brick* – 6 × 4.5 = 27 courses

Practice Exercise (2.4.2 and 2.4.3)

1. 0.5 in × 2.540 cm = 1.27 cm
2. 0.625 in × 2.540 cm = 1.59 cm
3. 0.5625 in × 2.540 cm = 1.43 cm
4. 5 gallons × 3.785 liters/gallon = 18.925 liters
5. 94 pounds × 0.4536 kg/lb = 42.64 kg
6. Your height to meters and centimeters

   _____ in × 2.540 in/cm = _____ cm

   _____ cm ÷ 100m/cm = _____ m

Practice Exercise (2.5.0)

1. $A = s^2$
   $A = (1.56 \text{ meters})^2 = 2.4336 \text{ m}^2$
   $A = 2.4336 \text{ m}^2 \times 10.76 \text{ ft}^2/\text{m}^2 = 26.19 \text{ ft}^2$

2. A = lw, l = 2w, w = 3 ft 8 in
   w = 3 ft (12 in/ft) + 8 in = 36 in + 8 in = 44 in
   l = 2w = 2(44 in) = 88 in
   A = 44 in × 88 in
   A = = 3,872 in^2/144 in^2/ft^2
   A = 26.89 ft^2
   A = 26.89 ft^2 × 0.0929 ft^2/m^2 = 2.498 m^2

3. A = bh ÷ 2
   h = 0.5 m × 100 cm/m = 50 cm
   A = 35 cm × 50 cm ÷ 2
   A = 1,750 cm ÷ 2 = 875 cm^2
   A = 1,750 cm ÷ 2 = 875 cm^2
   875 cm^2 × 0.155 cm^2/in^2 = 135.63 in^2

4. $A = \pi r^2$
   r = d ÷ 2 = 0.52 m ÷ 2 = 0.26 m
   A = 3.14 × (0.26 m)^2
   A = 3.14 × (0.676 m)^2 = 0.2123 m^2
   0.2123 m^2 × 10.76 ft^2/m^2 = 2.284 ft^2

5. $A = (S_1 + S_2 + S_3 + S_4 + S_5) \, r \div 2$
   A = (24 + 24 + 24 + 24 + 24) 28.5 ÷ 2
   A = (120 in) 28.5 in ÷ 2
   A = 3,420 in ÷ 2
   A = 1,710 in^2 ÷ 144 in^2/ft^2 = 11.875 ft^2
   11.875 ft^2 × 929 cm^2/ft^2 = 11,031.88 cm^2
   11,032 cm^2 × 0.0001 m^2/cm^2 = 1.1032 m^2

6. A = lw
   Area of hole = 3 ft × 6 ft = 18 ft^2
   8 in ÷ 12 in/ft = 0.667 ft
   16 in ÷ 12 in/ft = 1.333 ft
   Area of block face = 0.667 ft × 1.333 ft = 0.889 ft^2
   # blocks = area of hole ÷ area of bricks
   # blocks = 18 ft^2 ÷ 0.889 ft^2 = 20.25 rounded to 21

   Add 5% for breakage:
   20.25 × 1.05 = 21.26 rounded to 22 blocks

   Add one-third more for Flemish bond:
   22 × 1.33 = 29.26 rounded to 30 blocks

7. If you know how many blocks per square foot, then you can skip the calculation of the area of the block face. You would get the same answer.
   # blocks = 18 ft^2 × 1.125 blocks/ft^2 = 20.25 rounded to 21 blocks

8. Area of wall = total area of wall − area of window
   Total area of wall = lw
   Total area of wall = 4 m × 5 m = 20 m^2
   Area of window = πr^2
   r = d ÷ 2 = 1 m ÷ 2 = 0.5 m
   Area of window = πr^2 = 3.14 × (0.5 m)^2 = 0.785 m^2
   Area of wall = 20 m^2 − 0.785 m^2 = 19.215 m^2
   19.215 m^2 × 10.76 ft^2/m^2 = 206.753 ft^2

Practice Exercise (2.6.0)

1. V = l × w × H
   8 feet = 8 feet ÷ 3 ft/yd = 2.67 yds
   20 feet = 20 feet ÷ 3 ft/yd = 6.67 yds
   4 inches = 4 in ÷ 36 in/yd = 0.11 yd
   V = 2.67 yds × 6.67 yds × 0.11 yds = 1.96 yds^3

2. Total volume = volume of core × number of cores
   Volume of core = l × w × H
   = 2 in × 4 in × 8 in = 64 in^3

   # of blocks per row = length of row ÷ length of block
   = 26 ft 8 in ÷ 16 in
   = [(26 ft × 12 in/ft) + 8 in] ÷ 16 in
   = (312 in + 8 in) ÷ 16 in
   = 320 in ÷ 16 in
   = 20

   # of cores = # blocks per row × 2 cores per block × number of courses in wall
   = 20 × 2 × 9
   = 360

   Volume = 64 in^3 × 360 = 23,040 in^3
   = 23,040 in^3 ÷ 46,656 in^3/yd^3
   = 0.4938 yd^3
   = 0.5 yd^3

APPENDIX B

Figures and Mathematical Formulas

AREAS OF PLANE FIGURES

NAME	FORMULA	SHAPE
	(A = Area)	
Parallelogram	$A = b \times h$	
Trapezoid	$A = \dfrac{b+c}{2} \times h$	
Triangle	$A = \dfrac{b \times h}{2}$	
Trapezium	(Divide into 2 triangles) A = Sum of the 2 triangles (See above)	
Regular Polygon	$A = \dfrac{\text{Sum of sides (s)}}{2} \times \text{inside radius (r)}$	
Circle	$\pi = 3.14$ (1) πr^2 $A = (2)\ 0.784 \times d^2$	
Sector	(1) $\dfrac{a^2}{360°} \times \pi r^2$ $A = (2)$ Length of arc $\times \dfrac{r}{2}$ ($\pi = 3.14$)	
Segment	A = Area of sector minus triangle (see above)	
Ellipse	$A = M \times m \times 0.7854$	
Parabola	$A = b \times \dfrac{2h}{3}$	

VOLUMES OF SOLID FIGURES

NAME	FORMULA	SHAPE
	(V = volume)	
Cube	$V = a^3$ (in cubic units)	
Rectangular Solids	$V = l \times w \times h$	
Prisms	$V(1) = \dfrac{b \times a}{2} \times h$ $V(2) = \dfrac{s \times r}{2} \times 6 \times h$ V = Area of end $\times$ h	
Cylinder	$V = \pi r^2 \times h$ ($\pi = 3.14$)	
Cone	$V = \dfrac{\pi r^2 \times h}{3}$ ($\pi = 3.14$)	
Pyramids	$V(1) = l \times w \times \dfrac{h}{3}$ $V(2) = \dfrac{b \times a}{2} \times \dfrac{h}{3}$ V = Area of base $\times \dfrac{h}{3}$	
Sphere	$V = \dfrac{1}{6}\pi d^3$	
Circular Ring (Torus)	$V = 2\pi^2 \times Rr^2$ V = Area of section $\times 2\pi R$	

APPENDIX C

ASTM Standards for Masonry Construction

Clay Masonry Units

ASTM C27, *Fire Clay and High Alumina Refractory Brick*

ASTM C32, *Sewer and Manhole Brick*

ASTM C34, *Structural Clay Loadbearing Wall Tile*

ASTM C43, *Terminology Relating to Structural Clay Products*

ASTM C56, *Structural Clay Non-Loadbearing Tile*

ASTM C62, *Building Brick*

ASTM C106, *Fire Brick Flue Lining for Refractories and Incinerators*

ASTM C126, *Ceramic Glazed Structural Clay Facing Tile, Facing Brick and Solid Masonry Units*

ASTM C155, *Insulating Fire Brick*

ASTM C212, *Structural Clay Facing Tile*

ASTM C216, *Facing Brick*

ASTM C279, *Chemical Resistant Brick*

ASTM C315, *Clay Flue Linings*

ASTM C410, *Industrial Floor Brick*

ASTM C416, *Silica Refractory Brick*

ASTM C530, *Structural Clay Non-Loadbearing Screen Tile*

ASTM C652, *Hollow Brick*

ASTM C902, *Pedestrian and Light Traffic Paving Brick*

ASTM C1261, *Firebox Brick for Residential Fireplaces*

ASTM C1272, *Heavy Vehicular Paving Brick*

Cementitious Masonry Units

ASTM C55, *Concrete Building Brick*

ASTM C73, *Calcium Silicate Face Brick (Sand-Lime Brick)*

ASTM C90, *Loadbearing Concrete Masonry Units*

ASTM C129, *Non-Loadbearing Concrete Masonry Units*

ASTM C139, *Concrete Masonry Units for Construction of Catch Basins and Manholes*

ASTM C744, *Prefaced Concrete and Calcium Silicate Masonry Units*

ASTM C936, *Solid Concrete Interlocking Paving Units*

ASTM C1319, *Concrete Grid Paving Units*

Natural Stone

ASTM C119, *Terminology Relating to Building Stone*

ASTM C503, *Marble Building Stone*

ASTM C568, *Limestone Building Stone*

ASTM C615, *Granite Building Stone*

ASTM C616, *Sandstone Building Stone*

ASTM C629, *Slate Building Stone*

Mortar and Grout

ASTM C5, *Quicklime for Structural Purposes*

ASTM C33, *Aggregates for Concrete*

ASTM C91, *Masonry Cement*

ASTM C144, *Aggregate for Masonry Mortar*

ASTM C150, *Portland Cement*

ASTM C199, *Pier Test for Refractory Mortar*

ASTM C207, *Hydrated Lime for Masonry Purposes*

ASTM C270, *Mortar for Unit Masonry*

ASTM C330, *Lightweight Aggregates for Structural Concrete*

ASTM C331, *Lightweight Aggregates for Concrete Masonry Units*

ASTM C404, *Aggregates for Masonry Grout*

ASTM C476, *Grout for Reinforced and Nonreinforced Masonry*

ASTM C658, *Chemical Resistant Resin Grouts*

ASTM C887, *Packaged, Dry, Combined Materials for Surface Bonding Mortar*

ASTM C1142, *Extended Life Mortar for Unit Masonry*

ASTM C1329, *Mortar Cement*

Reinforcement and Accessories

ASTM A82, *Cold Drawn Steel Wire for Concrete Reinforcement*

ASTM A153, *Zinc Coating (Hot-Dip) on Iron or Steel Hardware*

ASTM A165, Electro-Deposited Coatings of Cadmium on Steel

ASTM A167, Stainless and Heat Resisting Chromium-Nickel Steel Plate, Sheet and Strip

ASTM A185, Welded Steel Wire Fabric for Concrete Reinforcement

ASTM A496, Deformed Steel Wire for Concrete Reinforcement

ASTM A615, Deformed and Plain Billet-Steel Bars for Concrete Reinforcement

ASTM A616, Rail-Steel Deformed and Plain Bars for Concrete Reinforcement

ASTM A617, Axle-Steel Deformed and Plain Bars for Concrete Reinforcement

ASTM A641, Zinc Coated (Galvanized) Carbon Steel Wire

ASTM A951, Joint Reinforcement

ASTM B227, Hard-Drawn Copper-Covered Steel Wire, Grade 30HS

ASTM C1242, Guide For Design, Selection and Installation of Exterior Dimension Stone Anchors and Anchoring Systems

Sampling and Testing

ASTM C67, Sampling and Testing Brick and Structural Clay Tile

ASTM C97, Absorption and Bulk Specific Gravity of Natural Building Stone

ASTM C109, Compressive Strength of Hydraulic Cement Mortars

ASTM C140, Sampling and Testing Concrete Masonry Units

ASTM C170, Compressive Strength of Natural Building Stone

ASTM C241, Abrasion Resistance of Stone

ASTM C267, Chemical Resistance of Mortars, Grouts and Monolithic Surfacings

ASTM C426, Drying Shrinkage of Concrete Block

ASTM C780, Preconstruction and Construction Evaluation of Mortars for Plain and Reinforced Unit Masonry

ASTM C880, Flexural Strength of Natural Building Stone

ASTM C952, Bond Strength of Mortar to Masonry Units

ASTM C1006, Splitting Tensile Strength of Masonry Units

ASTM C1019, Sampling and Testing Grout

ASTM C1072, Method of Measurement of Masonry Flexural Bond Strength

ASTM C1093, Accreditation of Testing Agencies for Unit Masonry

ASTM C1148, Measuring the Drying Shrinkage of Masonry Mortar

ASTM C1194, Compressive Strength of Architectural Cast Stone

ASTM C1195, Absorption of Architectural Cast Stone

ASTM C1196, In Situ Compressive Stress Within Solid Unit Masonry Estimated Using Flatjack Method

ASTM C1197, In Situ Measurement of Masonry Deformability Using the Flatjack Method

ASTM C1262, Evaluating the Freeze-Thaw Durability of Manufactured Concrete Masonry and Related Concrete Units

ASTM C1314, Method of Constructing and Testing Masonry Prisms Used to Determine Compliance with Specified Compressive Strength of Masonry

ASTM C1324, Examination and Analysis of Hardened Masonry Mortar

ASTM D75, Sampling Aggregates

ASTM E72, Conducting Strength Tests of Panels for Building Construction

ASTM E447, Compressive Strength of Masonry Prisms

ASTM E488, Strength of Anchors in Concrete and Masonry Elements

ASTM E514, Water Permeance of Masonry

ASTM E518, Flexural Bond Strength of Masonry

ASTM E519, Diagonal Tension in Masonry Assemblages

ASTM E754, Pullout Resistance of Ties and Anchors Embedded in Masonry Mortar Joints

Assemblages

ASTM C901, Prefabricated Masonry Panels

ASTM C946, Construction of Dry Stacked, Surface Bonded Walls

ASTM E835, Guide for Dimensional Coordination of Structural Clay Units, Concrete, Masonry Units, and Clay Flue Linings

ASTM C1283, Practice for Installing Clay Flue Lining

ASTM E1602, Guide for the Construction of Solid Fuel-Burning Masonry Heaters

REFERENCES & ACKNOWLEDGMENTS

Additional Resources

This module is intended to be a thorough resource for task training. The following reference works are suggested for further study. These are optional materials for continued education rather than for task training.

Building Block Walls—A Basic Guide, 1988. Herndon, VA: National Concrete Masonry Association.

Masonry Design and Detailing—For Architects, Engineers and Contractors, Fourth Edition. Christine Beall. New York, NY: McGraw-Hill.

The ABC's of Concrete Masonry Construction, Videotape. Skokie, IL: Portland Cement Association.

Figure Credits

Bon Tool Company	103F04
Topaz Publications, Inc.	103F11, 103F14
U.S. Department of Housing and Urban Development	103F26
American Concrete Institute	103F28

CONTREN® LEARNING SERIES — USER UPDATES

The NCCER makes every effort to keep these textbooks up-to-date and free of technical errors. We appreciate your help in this process. If you have an idea for improving this textbook, or if you find an error, a typographical mistake, or an inaccuracy in NCCER's Contren® textbooks, please write us, using this form or a photocopy. Be sure to include the exact module number, page number, a detailed description, and the correction, if applicable. Your input will be brought to the attention of the Technical Review Committee. Thank you for your assistance.

Instructors – If you found that additional materials were necessary in order to teach this module effectively, please let us know so that we may include them in the Equipment/Materials list in the Instructor's Guide.

Write: Product Development and Revision
National Center for Construction Education and Research
P.O. Box 141104, Gainesville, FL 32614-1104

Fax: 352-334-0932

E-mail: curriculum@nccer.org

Craft

Module Name

Copyright Date

Module Number

Page Number(s)

Description

(Optional) Correction

(Optional) Your Name and Address

Module 28104-04

Mortar

COURSE MAP

This course map shows all of the modules in the first level of the *Masonry* curriculum. The suggested training order begins at the bottom and proceeds up. Skill levels increase as you advance on the course map. The local Training Program Sponsor may adjust the training order.

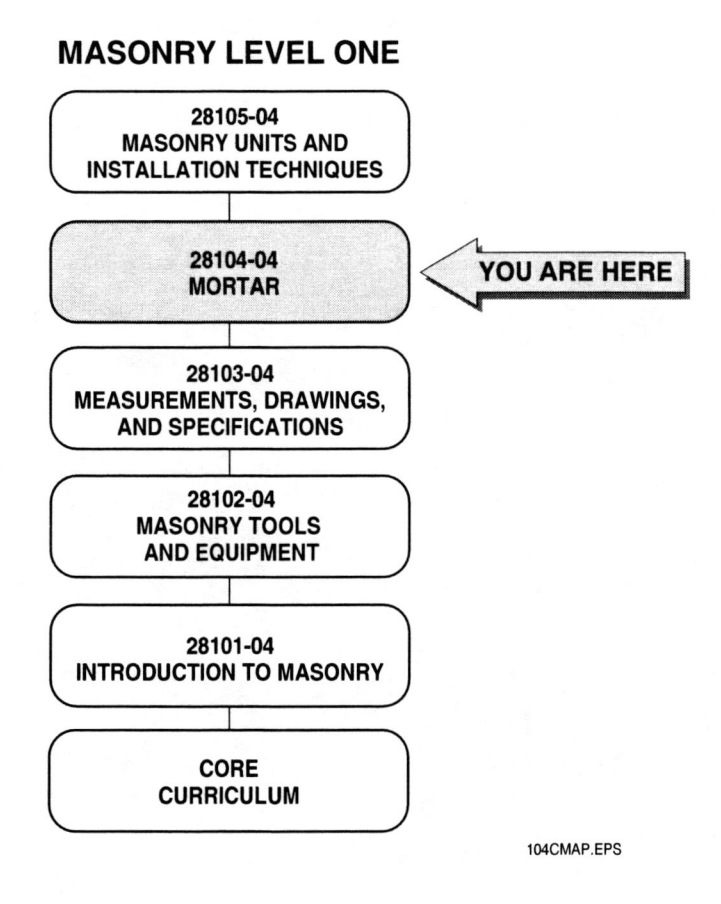

MASONRY LEVEL ONE

28105-04
MASONRY UNITS AND INSTALLATION TECHNIQUES

28104-04
MORTAR ⬅ YOU ARE HERE

28103-04
MEASUREMENTS, DRAWINGS, AND SPECIFICATIONS

28102-04
MASONRY TOOLS AND EQUIPMENT

28101-04
INTRODUCTION TO MASONRY

CORE CURRICULUM

Copyright © 2004 National Center for Construction Education and Research, Gainesville, FL 32614-1104. All rights reserved. No part of this work may be reproduced in any form or by any means, including photocopying, without written permission of the publisher.

MODULE 28104-04 CONTENTS

1.0.0 **INTRODUCTION** 4.1
2.0.0 **MORTAR MATERIALS** 4.2
    2.1.0   Portland Cement 4.3
    2.2.0   Hydrated Lime 4.3
    2.3.0   Masonry Cement 4.4
    2.4.0   Sand 4.5
    2.5.0   Water 4.6
    2.6.0   Admixtures 4.7
3.0.0 **MORTAR TYPES** 4.9
4.0.0 **MORTAR PROPERTIES** 4.10
    4.1.0   Plastic Properties of Mortar 4.11
    4.1.1   *Workability* 4.11
    4.1.2   *Water Retention* 4.11
    4.1.3   *Water Content* 4.11
    4.1.4   *Consistent Rate of Hardening* 4.12
    4.2.0   Properties of Hardened Mortar 4.12
    4.2.1   *Durability* 4.13
    4.2.2   *Compressive Strength* 4.13
    4.2.3   *Mechanical Bond* 4.13
    4.2.4   *Volume Change* 4.14
    4.2.5   *Appearance* 4.14
5.0.0 **SETTING UP, STORING, AND MEASURING** 4.14
    5.1.0   Setting Up the Mortar Area 4.15
    5.2.0   Storing Mortar Materials 4.15
    5.3.0   Measuring Mortar Materials 4.16
6.0.0 **MIXING BY MACHINE** 4.17
    6.1.0   Machine Mixing Steps 4.19
    6.2.0   Safety Tips for Machine Mixers 4.20
7.0.0 **MIXING BY HAND** 4.20
8.0.0 **PROBLEMS MIXING MORTAR** 4.21
    8.1.0   Proportioning Materials 4.22
    8.2.0   Poor Quality Materials 4.22
    8.3.0   Cold Weather 4.22
    8.4.0   Retempering 4.23
    8.5.0   Efflorescence 4.23
SUMMARY 4.24
REVIEW QUESTIONS 4.24
PROFILE IN SUCCESS 4.26
GLOSSARY 4.27
REFERENCES & ACKNOWLEDGMENTS 4.28

Figures

Figure 1	Mortar is twenty percent of the building's surface	4.2
Figure 2	Commercial hydrated lime	4.3
Figure 3	Pre-mixed mortar	4.5
Figure 4	Commercial sand for mortar	4.5
Figure 5	Siltation test jar	4.7
Figure 6	Multipurpose admixture for faster set and increased strength	4.7
Figure 7	Design effects using white mortar	4.8
Figure 8	Mortar with good cohesion	4.11
Figure 9	Slump test	4.12
Figure 10	Slump test comparison	4.12
Figure 11	Mortar must be able to support loads	4.13
Figure 12	Mortar bonding to a porous masonry surface	4.14
Figure 13	Storing cement	4.15
Figure 14	Sand pile	4.16
Figure 15	Bag of portland cement	4.17
Figure 16	Cubic foot box	4.17
Figure 17	Mortar mixer	4.17
Figure 18	Adding water to wet the mortar mixer	4.19
Figure 19	Dumping mortar into wheelbarrow	4.19
Figure 20	A cubic foot box can be used to measure sand	4.20
Figure 21	Portland cement is added to the mortar mix	4.20
Figure 22	Dry ingredients are moved to one end before water is added	4.21
Figure 23	Mortar consistency test	4.21
Figure 24	Transferring mortar to mortar board	4.21
Figure 25	Retempering mortar	4.23

Tables

Table 1	ASTM Standards for Mortar Materials	4.2
Table 2	Five Types of Portland Cement	4.3
Table 3	Recommended Sand Gradation Limits	4.6
Table 4	Admixtures and Their Effects	4.8
Table 5	Pigments Needed to Create Certain Colors	4.8
Table 6	Proportion Specifications for Mortar	4.9
Table 7	Mortar Types for Classes of Construction	4.10

MODULE 28104-04

Mortar

Objectives

When you have completed this module, you will be able to do the following:

1. Name and describe the primary ingredients in mortar and their properties.
2. Identify the various types of mortar used in masonry work.
3. Describe the common admixtures and their uses.
4. Identify the common problems found in mortar application and their solutions.
5. Properly set up the mortar mixing area.
6. Properly mix mortar by hand.
7. Properly mix mortar with a mechanical mixer.

Recommended Prerequisites

Core Curriculum; Masonry Level One, Modules 28101-04 through 28103-04

Required Trainee Materials

1. Pencil and paper
2. Appropriate personal protective equipment

1.0.0 ♦ INTRODUCTION

Mortar bonds masonry units. It is one of the basic building materials used by the mason. This module describes the materials used to make mortar, their characteristics, the types of mortar used, and how to mix mortar.

Masonry structures are a combination of mortar and masonry units. Mortar is twenty percent of the surface of the structure, as shown in *Figure 1*. Good mortar is critical to building a solid structure.

> **DID YOU KNOW?**
> ### New Additives for an Ancient Recipe
> Mortar was used to bond masonry units in ancient times. People used naturally available materials to make mortar. Primitive mortar was made from clay, chopped straw, and sand.
> Ancient Egyptians first used mud from the Nile River. In some areas, volcanic ash or ground pumice was added to mortar to increase strength. Later, burned gypsum and sand were used to make mortar.
> Lime and sand were used in the United States before the 1900s. In the 20th century, portland cement and additives were developed to increase workability and strength.

Because mortar joins the masonry units together, it must be strong and durable. The purpose of mortar is to do the following:

- Bond the units and seal the spaces between them
- Make up the differences in the size of the units.
- Bond metal ties, grids, and anchors
- Make the structure airtight and watertight
- Create a neat, attractive, and uniform appearance

The materials can be mixed to make different types of mortar. As a mason, you must understand how each of the ingredients contributes to the mix. You need to know how to mix different types and how to tell the difference between good and poor mortar.

MORTAR 4.1

2.0.0 ◆ MORTAR MATERIALS

Mortar is made of the following basic materials:

- Portland cement
- Hydrated lime
- Sand
- Water

Each material affects mortar quality. Generally, cement gives mortar strength. Sand adds bulk. Lime reacts with the water and air and hardens in a process called **hydration**.

ASTM (American Society for Testing and Materials) International created standards for each of the materials used in mortar. These standards are listed in *Table 1*. Different amounts of these materials can be used to mix different types of mortar. Other materials, known as **admixtures**, can be added to the mortar to change its physical or chemical properties. Color is a common admixture.

Figure 1 ◆ Mortar is twenty percent of the building's surface.

Table 1 ASTM Standards for Mortar Materials

Material	ASTM Standard
Portland cement (Types I, II, III)	C150
Air-entraining portland cement (Types IS, ISA, IP, IPA, S, SA)	C150
Masonry cement	C595
Quicklime	C5
Hydrated lime	C207
Aggregate	C144

Mortar Versus Concrete

Concrete and mortar contain the same principal ingredients. This does not mean that the two are mixed and handled in the same way.

Both mortar and concrete achieve their strength by the same chemical process of hydration. Both have a limited workable life span and need similar safety precautions.

However, mortar and concrete differ in several ways. The differences are listed here. For these reasons, mortar and concrete are mixed, treated, and used in vastly different ways.

Mortar
- Mortar binds masonry units into a single component.
- It is often mixed by hand.
- It is often placed by hand and therefore must remain pliable longer.
- It is usually placed between absorbent masonry units. Mortar loses water upon contact with the units.
- The water/cement ratio is less important. Mortars have a high water/cement ratio when mixed. This ratio decreases when the mortar contacts the absorbent units.

Concrete
- Concrete is usually a structural element.
- It is usually mixed by a machine or is brought to the job site already mixed.
- It is usually poured into forms by machine.
- It is usually placed in non-absorbent metal or wooden forms that absorb little water.
- The water/cement ratio for concrete is significant.

2.1.0 Portland Cement

The main ingredient in mortar is portland cement. It gives mortar the typical gray color and many of its properties. It adds to durability, high strength, and early setting of the mortar.

ASTM specifications list several types of portland cement that may be used for mortar. *Table 2* compares the five types of portland cement. Generally, only Types I, II, and III are used for mortar. Type I is most often used in mortar. The other two are usually used only for concrete.

Air-entraining admixtures can be added to the various types of portland cement. These mixtures are labeled with an A. For example, Type IA is Type I portland cement with air-entraining admixtures.

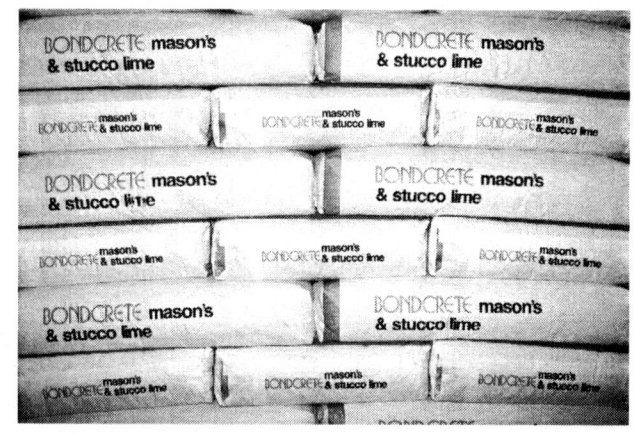

2.2.0 Hydrated Lime

The second ingredient in mortar is hydrated lime (*Figure 2*). It sets only upon contact with the air. Thus, hardening occurs slowly over a long period of time. Lime gives the following properties to mortar:

- Bond
- **Workability**
- **Water retention**
- Elasticity

Hydrated lime in *ASTM C207 Specification for Hydrated Lime for Masonry Purposes* is available in four types: S, N, NA, and SA. Only Type S hydrated lime should be used in mortar. Type N hydrated lime contains no limits on the quantity of unhydrated oxides. Types NA and SA are not recommended for mortar. They contain air-entraining additives that reduce the bond between the mortar and masonry units or reinforcement.

Type S hydrated lime adds several desirable characteristics to mortar:

- *Workability* – Mortar must be smooth and pliable. Workability is a measure of these features. Lime helps mortar to be very workable. This allows the mason to butter joints easily, and results in better workmanship and more economical construction. Mortar with the correct lime content spreads easily with a trowel. Cement that is not mixed with lime is more difficult to spread.

Figure 2 ♦ Commercial hydrated lime.

Table 2 Five Types of Portland Cement

TYPE	DESCRIPTION	USE
I	General Purpose	This is a general-purpose cement and is the one masons use most often. It may be used in pavements, sidewalks, reinforced concrete bridge culverts, and masonry mortar.
II	Modified	This cement hydrates at a lower temperature than Type I and generates less heat. It has better resistance to sulfate than Type I. It is usually specified for use in such places as large piers, heavy abutments, and heavy retaining walls.
III	High Early Strength	Although this cement requires as long to set as Type I, it achieves its full strength much sooner. Generally, this cement is recommended when high strength is required in one to three days. Type III is often used in cold weather, when protection from freezing weather is important.
IV	Low Heat	This is a special cement for use where the amount and rate of heat generated must be kept to a minimum. Since the concrete cures slowly, strength also develops at a slower rate. Low heat portland cement is used in areas where there are large masses of concrete, such as dams or large bridges, and where Type I or Type II cement would generate too much heat.
V	Sulfate Resistant	This is intended for use only in construction exposed to severe sulfate actions. It also gains strength at a slower rate than normal portland cement.

MORTAR

Building Blocks

How Portland Cement Is Made

To make portland cement, first limestone, marl, shale, iron ore, clay, and fly ash are crushed and screened. These materials contain the right amounts of calcium compounds, silica, alumina, and iron oxide. They are then placed in a rotating cement kiln.

The kiln looks like a large horizontal pipe. It has a diameter of 10' to 15' and a length of 300 feet or more. One end is slightly raised. The raw mix is placed in the high end. As the kiln rotates, the materials move slowly toward the lower end.

Flame jets are at the lower end. All the materials in the kiln are heated to temperatures between 2,700°F and 3,000°F. This high heat drives off the water and carbon dioxide from the raw materials. This process forms new compounds (tricalcium silicate, dicalcium silicate, tricalcium aluminate, and tetracalcium aluminoferrite).

Marble-sized pellets, called clinker, come out of the lower end of the kiln. The clinker is then very finely ground to produce portland cement. A small amount of gypsum is added during the grinding process to control the cement's set, or rate of hardening.

104SA01.EPS

- *Water retention* – Lime increases the water-holding capacity of the mortar, decreases the loss of water from the mortar (bleeding), and reduces the separation of sand. Mortar with a low lime content loses its moisture and sets prematurely. When you use mortar that is too dry, you will get a poor bond. This seriously weakens the structure.
- *Bond strength* – Bond is the most important property of mortar. Lime decreases the strength of the cement in mortar but helps the mortar fill gaps and adhere to the structure. This results in an overall higher bond strength.
- *Flexibility* – Mortar must be flexible. This will prevent cracking during strong winds, lateral pressures, and hard jolts. Tall masonry structures, such as high chimneys, are likely to sway. Lime increases the **plasticity** and flexibility of the mortar.
- *Economy* – Lime is inexpensive. In addition, lime helps make a smoother, more uniform mortar. Lime-based mortars have a slower set time. This reduces the amount of retempering and allows the mason to work efficiently.
- *Minimal shrinkage* – Very hard, high-strength mortar can crack because it shrinks after hardening. These cracks result in a loss of strength. Lime undergoes the second least loss in volume of all the ingredients in mortar.
- *Resistance to weather* – Mortar must be able to resist strong winds, freezing temperatures, and alternating wet and dry weather. Lime-based mortar actually increases in overall strength as it ages. In fact, when small cracks appear in the mortar joints, the action of lime combined with rainwater often reseals these cracks.
- *Autogenous healing* – The ability of lime to reseal itself when hairline cracks form in the mortar is called autogenous healing. Lime reacts with water and elements in the air to reseal these minute cracks through a chemical process called recarbonation.

2.3.0 Masonry Cement

Most mortars are made from portland cement mixed with other materials. Modern **masonry cement** mixes, developed in recent years, only require the addition of sand and water. These mixes contain all the ingredients except sand and water in one bag. Some mixes also include sand; you only need to add water. Most masonry cements are a combination of portland cement, lime, gypsum, and air-entraining admixtures. Masonry cement is available in bags from 40 to 80 pounds as shown in *Figure 3*.

The look and quality of the mix are very consistent. The materials are mixed and ground together before packaging to make uniform mixes. Masonry cement is less subject to batch variations than mortar mixed on the job, which can vary greatly between batches. Sand must be added to most masonry cement. However, sand is not uniform. Of all the materials in mortar, sand varies the most from batch to batch. Even using masonry cement does not guarantee consistent mortar.

Mixing masonry cement is much faster than mixing mortar. There are fewer materials to handle and measure, which reduces mixing time.

Masonry cement is made in grades M, S, N, and O. Masonry cement is convenient, but be careful. These products must meet *ASTM C91 Standard Specification for Masonry Cement*.

2.4.0 Sand

Sand is a main ingredient in mortar (*Figure 4*). Sand acts as a filler. It adds to the strength of the mix. It also decreases shrinkage during the hardening phase and, therefore, decreases cracking. The water, cement, and lime in the mix form a paste that coats the sand particles. This paste lubricates the sand to form a workable mix.

The amount of air contained in the mix is determined by the range of particle sizes and grades of sand used. *Table 3* shows the recommended gradation limits for sand. The percentages show how much of the sand must pass through a set of screens with progressively finer mesh sizes. Note that 35 to 70 percent of the sand must pass through the #30 screen, which has finer mesh than the #4 screen.

Gradation limits are given in *ASTM C144 Specification for Aggregates for Masonry Mortar*. Adding fine or coarse sands is an easy way to change the gradation. Sometimes it is more practical to proportion the mortar mix to suit the available sand. It is often difficult to obtain a particular gradation of sand. If the sand does not meet the grading requirement of *ASTM C 144*, it can only be used if the mortar meets the property specifications of *BIA M1 Specification for Portland Cement-Lime Mortar for Brick Masonry* or *ASTM C270*.

Figure 3 ◆ Pre-mixed mortar.

Figure 4 ◆ Commercial sand for mortar.

WARNING!
Dry cement dust is harmful. It can enter open wounds and cause blood poisoning. When cement dust comes into contact with body fluids, it can cause chemical burns. Protect your eyes, nose, and mouth from dust. Cement dust can cause a fatal lung disease called silicosis. Repeated contact with cement or wet concrete can also cause an allergic skin reaction known as cement dermatitis.

Table 3 Recommended Sand Gradation Limits

Sieve Size	PERCENT PASSING Natural Sand	PERCENT PASSING Manufactured Sand
No. 4	100	100
No. 8	95–100	95–100
No. 16	60–100	60–100
No. 30	35–70	35–70
No. 50	15–35	20–40
No. 100	2–15	10–25
No. 200	—	0–10

104T03.EPS

It is important to use good-quality sand in the production of mortar. Sand is one of the least expensive building materials on the job. Cutting costs by using poor-quality sand will lead to a variety of problems. Poor-quality sand has no place on the job, as it lowers both the quality of the final product and the mason's productivity.

Poor-quality sand may contain silt. If silt is present in mortar, the mortar will stick to the mason's trowel, causing a drop in production rate. Silt reduces the bond strength of the mortar by preventing the cementitious material from bonding to the sand. Silt will cause small mud pits and holes to form on the surface of the joints. The mortar will not be of a uniform color, and the amount of impurities will vary from batch to batch.

If impurities are suspected in sand, a simple siltation test can determine if the sand is satisfactory for use in masonry mortar. The siltation test will show if organics and other undesirable elements are present. The test is a simple process that can be done at the job site. These are the steps to test sand for excess silt:

Step 1   Fill a one-quart glass jar half full of the sand, and add water to a level 3 inches above the top of the sand. Place the cap on the jar, and shake it vigorously.

Step 2   Set the jar aside until the next morning. The sand will settle to the bottom of the jar, and any foreign material will come to rest on top of the sand. If the accumulated silt and organic material measures more than ⅛", the sand should not be used.

Figure 5 shows an unacceptable result of a sand siltation test.

2.5.0 Water

Clean water makes good mortar. Water must be free of chemicals and impurities. Impurities can weaken or discolor the mix. Impurities include alkalis, salts, acids, and organic matter. Generally, in mortar, you can use any water that is safe to drink. Water purification chemicals found in city water supplies usually do not affect mortar. Water should be used at a cool ambient temperature.

BUILDING BLOCKS

Slaking Quicklime

Hydrated lime is made from limestone. The limestone, as shown here, is processed through two chemical reactions to create calcium hydroxide. This process is known as slaking quicklime. The lime is mixed with water to form a putty. It takes several weeks to slake quicklime. Processing is very dangerous, as substantial heat is released by the lime/water reaction. Manufacturers slake the lime as part of producing hydrated lime for mortar.

104SA02.EPS

Figure 5 ◆ Siltation test jar.

2.6.0 Admixtures

Admixtures are additives put into mortar to change its appearance or properties (*Figure 6*). They can be used in mortar to increase workability, strength, and weather resistance. Admixtures to add color to mortar are also widely used.

Some admixtures are harmless. Others can be harmful to mortar and the resulting brickwork. The properties of mortar depend upon its ingredients. Admixtures should not be used unless you know their effect on the mortar. The effect of admixtures on the finished structure, masonry units, and items embedded in the wall must also be considered. The following are common admixtures:

- *Air-entraining admixtures* – These admixtures reduce water content in mortar. They increase workability and freeze-thaw resistance. Bond strength can be reduced. Use prepared cements that include an air-entraining agent.

Figure 6 ◆ Multipurpose admixture for faster set and increased strength.

- *Bonding admixtures* – Using bonding admixtures results in an increased bond between mortar and masonry units. Some organic modifiers provide an air-cure adhesive. This increases the bond strength of dry masonry.
- *Plasticizers* – Using plasticizers adds workability to mortar. Clay, clay-shale, and finely ground limestone are inorganic plasticizers. These promote workability and water release for cement hydration. Organic plasticizers also increase workability; however, mortar may stick to your tools more.
- *Set accelerators* – By increasing cement hydration, set accelerators decrease the time needed for the mortar to harden. They can increase the compressive strength of mortar. Nonchloride accelerators should be used. Chloride salts will deteriorate the cement in masonry.
- *Set retarders* – By delaying cement hydration, set retarders allow more working time for the mortar. This is useful during hot weather. Set retarders are also used in commercial ready-mixed mortars. The effect can last for up to 72 hours, but it dissipates when the mortar contacts masonry units.
- *Water reducers* – Used to lower the amount of water in the mortar, water reducers can increase the strength of the mortar. However, mortar rapidly loses water when it contacts absorptive masonry units. Water reducers may lower the water below the level needed for cement hydration. You must be sure to allow enough water for the curing process to take place.
- *Water repellents* – Water repellents modify masonry mortar during the early period after construction. However, they can reduce the rain penetration of the mortar for the long term.
- **Pozzolans** – Pozzolans increase the density and strength of mortar. They are fine particles that combine easily with lime. Ash, fly ash, and brick dust are examples of pozzolans.
- *Antifreeze admixtures* – These admixtures are not recommended for use in mortar. The amounts needed to prevent freezing also significantly lower the compressive and bonding strength of mortar.

Table 4 lists admixtures commonly used with cement and their effects on mortar.

> **NOTE**
> The addition of any admixture should be approved by the architect or project engineer, as it may affect structural stability and strength.

MORTAR 4.7

Table 4 Admixtures and Their Effects

ADMIXTURE	WORKABILITY	STRENGTH	WEATHER RESISTANCE
Air-entraining Agents	Increase	Decrease slightly	Increase
Bonding agents	—	Increase	—
Plasticizers	Increase	—	—
Set accelerators	Decrease	Increase	—
Set retarders	Increase	—	—
Water reducers	Decrease	Increase	Increase
Water repellants	—	—	Increase
Pozzolanic agents	Increase	Increase	—

Design effects can be created by using colored mortar to contrast or blend with the masonry units (*Figure 7*). White mortar is made by using white portland cement with lime and white sand. Light colors, such as cream, ivory, pink, and rose, are made using a white mortar base and pigments. Dark colors can be created by using colorants in normal gray mortar. Colors may not be as bright with a gray mortar base, but gray mortar is much less expensive.

Figure 7 ◆ Design effects using white mortar.

Color pigments are made of mineral oxide. They should not contain dispersants, which will slow or stop the hydration process. Mineral oxides should not exceed 5 percent by weight of cement for masonry cement or 10 percent by weight of cement for cement-lime mortars. You should use only the smallest amount needed to achieve the desired color. If pigment is in excess of 10 percent of the portland cement by weight, the mortar may have problems in strength and durability.

Iron, manganese, chromium, and cobalt oxides have also been used successfully. Avoid zinc and lead oxides because they may react with cement. You may use carbon black as a coloring agent to obtain dark gray or almost black mortar, but you should avoid using lamp black. Carbon black should be limited to 1 percent or 2 percent by weight of cement for masonry cement or cement-lime mortars, respectively. The color of carbon black mortar rapidly fades with exposure to weathering. Carbon black mortar has durability problems as well. Only use pigments that have been tested. A guide to the selection of pigments for colorants is shown in *Table 5*.

Table 5 Pigments Needed to Create Certain Colors

Color	Pigment
Red, yellow, brown, black, or gray	Iron oxide
Green	Chromium oxide
Blue	Cobalt oxide

4.8 MASONRY LEVEL ONE — TRAINEE MODULE 28104-04

BUILDING BLOCKS

Mixing Colorants

Pigments must be thoroughly mixed in the mortar. You can check the mortar to see if it is mixed thoroughly. Take a small sample of the mortar and flatten it out under a trowel. If there are streaks, it needs additional mixing.

Prepackaged colored masonry cement is the easiest and best way to achieve color results. Colored masonry cements are available in many areas as types M, S, and N.

3.0.0 ◆ MORTAR TYPES

There are five types of mortar. Each type is mixed differently, has different uses, and is assigned a letter: M, S, N, O, or K. These are the same as those used for masonry cement. *Table 6* gives the specifications for mixing each type of mortar.

The proportion specifications list the recipe for mixing mortar materials. A certain amount of each material is added to the mix. If more mortar is needed, each material must be added in proportion to the other materials. Mortar specified by this method does not need laboratory testing.

Table 6 Proportion Specifications for Mortar

MORTAR TYPE	PARTS BY VOLUME OF PORTLAND CEMENT	PARTS BY VOLUME OF HYDRATED LIME	SAND MEASURED IN A DAMP, LOOSE CONDITION
M	1	¼	Not less than 2¼ and not more than 3 times the sum of the volumes of the cement and lime used
S	1	½	
N	1	1	
O	1	2	
K	¼	1	

MORTAR

4.9

Several different types of mortar have been developed to meet the needs of different job requirements. Selecting the right mortar type will depend on the type of masonry work and its location in the structure. *Table 7* shows the type of construction suitable for each mortar.

- Type M mortar has the highest compressive strength. It is somewhat more durable than the other mortars. In laboratory tests, ASTM has found that a 2" cube of Type M mortar will stand up to at least 2,500 pounds per square inch (psi) of compression after it has cured for 28 days.
- Type S mortar has a medium compressive strength, good workability, and excellent durability. This type is often interchanged with Type M because it offers a relatively high compression strength but better workability, water retention, and flexibility. The ASTM specifications state that Type S mortar must stand up to at least 1,800 psi of compression.
- Type N mortar is a medium-strength mortar with a psi rating of at least 750. It has excellent workability. It does not have the strength of Types M or S. Type N is the most frequently used mortar because it is highly resistant to weather.
- Type O mortar is a low-strength mortar with very good workability. Because of its low durability, this type is not recommended for use where bad weather conditions exist. Type O mortar should have a compression strength of at least 350 psi.

> **DID YOU KNOW?**
> *Types of Mortar*
>
> There are five types of mortar: M, S, N, O, and K. These letters can be remembered by taking every other letter from the words *MaSoN wOrK*.

- Type K mortar is so low in compressive and bond strength that it is seldom used. ASTM requires that the testing strength be only 75 psi after 28 days of curing. Type K mortar made without cement is sometimes used for restoration of historic buildings.

4.0.0 ◆ MORTAR PROPERTIES

Properties of mortar can be divided into two categories related to its physical state. The first state begins immediately after it is mixed and is in use on the job. This is known as the plastic state. The second state begins after the mortar has cured into its final rigid condition. This is called the hardened state.

As a mason, you will be primarily concerned with the plastic properties of mortar. You will be working with the mortar in its plastic state. The mortar quickly moves to the hardened state and remains there for the life of the structure. You must also be familiar with the properties of mortar in its hardened state.

Table 7 Mortar Types for Classes of Construction

ASTM MORTAR TYPE DESIGNATION	CONSTRUCTION STABILITY	COMPRESSION STRENGTH (IN PSI)
M	Masonry subjected to high compressive loads, severe frost action, or high lateral loads from earth pressures, hurricane winds, or earthquakes; structures below grade, including manholes and catch basins	2,500
S	Structures requiring high flexural bond strength but subject only to normal compressive loads	1,800
N	General use in above-grade masonry; residential basement construction; interior walls and partitions; concrete masonry veneers applied to frame construction	750
O	Non-loadbearing walls and partitions; solid loadbearing masonry of allowable compressive strength not exceeding 100 psi	350
K	Interior non-loadbearing partitions where low compressive and bond strengths are permitted by building codes	75

> **Type M Mortar**
> Why don't masons always use Type M mortar when it is the strongest of all mortar types?

4.1.0 Plastic Properties of Mortar

Plastic mortar is freshly mixed and ready for use. Properly mixed mortar is soft enough to spread easily but thick enough to retain its shape. The plastic properties of mortar include the following:

- Workability
- Water retention
- Water content
- Rate of hardening

4.1.1 Workability

Several factors influence the workability of mortar. These include consistency, setting time, adhesion, and cohesion.

Consistency, or uniformity, measures how similar mortar is from one batch to the next or from one place to another in the same batch. Consistency is controlled and improved by closely following the mix recipe and the mixing procedure. The order in which the ingredients are added is important. Mixing the mortar for the same amount of time is also a key to consistency.

Setting time is the time the mason has to use the mortar before it hardens and is no longer workable. It is determined by the type of cement used in the mix. Setting times can be extended by using the maximum water allowable in the mix, by wetting the mortar board before placing mortar on it, and by retempering the mortar during use.

Adhesion is how well the mortar sticks to the masonry units. Greater adhesion means greater bonding strength.

Cohesion is how well mortar sticks to itself. Mortar with good cohesion will extrude between masonry units without smearing or dropping away (*Figure 8*).

4.1.2 Water Retention

Water retention is related to workability. Mortar with good water retention remains soft and plastic. It should remain plastic long enough for the mason to align, level, and plumb the blocks. Poor water retention may break the bond between the mortar and the units.

Figure 8 ♦ Mortar with good cohesion.

Good water retention is important to developing adhesion between the mortar and the masonry units. This will create strong, watertight joints. Water in the mortar allows the cement to hydrate. This is the chemical process that makes mortar harden.

Water retention is improved by adding more lime, finer sand, entrained air, and more water. These also increase the workability of the mortar. However, the proportions of the ingredients must continue to stay within the limits specified in the mix formula.

4.1.3 Water Content

Water content is possibly the most misunderstood aspect of masonry mortar. Mortar and concrete are made from the same materials. Some designers mistakenly assume that mortar requirements are similar to those for concrete. This is false. Mortar and concrete have very different water-cement ratios.

Mortar, concrete, and grout have different amounts of water. This can be demonstrated using the slump test shown in *Figure 9*. To test the materials, the cone is first filled and then inverted. The stiffer materials will keep the shape of the cone better than the more plastic materials. *Figure 10* shows a comparison of the typical stiffness for concrete, mortar, and grout.

MORTAR 4.11

Some mortar specifications incorrectly require the minimum amount of water consistent with workability. Often, retempering is prohibited. Retempering results in mortar with higher compressive strength but lower bond strength, which is incorrect. Mortar should be mixed with the maximum amount of water consistent with workability. This will provide maximum bond strength. Retempering is permitted, but only to replace water lost by evaporation. This can usually be controlled by requiring that mortar be used within 2½ hours after mixing, depending on conditions. If used within this time, it will need little retempering.

4.1.4 Consistent Rate of Hardening

Mortar hardens when cement reacts chemically with water. This process is called hydration. The rate of hardening is the speed at which it develops resistance to an applied load. If it hardens too quickly, the mason may have difficulty placing it. Very slow hardening may impede the progress of the work. The mortar will flow from the completed masonry. During winter construction, slow hardening may also subject mortar to early damage from frost action. A well-defined, consistent rate of hardening assists the mason in laying the masonry units and in tooling the joints at the same degree of hardness. Uniform joint color of masonry reflects proper hardening and consistent tooling times.

4.2.0 Properties of Hardened Mortar

The qualities needed in good mortar are different after it has hardened. You can control the final appearance and performance of the mortar. By using good-quality materials in the mortar mix, you are halfway there. Your workmanship provides the other half. The hardened properties of mortar include the following:

- Durability
- Compressive strength
- Mechanical bond
- Volume change
- Appearance

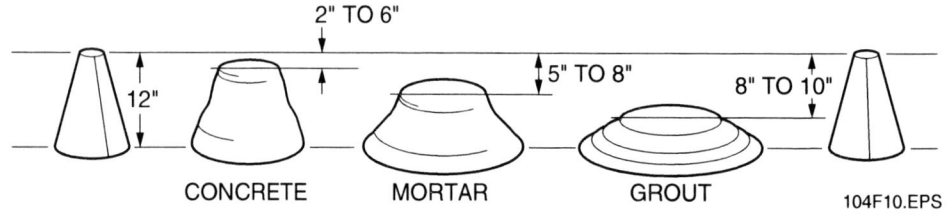

FILL THE CONE.

INVERT THE CONE.

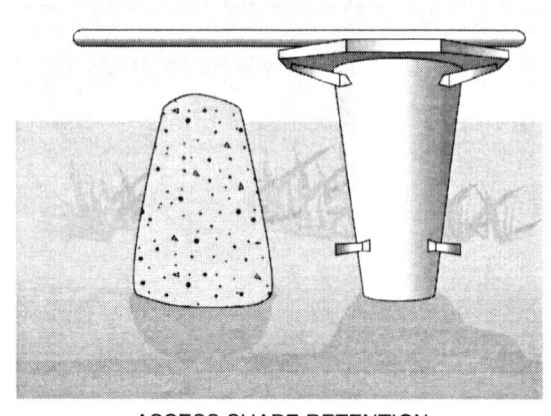

ASSESS SHAPE RETENTION

Figure 9 ◆ Slump test.

Figure 10 ◆ Slump test comparison.

4.2.1 Durability

Durability of mortar is measured by its ability to resist weather damage. Mortar must withstand repeated cycles of freezing and thawing. Freezing temperatures cause the water in mortar to expand. This causes cracking. Mortars with high-compressive strength are denser. They contain less water and are more durable. Air-entrained mortars withstand weathering better. Mortars made with masonry cement have higher air content than mortars made with portland cement and lime. Therefore, masonry-cement mortars resist freeze-thaw cycles well.

4.2.2 Compressive Strength

Compressive strength is the ability of the mortar not to crumble under pressure. High-strength mortar is needed for loadbearing walls or supports (*Figure 11*), masonry below grade, or masonry for paving. Compressive strength is measured in pounds per square inch, or psi. The compressive strength of mortar increases as the cement content increases. It decreases as more lime, air entrainment, and water are added. Type M mortar is the strongest. But there is more to the wall than the mortar. Masonry structures gain their strength from several factors, including the following:

- The compressive strength of the masonry units
- The compressive strength of the mortar
- The design of the structure
- The workmanship of the mason
- The amount of curing

If Type M mortar is used when Type S is specified, the structure may not be stronger. Brick walls are designed to move. They expand and contract with temperature changes. If the mortar is too hard, it will not move properly. This will weaken the final structure.

4.2.3 Mechanical Bond

Bond strength is the ability of the mortar to hold the masonry units together. This depends on the ability of the mortar to stick to the surface of the masonry units. Adhesion improves with greater contact between the mortar and the masonry unit.

The mortar must be completely spread over the unit face or shell. The mortar must have enough flow to wet the contact surfaces. Masonry units have surface irregularities, or micropores. This increases mechanical bond by increasing the area of the surface to be bonded. Both clay and concrete units will absorb wet mortar into their irregularities. *Figure 12* shows an enlargement of this process.

Mechanical bond strength in cured mortar depends on the mechanical interlocking of cement hydration crystals. Voids, smooth masonry units, and contamination can hinder this process. Voids are air holes in the mortar joint. These weaken the mechanical bond. Voids also allow water into the joint. Smooth masonry units such as molded brick or smooth stone do not have micropores. They provide less mechanical bond with the mortar. Loose sand particles, dirt, and other contamination also weaken the mechanical bond.

Figure 11 ◆ Mortar must be able to support loads.

BUILDING BLOCKS

Sulfate Resistance

Sulfates in the soil or water can also cause mortar to expand. The damage is similar to freeze-thaw cycles. Mortars that will contact soil, groundwater, seawater, or industrial processes must also withstand sulfate expansion. The damage can be minimized by using sulfate-resistant materials.

Masonry-cement mortars and Type V portland cement are sulfate-resistant. You can use either one, but Type V cement is slow to gain strength. Applying a protective coating to the finished work is another good option.

MORTAR

> **BUILDING BLOCKS**
>
> **Bonding**
>
> Skill is critical in bonding. Mortar joints must have complete contact with the masonry units. Once a unit is in place, the bond will break or weaken if it is moved. Wet mortar allows more time for placing units. Laboratory tests show that tapping the unit to level increases bond strength 50 to 100 percent over hand pressure alone.

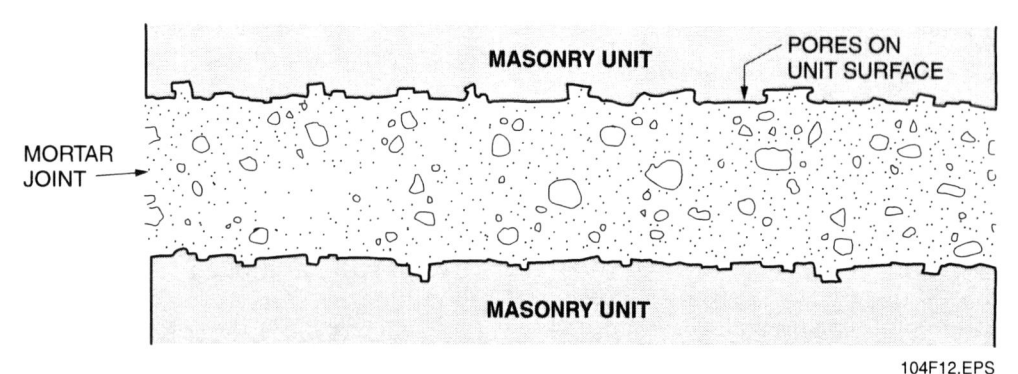

Figure 12 ◆ Mortar bonding to a porous masonry surface.

The strength of the bond depends upon the strength of the mortar itself. The main factors affecting mortar strength include the following:

- Ingredients, such as types and amounts of cement, retained water, and air content
- Surface texture and moisture content of the masonry units used
- Curing conditions, such as temperature, wind, relative humidity, and amount of retained water

Mortar strength is directly related to cement content. Both bond strength and mortar stiffness increase as cement content increases. Bond strength also increases as water content increases. The optimum bond strength is obtained by using a mortar with the highest water content compatible with workability.

4.2.4 Volume Change

As mortar cures, the volume changes as free water is lost. Shrinkage and expansion result from the volume changes in mortar. Shrinkage causes cracks to appear in joints. Expansion is generally due to poor-quality ingredients.

A mortar-joint made with properly proportioned mortar will shrink. However, it is usually very small and not of concern. Larger shrinkage cracks will appear when there is a large amount of water loss from the mortar.

The rate of water loss from mortar is determined by several factors. First is the amount of water in the mortar. Second is the rate at which it is absorbed into the masonry units. Air-entrained mortars need less water. However, workability and bond strength are directly related to the flow of the mortar and should be given priority when determining water content in the mixing operation.

Expansion in mortars is generally caused by using poor-quality material. Materials can become contaminated by impurities. Careful storage and handling of the material will eliminate much of this problem. There are tests that can be run to check for poor quality. The siltation test described earlier will ensure that the sand is clean. The autoclave expansion test (*ASTM C91*) measures the quality of the cementitious material.

4.2.5 Appearance

Mortar makes up twenty percent of the face of a masonry structure. The mortar should have a uniform color and shade. This affects the overall appearance of the finished structure. Several factors affect the color and shade of mortar joints:

- Atmospheric conditions
- Admixtures
- Moisture content of the masonry units
- Uniformity of proportions in the mortar mix
- Water content
- When the mortar joints are tooled

5.0.0 ◆ SETTING UP, STORING, AND MEASURING

Masons should be able to mix their own mortar, or mud, as it is known in the trade. They need to know how to modify the mix for different requirements of the work. On large jobs, mortar may not

> ### Mixing Area
> Keep the mixing area clean by measuring and loading materials carefully. Clear away spills with a shovel or brush. A clean mixing site is safer as well as easier to use.

be mixed on the site but brought in from an off-site mixing plant. On small jobs, however, it may be practical for masons to mix their own mortar either by hand or with a power mixer. The following sections present the steps needed to prepare for mixing mortar.

5.1.0 Setting Up the Mortar Area

Efficient placement of the materials and the mixer can save time and work. Because they need to be close together, the mixer site should be set at the same time material stockpile sites are set. Avoid a site that is downhill from the work area if the mortar will be moved in wheelbarrows.

The mixer should be located as close as possible to the main section of the work with mix ingredients arranged around it. The stockpiles should be on the side of the mixer away from the work so that the mortar-filled light trucks and other equipment do not have to pass around the stockpiles. If the ground is soft or damp, set up a plywood runway for equipment.

Use a hose to bring the water to a barrel near the mixer or to fill the water measure directly. Make sure the hose is not in the path of traffic. When setting up for a good-sized job, the mortar mixing area needs to be as follows:

- Away from materials and equipment movement paths
- Close to stockpiles, so materials will not have to be moved a long distance
- Close to the work site, so mortar will not have to be moved a long distance, but not so close as to interfere with the work
- Positioned so that mortar-filled wheelbarrows will not have to be pushed uphill to the work area

If the job requires only a small amount of mortar, the mortar can be mixed in a wheelbarrow and rolled directly to the work site.

5.2.0 Storing Mortar Materials

The materials used in making mortar should be stored on the job site in an area convenient to the mixing site but not where they will interfere with the work. Cement and lime are generally sold and delivered by the bag. Sand is ordered by weight, but it is usually delivered in loose form.

The cement and lime should be kept in a storage shed. If kept outside, they must be covered with plastic sheets or canvas tarpaulins (tarps) (see *Figure 13*). Any other materials stored outside should be covered in a similar manner. In either case, the bags should be kept off the ground by using wooden pallets to keep them from ground moisture.

Figure 13 ◆ Storing cement.

> ### Measuring and Mixing
> Careful measurement of mortar materials and thorough mixing are the first steps to making a good-looking masonry structure.

MORTAR 4.15

Mortar-Box Safety

Some mortar boxes have special fittings, so they can be moved by a forklift. Always place these boxes for easy access by the forklift operator. It is much easier to adjust the box before it is full of mortar. Never use a forklift to move a mortar box that does not have fittings for doing so. The wet mortar could shift, and the box will fall. This can cause serious injury and/or damage.

Sand can be dumped on the ground and, in most cases, does not need to be protected from moisture in warm weather (*Figure 14*). In winter, cover the stockpile to prevent wet sand from freezing. Another method to keep sand from freezing is to run pipes through the stockpile, then pump heated air or water through the pipes. If the sand does freeze, it must be thawed before it can be used.

Keep foreign material out of the stockpile. If the only location for sand at the job site is muddy or vegetation-covered ground, the sand can be dumped on plastic sheets or tarpaulins.

Figure 14 ◆ Sand pile.

5.3.0 Measuring Mortar Materials

Quality control of a mortar mix and consistency between batches start with the proper measuring of mortar ingredients. Proper measuring will ensure that the properties of the mortar, such as workability, color, and strength, will be the same from mix to mix. This ability to produce the same results in each batch will increase your productivity and lead to a quality, good-looking job.

Mortar mix specifications are expressed in terms of a proportion of cement, lime, and sand volumes. For example, for a Type N mortar, the proportions might be expressed as 1:1:6, or 1 part portland cement, 1 part lime, and 6 parts sand. For a Type O cement-lime mortar with a mix ratio of 1:2:9, the proportions would be 1 part portland cement, 2 parts lime, and 9 parts sand.

Portland cement is available in 94-pound bags containing one cubic foot of material (*Figure 15*). Hydrated lime comes in 50-pound bags containing one cubic foot. Masonry cement comes in 70-pound bags containing one cubic foot. Because sand comes in bulk form, it needs a practical and consistent measuring method. Many masons measure sand by the shovel. The number of shovels of sand in a cubic foot will vary, however, depending on the amount of water in the sand. One method that works well is to use a cubic foot box (*Figure 16*). This box measures 1' long × 1' wide × 1' deep and holds exactly 1 cubic foot of material.

> **BUILDING BLOCKS**
>
> *Cement Storage*
>
> In the United States, portland cement is normally sold in paper bags by volume. Bags must be stored off the ground and covered. This prevents the cement from absorbing moisture, which can cause lumps to form in the cement powder. Cement powder should be free flowing. Do not use any cement with lumps that cannot be broken up easily into powder by squeezing them in your hand.

The cubic foot box can be used to measure all the sand used in the mix. It is more commonly used to calibrate the shovel for loading the sand into the mixer. The mason determines the number of shovels of sand required for a cubic foot by filling the box. The box is no longer necessary for most of the mixing operations. You should recheck your shovel count twice each day to allow for factors such as moisture content of the sand, changing shovels, or changing people doing the shoveling.

Water should always be measured with a container. Never pour water directly into the mixer from a hose, as it is not possible to calculate the exact amount added. A plastic five-gallon bucket works well.

The volume of mortar obtained from a mix will only equal the volume of sand in the mix, even though you have also added a quantity of cement, lime, or masonry cement. This is because the cement, lime, and water occupy the voids between the grains of sand. The mix simply becomes denser as more cement and lime are added. For this reason, consider only the amount of sand added when you want to know how much mortar will be produced in your mix.

6.0.0 ♦ MIXING BY MACHINE

Machine mixing is the preferred method for preparing a batch of mortar. For larger jobs, a machine mixer (*Figure 17*) can produce 4 to 7 cubic feet of mortar at one time with the same quality and the same properties batch after batch. Machine mixing requires less human energy, which frees masons to use their energy where it counts most—in the construction of their project.

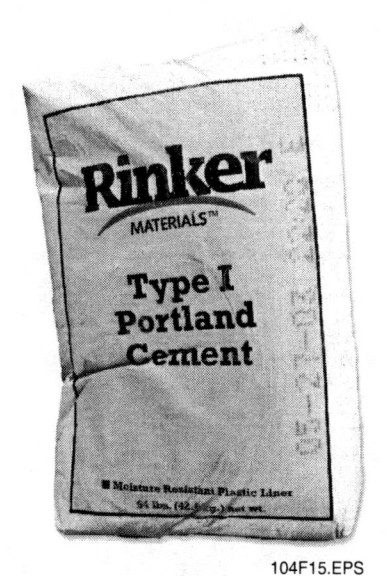

Figure 15 ♦ Bag of portland cement.

Figure 16 ♦ Cubic foot box.

Figure 17 ♦ Mortar mixer.

MORTAR

4.17

Calibrating Your Shovel

When you measure mortar materials with a shovel, you must first calibrate your shovel. You can do this by first using the shovel to fill a cubic foot box or other container of known volume.

Mortar Mixing

For large jobs, the dry mix is delivered to the site in a large hopper. The mixing machine is placed below the hopper. Batches of mortar can be mixed easily and rapidly. The materials must still be carefully measured to obtain a consistent mortar.

When using a power mixer, place the materials near the mixing area. The mixer should be at the center of these materials in such a way that all materials can be easily reached. Leave a clear path for the wheelbarrow or other equipment used to move the finished mortar. The mixer should be securely supported and the wheels blocked to avoid tire wear.

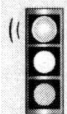

> **WARNING!**
> Wear eye protection and other appropriate personal protective equipment when using a power mixer. Never place any part of your body in the mixer.

6.1.0 Machine Mixing Steps

Review the operation manual for the machine used, and review safety procedures for working with power equipment. Have the mix formula written down, including the number of cubic feet of cement and lime as well as the cubic feet or shovelsful of sand needed for the size of the batch to be mixed.

Follow these steps for mixing mortar:

Step 1 With the blades turning, add a small amount of water (*Figure 18*). Add enough to wet the inside of the mixing drum to prevent the mortar from caking on the mixing paddles or the sides of the drum.

Step 2 Add one third to one half of the sand needed for the batch. Keep the paddles turning to prevent stress on the turning mechanism.

Step 3 Add all the necessary cement and lime or the masonry cement to the mixer. This is best done by the bag. Place the bag on the safety grate, and open the bag by cutting with a small knife or trowel or by pulling the bag opening. Some mixers have a metal tooth on the safety grate specifically designed for this purpose.

Step 4 Add the remaining sand at this time. Then add more water to bring the mixture to the desired consistency. Allow the mixing process to continue for three to five minutes in order to completely blend the materials. Do not mix any longer, however, as this will allow extra air to be trapped in the mix. This will cause the mortar to be spongy and weak.

Step 5 When the mixing is complete, with the drum still turning, grasp the drum handle and dump the mortar into the wheelbarrow or mortar box (*Figure 19*). The turning blades will clear most of the mortar out of the mixer. Next, take the blades out of gear and turn off the mixer. The remaining mortar can now be removed by hand.

Step 6 If more mortar is needed immediately, start the process over again by adding enough water to the drum to clean the blades as they are turning. If no more mortar will be needed for some time, leave the mixer turned off, and clean the inside and the blades with a water hose and a stiff bristle brush.

Figure 18 ◆ Adding water to wet the mortar mixer.

Figure 19 ◆ Dumping mortar into wheelbarrow.

MORTAR

4.19

6.2.0 Safety Tips for Machine Mixers

As with any piece of power equipment, a power mortar mixer should be treated with respect and caution. You should read the manufacturer's operation manual. Follow the procedures listed in the manual. Here are some basic safety tips:

- Wear safety glasses or goggles to protect your eyes. Wear a nose mask or face mask to avoid breathing any cement or lime dust.
- When adding water, shoveling sand, or pouring cement or lime into the drum, be extremely careful not to place the shovel or bucket into the mouth of the mixer where the blades are turning. Serious injury could result if equipment is caught by the mixing blades.
- Do not place your hands or arms into the turning mixer. Do not wear loose clothing; it can be caught in the turning mixer blades.
- If the mixer does not have a safety grate over the mixing chamber, consider adding one.
- Do not scrape the last mortar out of the drum while the blades are still turning. Disengage the blades, and turn off the mixer first.
- Turn off the mixer before doing final cleanup.

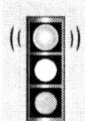

> **WARNING!**
> Keep your hands out of the mouth of the mixer. If a torn bag falls into the mixer, do not try to remove it while the mixer is turning. Turn the machine off and allow its blades to stop completely before reaching inside.

7.0.0 ◆ MIXING BY HAND

For small jobs, manual mixing of the mortar usually makes the most sense. The preliminary steps are the same as those for mechanical or power mixing. Position the material near the mortar box. Ensure that the mortar box is level and stable. Have the mix recipe written down. Make sure that there is enough room at the two ends of the mortar box for you to stand while mixing the mortar. There is no particular height required for the box. It should be at a height that will be convenient for the mason when using the hoe to mix the mortar.

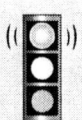

> **WARNING!**
> Cement dust is harmful to skin and mucous membranes. Always wear personal protective equipment when mixing mortar.

The basic steps for mixing mortar by hand are as follows:

Step 1 Place half of the sand in the box (*Figure 20*). Spread this sand out evenly across the bottom of the box.

Step 2 Place the desired amount of cement, lime, or masonry cement over the sand (*Figure 21*). For small batches of mortar, you may wish to use standard shovelsful of material rather than bags as your measuring device. Use a shovel to spread out the remaining sand across the cement and lime layer.

Step 3 Blend the dry ingredients together with the shovel or hoe. When they are thoroughly mixed, push them to one end of the box (*Figure 22*).

Figure 20 ◆ A cubic foot box can be used to measure sand.

Figure 21 ◆ Portland cement is added to the mortar mix.

Figure 22 ♦ Dry ingredients are moved to one end before water is added.

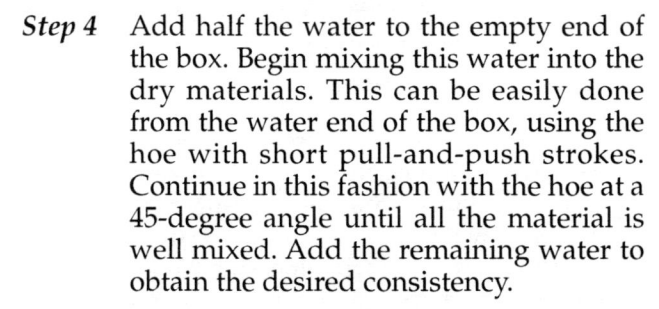

Figure 23 ♦ Mortar consistency test.

Step 4 Add half the water to the empty end of the box. Begin mixing this water into the dry materials. This can be easily done from the water end of the box, using the hoe with short pull-and-push strokes. Continue in this fashion with the hoe at a 45-degree angle until all the material is well mixed. Add the remaining water to obtain the desired consistency.

Step 5 After the mortar is mixed, pull it to one end of the box to prevent it from drying out.

Step 6 At this time, the mortar can be checked to see if it is of the proper consistency and workability. Pick up a small amount of mortar on a trowel, and set it firmly on the trowel by tapping it once on the side of the box. Turn the trowel upside down. If the mortar is the proper consistency, it will remain on the trowel (*Figure 23*). This is also a measure of the mortar's adhesion. If the mortar does not adhere, use the corrective techniques discussed in the next section.

Step 7 Transport the mortar to the work site with a wheelbarrow or other equipment, but wet the inside surface with water before loading the mortar. This will prevent the mortar from sticking to the sides of the container. When only a small amount of mortar is needed, it can be mixed directly in the wheelbarrow.

Step 8 Load the mortar from the wheelbarrow to a mortar board or pan with a shovel. The mortar should be sticky and not runny (*Figure 24*).

Figure 24 ♦ Transferring mortar to mortar board.

Step 9 The final step is to clean the mixing equipment of all mortar as soon as possible after the mixing box or wheelbarrow is no longer needed. A water hose with a spray nozzle and a stiff brush are the best tools to use when cleaning your masonry tools. You can also use a water barrel as a bath to keep the mortar from drying on your tools.

8.0.0 ♦ PROBLEMS MIXING MORTAR

Typically, the mason faces four general types of problems when mixing mortar:

- Improper proportioning of materials
- Poor quality of materials
- Working in cold weather
- Retempering

Efflorescence is another mortar problem in which salts seep out of the mortar and stain the surface. It does not show up until after the mortar has hardened.

8.1.0 Proportioning Materials

Improper proportioning of materials is a common problem in mixing mortar. It is caused primarily by poor work techniques. The first step in preparing the mix is to plan adequately. Ensure that the mix ratios are completely understood and the formula written down. The mason or helper preparing the mix should go through a mental checklist of the things needed and the steps required for the mixing operation. Planning ahead will bring out any missing pieces or shortages that would lead to problems during the mixing.

The four most common problems are as follows:

- *Adding too much water* – This problem is caused by using a hose to add water. Adding excessive amounts of water is known as drowning the mortar. This lowers the water-cement ratio, which leads to lower mortar strength. This problem can be corrected by adding dry ingredients in their proper proportions until the desired consistency is reached. The best way to avoid adding too much water is to pre-measure the water in a bucket. This is important because trying to correct this error can lead to new problems.

- *Adding too much sand* – This problem is caused by not measuring your materials. If you use a shovel to initially fill the power mixer or mortar box, calibrate it ahead of time. You should calibrate the shovel at least twice a day, once before you start in the morning and once midway through the day. If you continue to have problems with measuring the sand, using the cubic foot box may be the best answer. This is actually a faster way to load sand, but you will need a helper to help lift the box and empty it into the drum.

- *Adding too much sand to counteract the effects of too much water* – This practice results in weaker mortar and should be avoided. Overly sanded mortar is harsh, difficult to use, and forms weak bonds with the masonry units. The color of the mortar will also vary. Never add sand to a mix without adding the other ingredients in proportion.

- *Adding too much or too little cement* – Mortar that is too high in cement content is known as fat mortar. It is sticky and hard to remove from the drum. Mortar that is low in cement content is called lean mortar. It will display all the characteristics of weak mortar, such as poor bonding, lack of cohesion and adhesion, and lack of compressive and tensile strength. Both of these conditions can be avoided by always using a one-bag mix. That is, mix a mortar batch of a size that requires one bag each of cement and lime, or one bag of masonry cement, and the correct amount of sand.

All of these problems can be avoided by proper planning, thinking ahead to anticipate problems, using calibrated containers, and approaching the mixing process in a methodical manner.

8.2.0 Poor Quality Materials

Problems with poor quality or spoiled materials can be solved by properly storing, inspecting, and testing materials prior to use. Do not use cement that has knots or lumps; it will not mix well, and taking the lumps out will waste time. You can avoid spoiling materials by planning to minimize spills onto raw materials.

Lay out the mixing area in an organized way. This will prevent good material from being spoiled. It will also increase your efficiency. For example, place the mixer downhill from the cement pallets, to avoid getting them wet from possible water spills. Leave a pathway through the mixing site for wheelbarrows or forklifts. Mortar can be easily spilled during transfers. A spill into your materials piles will ruin future batches.

8.3.0 Cold Weather

Cold weather presents special problems in mixing mortar. Certain admixtures can be added to improve the performance of mortar in cold weather.

Air-entraining admixtures will increase the workability and freeze-thaw durability of mortar. Adding too much will reduce the strength of the mortar. Adding an air-entraining admixture separately at the mixer is not recommended. This is due to the lack of control over the final air content. If air-entraining is needed, use factory-prepared cements and lime that include an air-entraining agent.

At low temperatures, performance can be improved by using Type III (high early strength) cement. Also, mortar made with lime in dry form is preferred for winter use over **slaked lime** or lime putty because it requires less water.

Another option for cold weather is to heat the material used to make the mortar. The sand and the water can both be heated. It is easier and safer to heat the water. Consider heating the materials when the temperature falls below 40°F. After mixing the heated ingredients, the temperature should be between 70°F and 105°F.

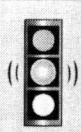

> **CAUTION**
>
> Do not heat the water over 160°F. If the water is too hot, the mix may flash set when the cement is added. A flash set occurs when the cement sets prematurely due to excessive heat.

Some admixtures are not recommended due to their negative effect on mortar. Any antifreeze admixtures, including several types of alcohol, would have to be used in such large quantities that the strength of the mortar would be greatly reduced. Do not use calcium chloride. It is commonly used as an accelerator in concrete, but it produces increased shrinkage, efflorescence, and metal corrosion in mortar.

Figure 25 ◆ Retempering mortar.

8.4.0 Retempering

Fresh mortar should be prepared as it is needed. That way its workability will remain about the same throughout the day. Mortar that has been mixed but not used immediately tends to dry out and stiffen. However, loss of water on a dry day can be reduced by wetting the mortar board. Covering the mortar in the mortar box, wheelbarrow, or tub will also help keep it fresh.

If necessary to restore workability, mortar may be retempered by adding water (*Figure 25*). Thorough remixing is then necessary. Remixing can be done in the wheelbarrow or on a mortar board. Small additions of water may slightly reduce the compressive strength of the mortar; however, the end effect is acceptable. Masonry built using a workable plastic mortar has a better bond strength than masonry built using dry, stiff mortar. Mortar used within one hour after mixing should not need retempering unless the weather is very hot, and evaporative conditions prevail.

Colored mortar is very sensitive to retempering; additional water may cause a noticeable lightening of the color. Mortar should be retempered with caution to avoid variations in the color of the hardened mortar. Mixing smaller batches can lessen the need to retemper.

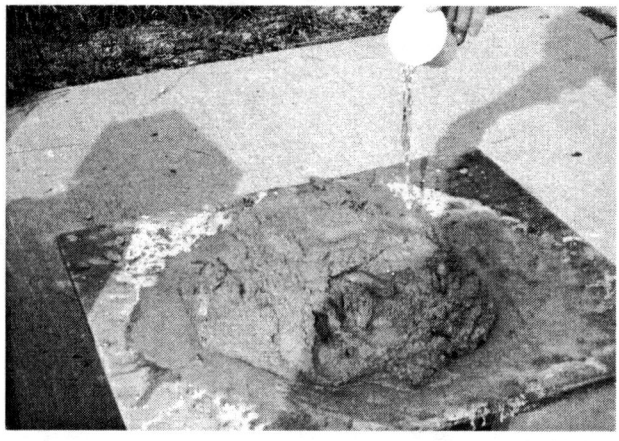

8.5.0 Efflorescence

Efflorescence is a deposit of water-soluble salts on the surface of a masonry wall. It is usually white and generally appears soon after the wall has been built. Of course, this is when the owner and architect tend to be most concerned with the structure's appearance.

Efflorescence occurs when soluble salts and moisture are both present in the wall. If either of those elements is missing or below a certain concentration, efflorescence will not occur.

The salts usually originate in the walls, in either the masonry units or the mortar itself. In some cases, the salts may come from ground moisture behind a basement wall. For this reason, basement walls should be protected with a moisture barrier.

Efflorescence can be prevented in several ways:

- Reduce salt content in the mortar materials by using washed sand.
- Use clean water for making the mortar.
- Keep materials properly stored and off the ground prior to use.
- Protect newly built masonry walls with a canvas or suitable waterproofing material.
- Use quality workmanship to ensure strong, tight bonds between mortar and masonry units.

BUILDING BLOCKS

Mortar Hydration Hardening

Mortar that has stiffened because of hydration hardening should be discarded. It is difficult to tell by sight or feel whether mortar stiffening is due to evaporation or hydration. The most practical method is to consider the time elapsed after mixing. Mortar should be used within 2½ hours after the initial mixing to avoid hydration hardening.

Building Blocks

Efflorescence

Efflorescence can be removed by washing the wall with water or a muriatic acid-water solution. Use a diluted solution of one part muriatic acid to nine parts water. This washing will remove the efflorescence, but it may return after a period of time. The best solution is to reduce the amount of moisture entering the wall and to lower the amount of water-soluble salts in the mortar materials.

Summary

Mixing quality mortar is one of the primary skills a mason needs. Good mortar adds to the strength and stability of the structure. Although you can purchase premixed mortar, a mason must know the ingredients in mortar. Each of the materials affects the overall properties and appearance of the final product. Knowing these properties will help you mix strong, effective mortar under different conditions.

Review Questions

1. Which type of cement is used most often by masons to mix mortar?
   a. Type I
   b. Type II
   c. Type III
   d. Type IV

2. One of the advantages of using lime in mortar is its characteristic of autogenous healing, or the ability to _____.
   a. reseal the effects of efflorescence
   b. resist the effects of airborne acids and alkalis
   c. reseal small cracks in mortar
   d. resist the strength-weakening effects of air-entraining admixtures

3. The siltation test for sand is meant to determine _____.
   a. acceptable limits for sodium chloride
   b. how well color will adhere to the sand
   c. the presence of undesirable elements
   d. the strength of the sand

4. Air-entraining admixtures reduce _____ in mortar.
   a. workability
   b. water content
   c. freeze-thaw resistance
   d. lime content

5. If too much color pigment is added to the mix, the mortar will lose _____.
   a. workability and water retention
   b. strength and durability
   c. water retention and consistency
   d. durability and autogenous healing

6. Of the following types of mortar, _____ is the strongest.
   a. Type K
   b. Type N
   c. Type M
   d. Type O

7. Water retention in mortar is desired to _____.
   a. increase the workability of the mortar
   b. keep the mortar cool
   c. increase the mortar's ability to retain color
   d. decrease the chance of efflorescence

8. Mortar gains compressive strength as _____ is added to the mix.
   a. cement
   b. lime
   c. sand
   d. water

9. Bond strength refers to the mortar's _____.
   a. ability to hold the masonry units together
   b. compressive strength
   c. freeze-thaw tolerance
   d. percentage of cement

10. Mortar is commonly referred to in the trade as _____.
    a. concrete
    b. gypsum
    c. cement
    d. mud

11. You should never use a hose to add water directly to a mechanical mixer because _____.
    a. you can trip over the hose
    b. the hose will get caught in the mixer blades
    c. you cannot accurately measure the water
    d. the water should be heated first

12. Water should be placed first into the power mixing drum in order to _____.
    a. cool it off on hot days
    b. prevent cement from sticking to its sides
    c. increase the workability of the mix
    d. help speed up the mixing process

13. Typically, the time for mixing mortar in a power mixer is _____ after all ingredients are added.
    a. 2 minutes
    b. 3 to 5 minutes
    c. 6 to 8 minutes
    d. 10 to 12 minutes

14. When mixing mortar by hand, the first thing you put in the mixing box is _____.
    a. all of the sand
    b. all of the lime
    c. half of the sand
    d. half of the lime

15. A mortar mix with insufficient cement produces a weak, harsh mix that lacks strength and adherence. This type of mortar is called _____ mortar.
    a. lightweight
    b. sharp
    c. lean
    d. fat

16. If too much water has been added to the mix, you should _____.
    a. add more sand
    b. add more cement
    c. pour off the excess water
    d. add all dry ingredients in proportion

17. For cold weather masonry work, at what temperature should you begin heating the mortar ingredients?
    a. 32°F
    b. 40°F
    c. 45°F
    d. 50°F

18. Calcium chloride should be used to temper mortar in cold weather.
    a. True
    b. False

19. Retempering mortar is done to extend its workability after the mortar _____.
    a. freezes and then thaws out
    b. was mixed too lean
    c. stiffens due to evaporation
    d. stiffens due to hydration

20. Efflorescence can be reduced by _____.
    a. adding muriatic acid to the mortar
    b. using washed sand
    c. boiling the water used in mixing the mortar
    d. using wet sand to make the mortar

PROFILE IN SUCCESS

Sam McGee, President
McGee Brothers Masonry
Charlotte, NC

Sam McGee had a good management job with a large corporation, but it kept him away from his growing family. He decided to learn the masonry trade by working with his brothers and his in-laws and then went on to found the largest masonry contracting company in the United States. Today, McGee Brothers employs 1,200 people and has facilities at many locations in North Carolina and South Carolina.

Sam had worked in management for two international companies and had also managed a retail store. His job required him to spend a lot of time traveling. His brothers and his in-laws were masons, and he envied the fact that they were able to be home with their families every night. He decided right then, at the age of 30, to make a life change. He started working as a laborer at $2.75 an hour, then gradually learned to lay brick.

After laying brick for a while, Sam and his brother decided to go out on their own. They rented a tractor and developed a lifting device for the tractor so that they could move brick and other materials.

That was over 30 years ago. Today, Sam oversees the business and deals with special problems. He spends a lot of time visiting company facilities in North Carolina and South Carolina. He also manages the financial side of the business. A great deal of his time is spent in promoting the masonry industry by lecturing, and by participating in masonry associations. He is active in groups like NCCER that promote masonry training, and he is one of the subject matter experts who helped develop the masonry curriculum.

Sam believes the key to the company's success lies in their approach to the business. When the company was founded, they started by hiring people they knew and trusted, then taught them the masonry trade. Today, they generally hire people who have been referred to them by employees. The employees tend to refer people they know and trust because they know the company management has high standards. They also know that their success is directly linked to the success of the company, so they are not likely to recommend someone they are uncertain about.

Sam's view of the business is straightforward: "In the masonry trade, it's all about teamwork. We treat each crew as though they were an independent business. It's up to them to succeed or fail. If they can't work together as a team, they will not succeed." He takes a great deal of pleasure in working with young people and watching them grow in the business. He feels fortunate to have the opportunity to share what he has learned with people around the country.

Sam believes there are many opportunities in residential masonry, but it requires that you work hard, have a good attitude, and learn everything you can. He points out that the company employs many young masons who have beautiful homes and earn incomes that most people only dream about, simply because they have been willing to invest the time to learn the trade and have a great work ethic.

GLOSSARY

Trade Terms Introduced in This Module

Air-entraining: A type of admixture added to mortar to increase microscopic air bubbles in mixed mortar. The air bubbles increase resistance to freeze-thaw damage.

Hydration: A chemical reaction between cement and water that hardens the mortar. Hydration requires the presence of water and an air temperature between 40°F and 80°F.

Masonry cement: Cement that has been modified by adding lime and other materials.

Plasticity: The ability of mortar to flow like a liquid and not form cracks or break apart.

Pozzolan: A finely powdered material that can be added to mortar to increase durability and provide a positive set.

Slaked lime: Lime reduced by mixing with water to a safe form that can be used in the production of mortar.

Water retention: The ability of mortar to keep sufficient water in the mix to enhance plasticity and workability.

Workability: The property of mortar to remain soft and plastic long enough to allow the mason to place and align masonry units and strike off the mortar joints before the mortar hardens completely.

REFERENCES & ACKNOWLEDGMENTS

Additional Resources

This module is intended to be a thorough resource for task training. The following reference works are suggested for further study. These are optional materials for continued education rather than for task training.

Masonry Construction. David L. Hunter, Sr. Upper Saddle River, NJ: Prentice Hall.

Building Block Walls—A Basic Guide, 1988. Herndon, VA: National Concrete Masonry Association.

The ABCs of Concrete Masonry Construction, Videotape. 1980. Skokie, IL: Portland Cement Association.

Figure Credits

Topaz Publications, Inc.	104F01, 104F07, 104F11, 104F14–104F16, 104F21–104F24, 104SA06
Graymont Dolime (OH), Inc.	104F02, 104SA02
Quikrete	104F03, 104F04, 104F06, 104SA03
Bon Tool Company	104SA04
Portland Cement Association	104F08, 104F13, 104F18–104F20, 104F25, 104SA01, 104SA05
Granite City Tool Co.	104F17

CONTREN® LEARNING SERIES — USER UPDATES

The NCCER makes every effort to keep these textbooks up-to-date and free of technical errors. We appreciate your help in this process. If you have an idea for improving this textbook, or if you find an error, a typographical mistake, or an inaccuracy in NCCER's Contren® textbooks, please write us, using this form or a photocopy. Be sure to include the exact module number, page number, a detailed description, and the correction, if applicable. Your input will be brought to the attention of the Technical Review Committee. Thank you for your assistance.

Instructors – If you found that additional materials were necessary in order to teach this module effectively, please let us know so that we may include them in the Equipment/Materials list in the Instructor's Guide.

Write: Product Development and Revision
National Center for Construction Education and Research
P.O. Box 141104, Gainesville, FL 32614-1104

Fax: 352-334-0932

E-mail: curriculum@nccer.org

Craft _____ Module Name _____

Copyright Date _____ Module Number _____ Page Number(s) _____

Description

_____

_____

_____

_____

(Optional) Correction

_____

_____

_____

(Optional) Your Name and Address

_____

_____

_____

Module 28105-04

Masonry Units and Installation Techniques

COURSE MAP

This course map shows all of the modules in the first level of the *Masonry* curriculum. The suggested training order begins at the bottom and proceeds up. Skill levels increase as you advance on the course map. The local Training Program Sponsor may adjust the training order.

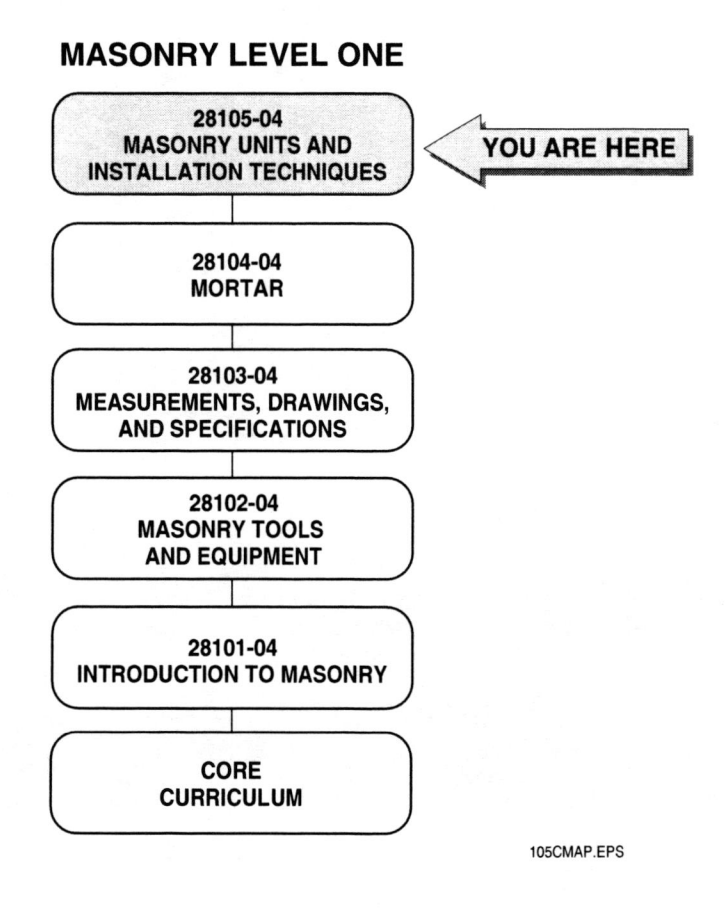

MASONRY LEVEL ONE

- 28105-04 MASONRY UNITS AND INSTALLATION TECHNIQUES ← YOU ARE HERE
- 28104-04 MORTAR
- 28103-04 MEASUREMENTS, DRAWINGS, AND SPECIFICATIONS
- 28102-04 MASONRY TOOLS AND EQUIPMENT
- 28101-04 INTRODUCTION TO MASONRY
- CORE CURRICULUM

Copyright © 2004 National Center for Construction Education and Research, Gainesville, FL 32614-1104. All rights reserved. No part of this work may be reproduced in any form or by any means, including photocopying, without written permission of the publisher.

MODULE 28105-04 CONTENTS

1.0.0 INTRODUCTION 5.1
2.0.0 CONCRETE MASONRY MATERIALS 5.2
    2.1.0 ASTM Specifications 5.2
    2.1.1 Compressive Strength 5.2
    2.1.2 Moisture Absorption and Content 5.2
    2.2.0 Contraction and Expansion Joints 5.2
    2.3.0 Concrete Block Characteristics 5.3
    2.4.0 Concrete Brick 5.8
    2.5.0 Other Concrete Units 5.8
    2.5.1 Pre-faced Units 5.9
    2.5.2 Calcium Silicate Units 5.9
    2.5.3 Catch Basins 5.10
3.0.0 CLAY AND OTHER MASONRY MATERIALS 5.10
    3.1.0 Clay Masonry Units 5.10
    3.2.0 Stone 5.11
    3.3.0 Other Masonry Materials 5.12
    3.3.1 Metal Ties 5.13
    3.3.2 Veneer Ties 5.13
    3.3.3 Reinforcement Bars 5.13
    3.3.4 Joint Reinforcement Ties 5.13
    3.3.5 Flashing 5.13
    3.3.6 Joint Fillers 5.14
    3.3.7 Anchors 5.14
4.0.0 SETTING UP AND LAYING OUT 5.14
    4.1.0 Setting Up 5.14
    4.2.0 Job Layout 5.16
    4.2.1 Planning 5.16
    4.2.2 Locating 5.17
    4.2.3 Dry Bonding 5.18
5.0.0 BLOCK HEAD JOINTS 5.18
    5.1.0 Buttering Block 5.18
    5.2.0 Block Bed Joints 5.19
    5.3.0 General Rules 5.20
6.0.0 BONDING MASONRY UNITS 5.20
    6.1.0 Mechanical or Mortar Bond 5.20
    6.2.0 Pattern Bond 5.21
    6.3.0 Structural Bond and Structural Pattern Bond 5.22
    6.4.0 Block Bond Patterns 5.23
7.0.0 CUTTING MASONRY UNITS 5.25
    7.1.0 Brick Cuts 5.25
    7.2.0 Block Cuts 5.25

MODULE 28105-04 CONTENTS (Continued)

7.3.0	Cutting with Hand Tools	5.26
7.3.1	*Cutting with Chisels and Hammers*	*5.26*
7.3.2	*Cutting with Masonry Hammers*	*5.27*
7.3.3	*Cutting with Trowels*	*5.27*
7.4.0	Cutting with Saws and Splitters	5.28
7.4.1	*Saws and Splitters*	*5.28*
7.4.2	*Units and Cuts*	*5.29*

8.0.0 LAYING MASONRY UNITS 5.29
   8.1.0 Laying Brick in Place 5.29
     8.1.1 Placing Brick *5.29*
     8.1.2 Checking the Height *5.30*
     8.1.3 Checking Level *5.30*
     8.1.4 Checking Plumb *5.30*
     8.1.5 Checking Straightness *5.31*
     8.1.6 Laying the Closure Unit *5.31*
   8.2.0 Placing Block 5.32
   8.3.0 Laying to the Line 5.33
     8.3.1 Setting Up the Line Using Corner Poles *5.33*
     8.3.2 Setting Up the Line Using Line Blocks and Stretchers ... *5.33*
     8.3.3 Setting Up the Line Using Line Pins *5.34*
     8.3.4 Setting Up the Line Using Line Trigs *5.34*
     8.3.5 Laying Brick to the Line *5.34*
     8.3.6 Laying Block to the Line *5.36*
   8.4.0 Building Corners and Leads 5.36
     8.4.1 Rackback Leads *5.37*
     8.4.2 Brick Rackback Corners *5.37*
     8.4.3 Block Rackback Corners *5.39*

9.0.0 MORTAR JOINTS 5.40
   9.1.0 Joint Finishes 5.40
   9.2.0 Striking the Joints 5.41
     9.2.1 Testing the Mortar *5.41*
     9.2.2 Striking *5.41*
     9.2.3 Cleaning Up Excess Mortar *5.42*

10.0.0 PATCHING MORTAR 5.42
   10.1.0 Pointing 5.42
   10.2.0 Tuckpointing 5.43

11.0.0 CLEANING MASONRY UNITS 5.44
   11.1.0 Clean Masonry Checklist 5.44
   11.2.0 Bucket Cleaning for Brick 5.44
   11.3.0 Cleaning Block 5.46

MODULE 28105-04 CONTENTS (Continued)

SUMMARY 5.46
REVIEW QUESTIONS 5.46
PROFILE IN SUCCESS 5.48
GLOSSARY 5.49
REFERENCES 5.50
ACKNOWLEDGMENTS 5.51

Figures

Figure 1	Locations of control joints	5.3
Figure 2	Block wall with reinforcing rods	5.3
Figure 3	Vertical control joint with plastic filler	5.4
Figure 4	Parts of blocks	5.4
Figure 5	Common concrete block shapes	5.6–5.7
Figure 6	An application of loadbearing block	5.7
Figure 7	Concrete brick	5.8
Figure 8	A wall made from slump brick	5.9
Figure 9	Concrete pre-faced units	5.9
Figure 10	Calcium silicate units	5.9
Figure 11	Manhole and vault units	5.11
Figure 12	Brick positions	5.11
Figure 13	Imitation stone made of concrete	5.12
Figure 14	Metal ties	5.13
Figure 15	Veneer ties	5.13
Figure 16	Joint reinforcement ties	5.13
Figure 17	Flashing applications	5.13
Figure 18	Joint fillers	5.14
Figure 19	Shaped anchors	5.15
Figure 20	Door and window openings	5.16
Figure 21	Foundation plan	5.17
Figure 22	Example of a dry bond	5.18
Figure 23	Block corner layouts	5.19
Figure 24	Placing block	5.19
Figure 25	Buttering block	5.20
Figure 26	Stack bond	5.21
Figure 27	Brick positions in walls	5.22
Figure 28	Running bonds	5.22
Figure 29	English and Flemish bonds	5.23
Figure 30	Common bond	5.23

Figure 31	Dutch bond	5.23
Figure 32	Garden wall bonds	5.24
Figure 33	Block bonds	5.24
Figure 34	Common brick cuts	5.25
Figure 35	Horizontal and vertical face cuts	5.25
Figure 36	End and web block cuts	5.26
Figure 37	Bond beam cut	5.26
Figure 38	Cutting with a hammer and chisel	5.27
Figure 39	Cutting with a brick hammer	5.27
Figure 40	Portable masonry saw	5.28
Figure 41	Brick splitter	5.28
Figure 42	Proper hand position	5.29
Figure 43	Mason's rules	5.30
Figure 44	Leveling the course	5.30
Figure 45	Checking plumb	5.31
Figure 46	Plumb and out-of-plumb bricks	5.31
Figure 47	Checking for straightness	5.31
Figure 48	Placing a closure unit	5.31
Figure 49	Placing a block	5.32
Figure 50	Line block	5.33
Figure 51	Using a line stretcher	5.34
Figure 52	Using a line pin	5.35
Figure 53	Using a line trig	5.35
Figure 54	Laying brick to the line	5.36
Figure 55	Building corner leads	5.37
Figure 56	Rackback lead	5.37
Figure 57	Checking alignment	5.37
Figure 58	Tailing the diagonal	5.38
Figure 59	Leveling the diagonal	5.38
Figure 60	Mortar joint finishes	5.40
Figure 61	Extruded or weeping joint	5.41
Figure 62	Using a jointer	5.41
Figure 63	Using the convex sled	5.42
Figure 64	Skate raker	5.42
Figure 65	Trimming and dressing mortar burrs	5.42
Figure 66	Preparation of joints for tuckpointing	5.43
Figure 67	Tuckpointing	5.43

Table

Table 1	Cleaning Guide for New Masonry	5.45

MODULE 28105-04

Masonry Units and Installation Techniques

Objectives

When you have completed this module, you will be able to do the following:

1. Describe the most common types of masonry units.
2. Describe and demonstrate how to set up a wall.
3. Lay a dry bond.
4. Spread and furrow a bed joint, and butter masonry units.
5. Describe the different types of masonry bonds.
6. Cut brick and block accurately.
7. Lay masonry units in a true course.

Recommended Prerequisites

Core Curriculum; Masonry Level One, Modules 28101-04 through 28104-04

Required Trainee Materials

1. Pencil and paper
2. Appropriate personal protective equipment

1.0.0 ♦ INTRODUCTION

This module contains detailed information on masonry materials. It also provides instructions for building a single-wythe masonry wall. This module gives you details about the following topics:

- Cement, clay, and stone masonry units
- Setting up and laying out a wall
- Spreading mortar
- Bonding masonry units
- Cutting masonry units
- Laying masonry units
- Mortar and other joints
- Patching, pointing, and tuckpointing
- Cleaning masonry units

Safety is a continuous effort. Being a mason calls for following safe work practices and procedures. This includes performing the following work activities:

- Inspecting tools and equipment before use
- Using tools and equipment properly
- Keeping tools and equipment clean and properly maintained
- Keeping your hands out of the mortar
- Using the right tools for the job
- Assembling and using scaffolding and foot boards properly
- Using caution and common sense when working on elevated surfaces

Accidents also happen when tools and equipment are left in the way of other workers. Store tools safely, where other people cannot trip over them and where the tools cannot get damaged. Clean, well-kept tools make for safe work as well as good work.

Do not drop or temporarily store tools or masonry units in pathways or around other workers. Stack masonry units neatly so the stack will be less likely to topple over. Stack materials by reversing the direction of the units on every other layer, so they will be less prone to tip. Keep the pile neat and vertical to avoid snagging clothes. As a rule, do not stack masonry units higher than your chest. A stack that is too high is more likely to tip over.

Various materials used to build masonry structures have properties that can be harmful. Mortar and grout, for example, can be very caustic. This means that continued contact with the skin can cause burns and irritation. It is important to protect your skin and eyes when mixing and working with these materials.

Cutting masonry units can be a dangerous job. When sawing or chiseling a masonry unit, chips can fly off and hit your eyes.

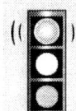

> **WARNING!**
> Always wear safety glasses or goggles and respiratory protection when cutting masonry units.

The process of using a saw or chisel is dangerous. Always keep your hands away from the blade. Do not operate power saws when you are tired, sick, or otherwise unable to give the process your full attention and effort.

2.0.0 ◆ CONCRETE MASONRY MATERIALS

Most masonry materials are made of clay, concrete, or stone. The masonry unit most commonly used in the United States is the concrete block. The term CMU stands for concrete masonry unit. CMUs are classified into six types:

- Loadbearing concrete block
- Nonbearing concrete block
- Concrete brick
- Calcium silicate units
- Pre-faced or prefinished concrete facing units
- Concrete units for manholes and catch basins

2.1.0 ASTM Specifications

Each of the six types of CMUs has its own ASTM (American Society for Testing and Materials) International standards for performance characteristics. The standards describe the expected performance of the CMU in compressive strength, water absorption, loadbearing, and other characteristics. CMUs are valuable for their fire resistance, sound absorption, and insulation value.

2.1.1 Compressive Strength

The compressive strength of a CMU measures how much weight it can support without collapsing. These figures are set according to ASTM test results. The tests are performed on a specific shape, size, and weight of CMU. The CMU tested then becomes the standard for that particular compressive strength. Compressive strength is measured in pounds of pressure per square inch (psi). This is the weight that the unit, and the structure made of the units, can support. A structure supports less imposed weight than the sum of its unit strength because the structure has to support itself as well.

The quality of a mason's work is an important part of how a masonry structure performs and whether it meets its compressive strength specification. Tests have shown that the compressive strength of a loaded wall is about 42 percent of the compressive strength of a single CMU when the mason uses a face shell mortar bedding. When the mason uses a full mortar bedding, the compressive strength increases to about 53 percent. The engineer factors these components into the equation for picking the CMU. The factoring is contained in the job specifications. This is another reason that specifications are important.

2.1.2 Moisture Absorption and Content

Moisture in the CMU has an effect on shrinking and cracking in the finished structure. Generally, the lower the moisture, the less likely the units are to shrink after they are set. Acceptable moisture content and absorption rates are set by ASTM standards and local codes. Manufacturers specify that their units meet ASTM or other standards. Unit tests are made to make sure they do. For the mason, this means keeping CMUs dry. Never wet CMUs immediately before or during the time they are to be laid. Stockpile them on planks or pallets off the ground. Use plastic or tarpaulin covers for protection against rain and snow.

When stopping work, cover the tops of masonry structures to keep rain or snow off. Be sure moisture does not get into cavities between wythes. When laying CMUs for interior use, dry them before laying. They should be dried to the average condition to which the finished wall will be exposed.

2.2.0 Contraction and Expansion Joints

CMUs have one major difference from other masonry units: like concrete slabs and sidewalks, CMU construction is prone to cracking from shrinkage. Shrinkage cracking occurs as the concrete slowly finishes drying. As the concrete shrinks, it moves slightly. The movement causes cracks in a rigid slab or wall. The shrinkage cracking in a CMU structure is controlled in the same way as for concrete slabs. This is done by using reinforcement in combination with contraction or control joints.

Figure 1 shows some typical locations for contraction joints. Walls are likely to crack at abrupt changes in wall thickness or heights, at openings, over windows, and over doors. Contraction joints are also used at intersections between loadbearing walls and partition walls.

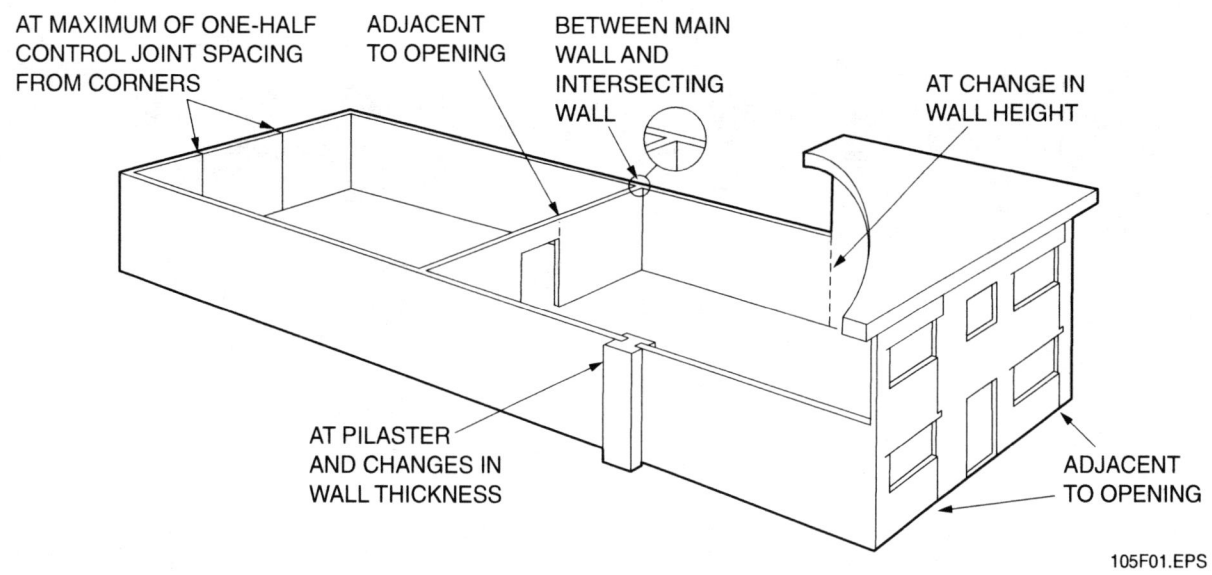

Figure 1 ♦ Locations of control joints.

CMU walls use two kinds of reinforcement: grout and steel with grout. Either the grout is poured into CMU cores, or steel rods are inserted in CMU cores, and grout is poured around the steel (*Figure 2*). The reinforcement gives rigidity and strength to the wall. CMU walls need reinforcement of either kind on both sides of a contraction joint.

Contraction joints control cracking by weakening the structure. Cracks occur at the weakened area instead of randomly. In a concrete slab, these control joints are made by cutting grooves. In a CMU wall, this is done by breaking the contact between two columns of units, as shown in *Figure 3*.

The control joints replace a standard mortar joint every 20' or so. Control joints must be no more than 30' apart. The control joints in CMU walls have no mortar; they are filled with silicon or another flexible material. The contraction joint filler keeps the rain out but allows slight movement. The reinforcement keeps the edges of the contraction joints aligned as they move.

Clay masonry units do not contract and shrink as they age, but they do change size very slightly with temperature and moisture changes in the air. Clay masonry structures need expansion joints to handle this type of movement. The control joints in CMU walls take care of expansion as well as contraction. Clay masonry walls must include expansion joints. These are usually soft, mortarless joints filled with foam and covered with a layer of silicon paste. As with contraction joints, the filler keeps the rain out and allows slight movement of the masonry.

2.3.0 Concrete Block Characteristics

Block is produced in four classes: solid loadbearing, solid nonbearing, hollow loadbearing, and hollow nonbearing. Block comes in two weights: normal and lightweight. Lightweight block is made with fly ash, pumice, and scoria or other lightweight aggregate. Loadbearing and appearance qualities of the two weights are similar. The major difference is that lightweight block is easier and faster to lay.

Solid blocks are for special needs, such as structures with unusually high loads, drainage catch basins, manholes, and firewalls. Like solid clay products, 25 percent or less of the surface of a solid concrete unit is hollow. Less than 75 percent of a hollow unit is solid. *Figure 4* shows the names of different parts of blocks.

Figure 2 ♦ Block wall with reinforcing rods.

MASONRY UNITS AND INSTALLATION TECHNIQUES

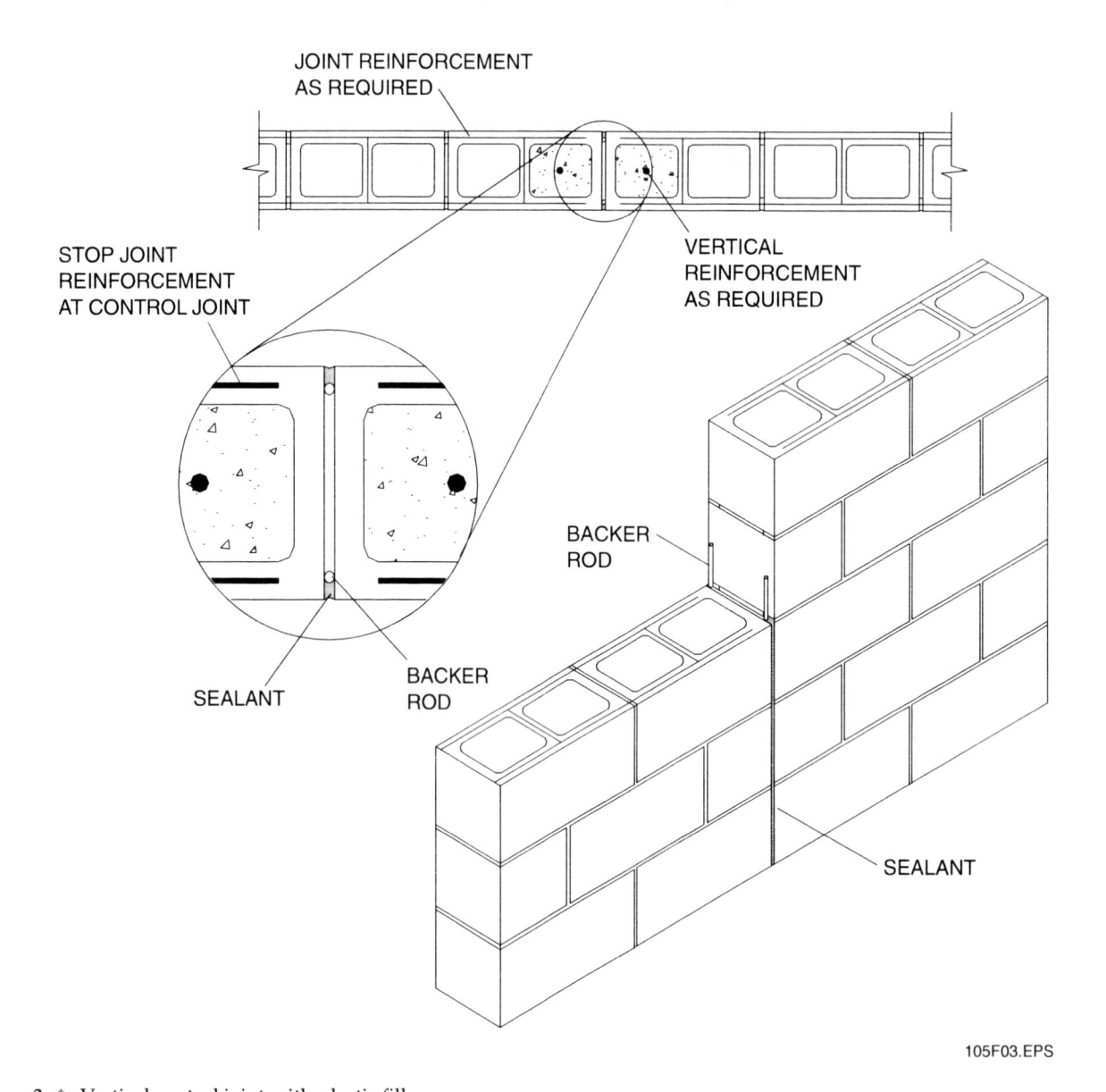

Figure 3 ♦ Vertical control joint with plastic filler.

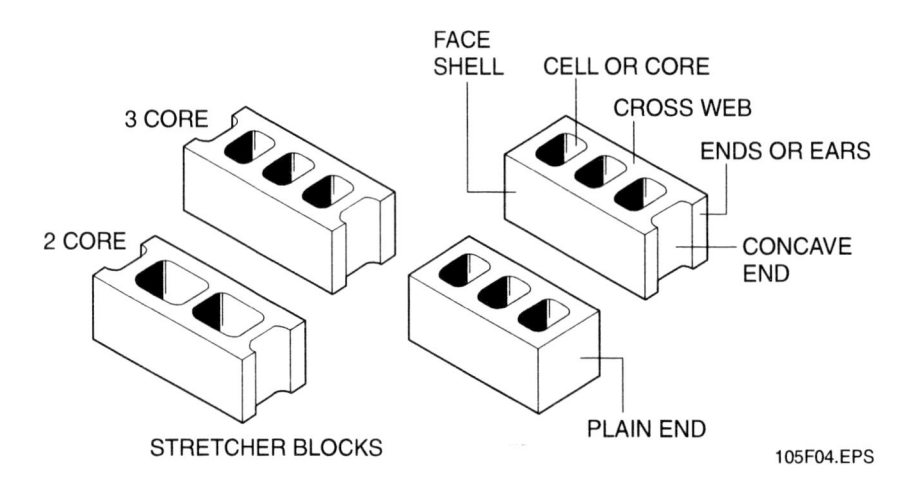

Figure 4 ♦ Parts of blocks.

5.4　　MASONRY LEVEL ONE — TRAINEE MODULE 28105-04

BUILDING BLOCKS

Reinforcing Block Walls

Grout is a mixture of cementitious material and aggregate with enough liquid content to make it flow readily. Grout is often pumped into block wall cavities with special pumps.

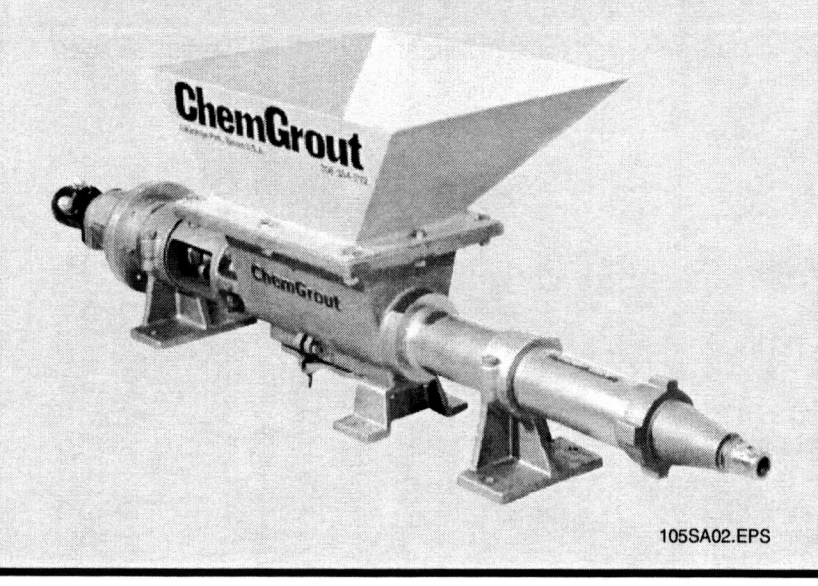

Most block is governed by *ASTM C90*, which covers hollow loadbearing block. At one time, *ASTM C90* specified Grades N and S. Grade S has been discontinued for this block, and *ASTM C90* block is now ungraded. The ASTM does specify a minimum compressive strength of 800 psi, however. Solid loadbearing block falls under *ASTM C145*, which specifies Grades N and S. Grade N has a compressive strength of 1,500 psi, and the compressive strength of Grade S is 1,000 psi.

Block comes in modular sizes, with colors determined by the cement ingredients, the aggregates, and any additives. The basic block is called a stretcher. It has a nominal face size of 8" × 16" with a standard ⅜" mortar joint. The most commonly used block has a nominal width of 8", but 4", 6", 10", and 12" widths are also common. Three modular bricks have the same nominal height as one nominal 8" block. Two nominal brick lengths equal one nominal block length. So, laying one stretcher covers the same area as laying six modular bricks.

Block comes in a variety of shapes to fit common and special purposes. *Figure 5* shows a sampling of block sizes and shapes. Common hollow units can have two or three cores. Most cores are tapered slightly to provide a larger bed joint surface. Block edges may be flanged, notched, or smooth. There are local variations as well, with some shapes available only in specific parts of the country.

MASONRY UNITS AND INSTALLATION TECHNIQUES

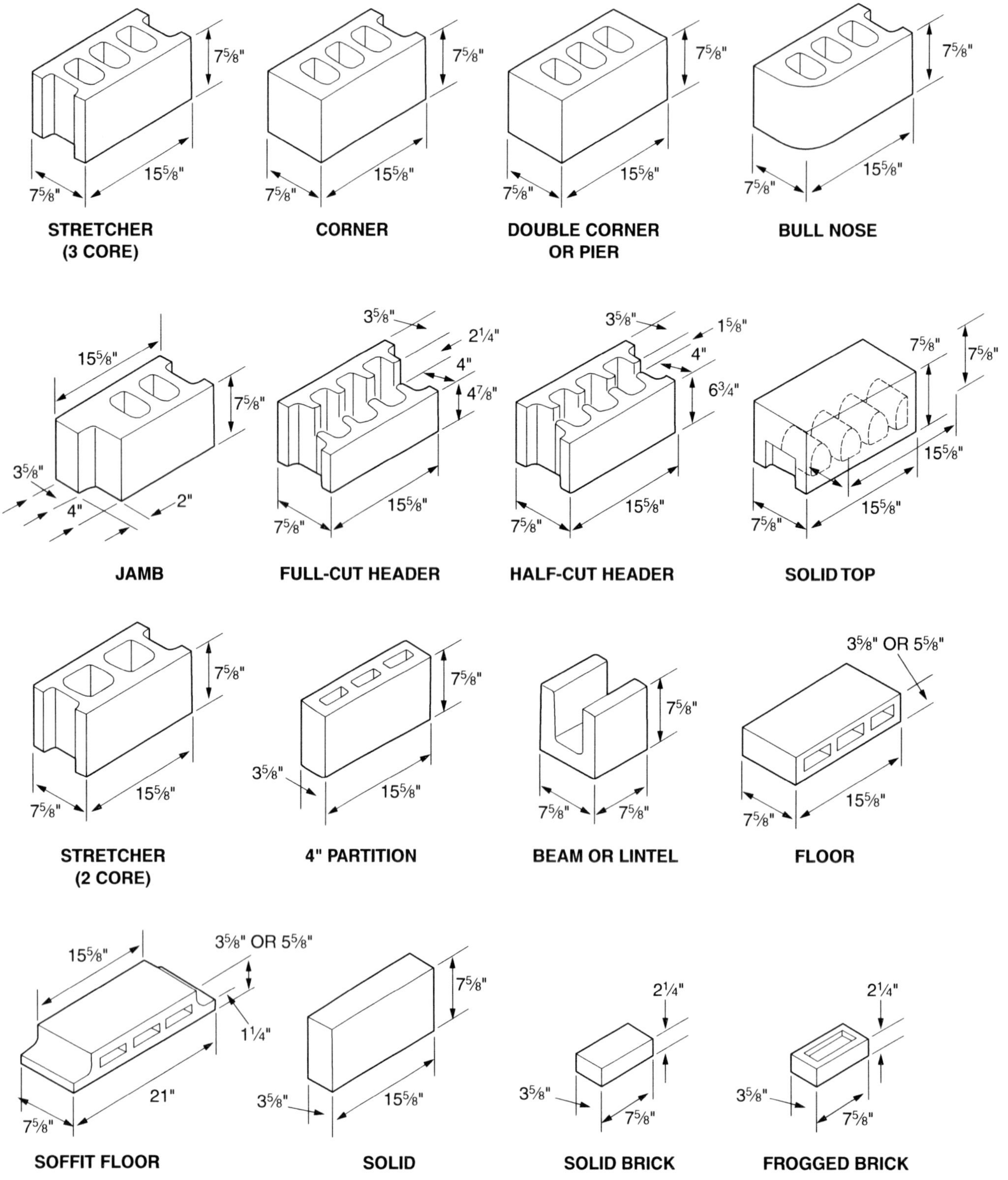
7⅝"
15⅝"
7⅝"
STRETCHER (3 CORE)
7⅝"
15⅝"
7⅝"
CORNER
7⅝"
15⅝"
7⅝"
DOUBLE CORNER OR PIER
7⅝"
15⅝"
7⅝"
BULL NOSE
15⅝"
7⅝"
3⅝"
4"
2"
JAMB
3⅝"
2¼"
4"
4⅞"
15⅝"
7⅝"
FULL-CUT HEADER
3⅝"
1⅝"
4"
6¾"
15⅝"
7⅝"
HALF-CUT HEADER
7⅝"
7⅝"
15⅝"
15⅝"
7⅝"
SOLID TOP
7⅝"
15⅝"
7⅝"
STRETCHER (2 CORE)
7⅝"
3⅝"
15⅝"
4" PARTITION
7⅝"
7⅝"
7⅝"
BEAM OR LINTEL
3⅝" OR 5⅝"
15⅝"
7⅝"
FLOOR
15⅝"
3⅝" OR 5⅝"
1¼"
21"
7⅝"
SOFFIT FLOOR
7⅝"
3⅝"
15⅝"
SOLID
2¼"
3⅝"
7⅝"
SOLID BRICK
2¼"
3⅝"
7⅝"
FROGGED BRICK

NOTE: Dimensions are actual unit sizes. A 7⅝" × 7⅝" × 15⅝" unit is an 8" × 8" × 16" nominal-size block.

105F05A.EPS

Figure 5 ◈ Common concrete block shapes. (1 of 2)

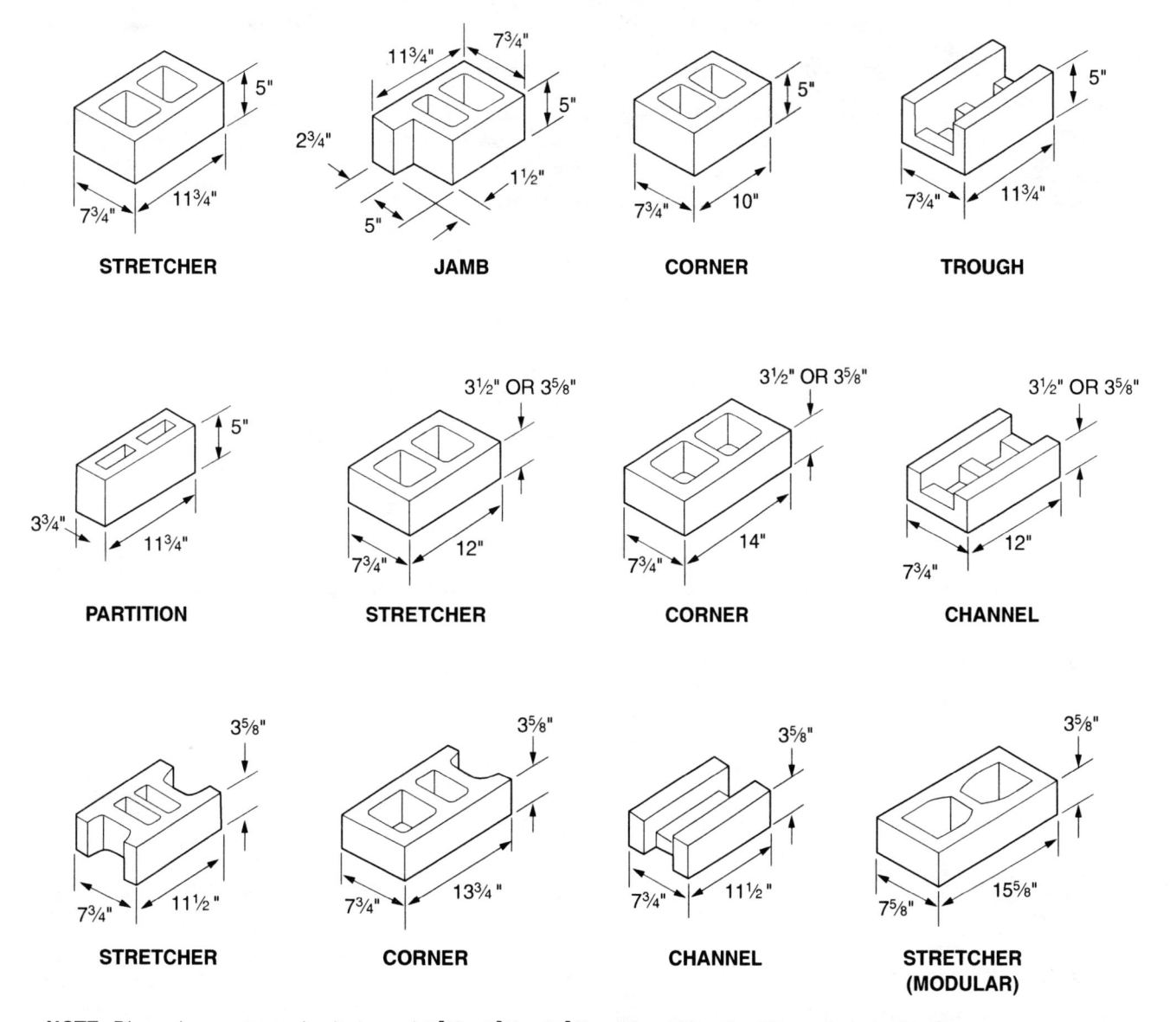

NOTE: Dimensions are actual unit sizes. A 7⅝" × 7⅝" × 15⅝" unit is an 8" × 8" × 16" nominal-size block.

Figure 5 ◆ Common concrete block shapes. (2 of 2)

A variety of surface and mixing treatments can give block varied and attractive surfaces. Newer finishing techniques give block face the appearance of brick, stone, ribbed columns, raised patterns, or architectural fabrics. Like clay masonry units, block can be laid in structural pattern bonds.

Loadbearing block is used as backing for veneer walls, bearing walls, and all structural assemblies. Both regular and specially shaped blocks are used for paving, retaining walls, and slope protection. Landscape architects call the newer, shaped-to-interlock blocks *hardscape* and use them as part of landscape design, as shown in *Figure 6*.

Figure 6 ◆ An application of loadbearing block.

MASONRY UNITS AND INSTALLATION TECHNIQUES

BUILDING BLOCKS

The Law of Gravity Applies

Cubes of brick and block must be placed on level ground. If a 500-pound load of bricks lets go, it can cause serious injuries and expensive damages.

Nonstructural block is specified under *ASTM C129*, listing a minimum compressive strength of 500 psi. This block is used for screening and non-bearing partition walls. Elegantly surfaced, solid nonstructural block is often used as a veneer wall for wood, steel, or other backing. Hollow nonstructural block is made with pattern cores much like clay tile. Pattern core blocks come in a variety of shapes and modular sizes and are commonly used for screen walls.

2.4.0 Concrete Brick

Concrete brick is a solid loadbearing unit, roughly brick size, used in the same way as clay brick. Concrete brick has no voids and may be frogged as shown in *Figure 7*. A frog is a depression in the head of a brick that lightens the weight of the brick. It also makes for a better mortar joint by increasing the area of mortar contact.

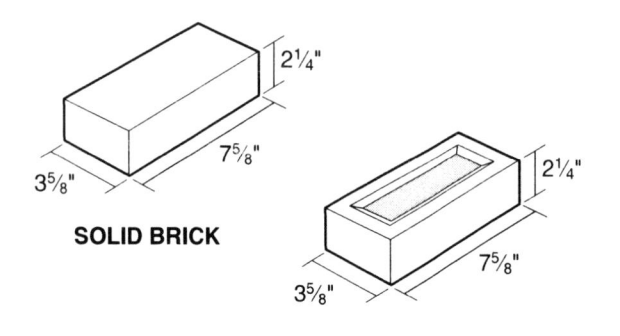

Concrete brick is designed to be laid with a ⅜" mortar joint. It comes in many sizes, with the most popular nominal dimensions of 4" × 8". This size gives three courses in a height of 8", like standard modular brick.

ASTM C55 specifies two grades of concrete brick:

- Grade N is used for architectural veneers and facing units in exterior walls. It has high resistance to moisture and frost penetration and has a compressive strength of 3,000 psi.
- Grade S is also used for architectural veneers and facing units. Grade S has moderate resistance to moisture and frost and is used in the southern region of the United States. Its compressive strength rating is 2,000 psi.

Slump brick is made from a wet mixture. The units sag or slump when removed from the molds. This gives an irregular face resembling stone, as shown in *Figure 8*. In other respects, slump brick meets concrete brick standards.

Solid block is also made from a slump mixture. Because of the greater surface area, the block face is very irregular. Its height, surface texture, and appearance resemble stone.

2.5.0 Other Concrete Units

CMUs include pre-faced units, calcium silicate CMUs, and catch basin units. ASTM specifications cover loadbearing, moisture retention, aggregate mix, and other characteristics of these units.

SOLID BRICK — 2¼" × 7⅝" × 3⅝"

FROGGED BRICK — 2¼" × 7⅝" × 3⅝"

Figure 7 ◆ Concrete brick.

Figure 8 ◆ A wall made from slump brick.

2.5.1 Pre-faced Units

Concrete pre-faced or pre-coated units are faced with colors, patterns, and textures on one or two face shells (*Figure 9*). The facings are made of resins, portland cement, ceramic glazes, porcelainized glazes, or mineral glazes. The slick facing is easily cleaned. These units are popular for use in gyms, hospital or school halls, swimming pools, and food processing plants. They come in a variety of sizes and special-purpose shapes, such as coving and bullnose corners.

2.5.2 Calcium Silicate Units

Calcium silicate units (*Figure 10*) are made of a mixture of sand, water, lime, and calcium silicate. The calcium silicate acts as a leavening agent and creates gas bubbles in the mix. The units are not fired or cured in a kiln but cured in an **autoclave** with pressurized live steam. In the autoclave, the lime reacts with the silica to bind the sand particles into a very lightweight, strong unit. ASTM performance specifications cover this type of brick and block with grading standards identical to those for traditional products.

The units are also called sand-lime brick or aerated block. They are used extensively in Europe, Australia, Mexico, and the Middle East. In the U.S., this brick is used mostly in flues, chimney stacks, and other high-temperature locations. They resist sulfates in soil, do not effloresce, and are not damaged by repeated freeze-thaw cycles. The block is now manufactured in the U.S. in a variety of sizes for commercial or home building.

Figure 10 ◆ Calcium silicate units.

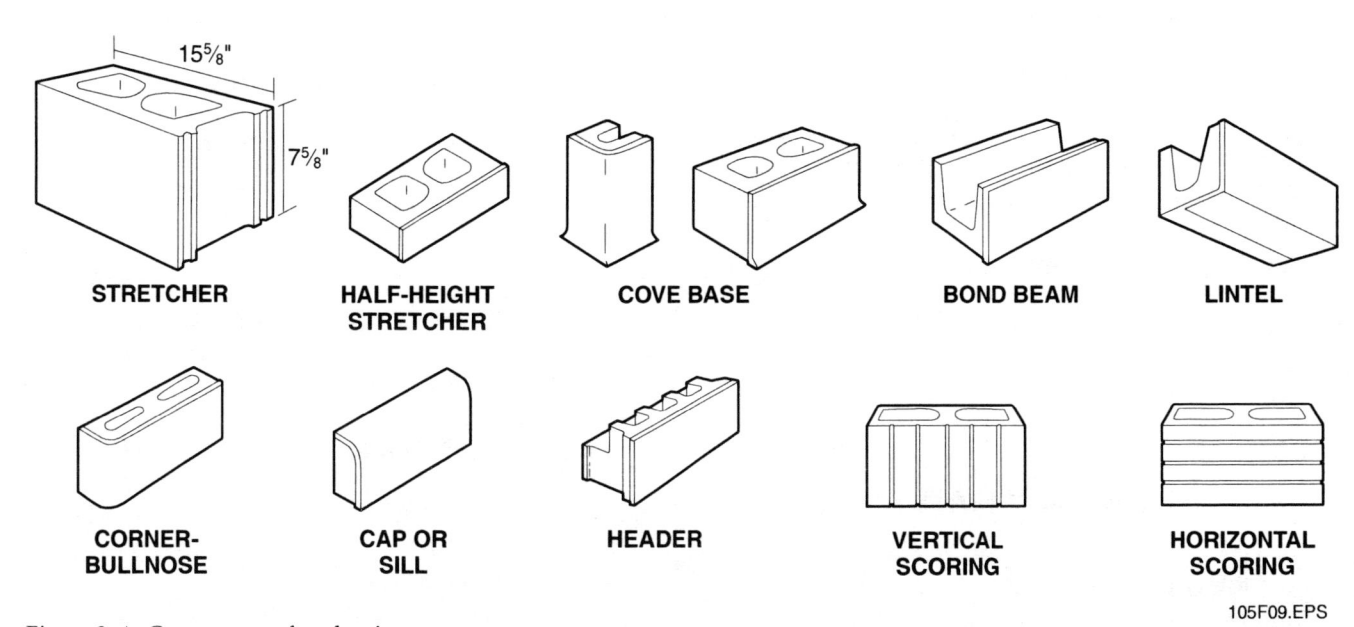

STRETCHER — 15 5/8" × 7 5/8"
HALF-HEIGHT STRETCHER
COVE BASE
BOND BEAM
LINTEL
CORNER-BULLNOSE
CAP OR SILL
HEADER
VERTICAL SCORING
HORIZONTAL SCORING

Figure 9 ◆ Concrete pre-faced units.

MASONRY UNITS AND INSTALLATION TECHNIQUES

Building Blocks

Concrete Pre-faced Units

Pre-faced concrete units like those shown here are often designed to imitate stone construction.

2.5.3 Catch Basins

Concrete manholes and catch basin units are specially made with high strength aggregates. They must resist the internal pressure generated by the liquid in the completed compartment. *Figure 11* shows the shaped units manufactured for the top of a catchment vault. These blocks are engineered to fit the vault shape and cast to specification. They are made with interlocking ends for further strength.

3.0.0 ◆ CLAY AND OTHER MASONRY MATERIALS

As you learned in *Introduction to Masonry*, clay masonry materials are the second-oldest building material. The following sections review clay and stone masonry units and introduce metal and plastic masonry materials.

3.1.0 Clay Masonry Units

Clay masonry units include the following:

- Solid masonry units or brick
- Hollow masonry units
- Architectural terra cotta units

Solid masonry units have 25 percent or less of their surface as a void or hole. Hollow masonry units have 75 percent or less of their surface as a void and are usually called tiles. ASTM standards cover all types of masonry units, loadbearing and nonbearing. Masonry units come in standard modular and non-modular sizes, in a wide range of colors, textures, and finishes. This module will focus on brick and laying brick.

Bricks can be installed in any of six positions. *Figure 12* shows each of these six positions and their names. The shaded part of the brick is the

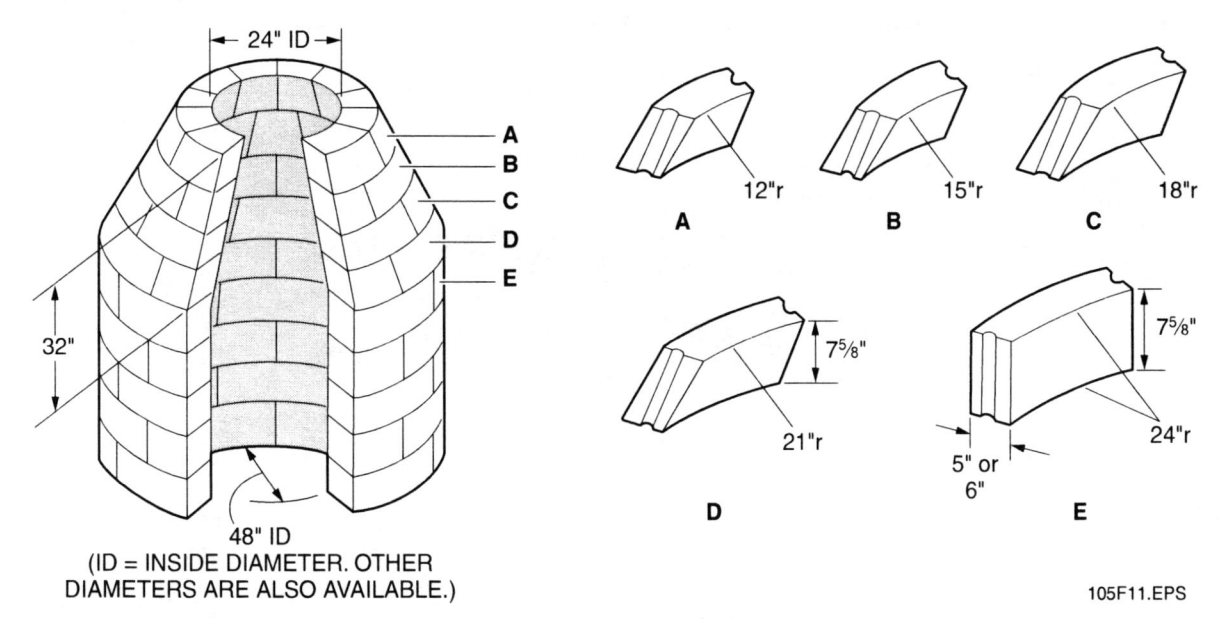

Figure 11 ◆ Manhole and vault units.

named part and the part that shows when the brick is laid in a pattern bond.

The most important characteristic of brick is its absorption capacity, or the amount of water it can soak up in a fixed length of time. The percentage of water present in brick affects the hardening of the mortar around the brick. If the brick contains a high percentage of moisture, the mortar will set more slowly than usual, and the bond will be poor. The brick will not absorb moisture and mortar into its microscopic irregularities. If the brick contains a low percentage of moisture, it will absorb too much moisture from the mortar. This will prevent the mortar from hardening properly because there will not be enough water left for good hydration.

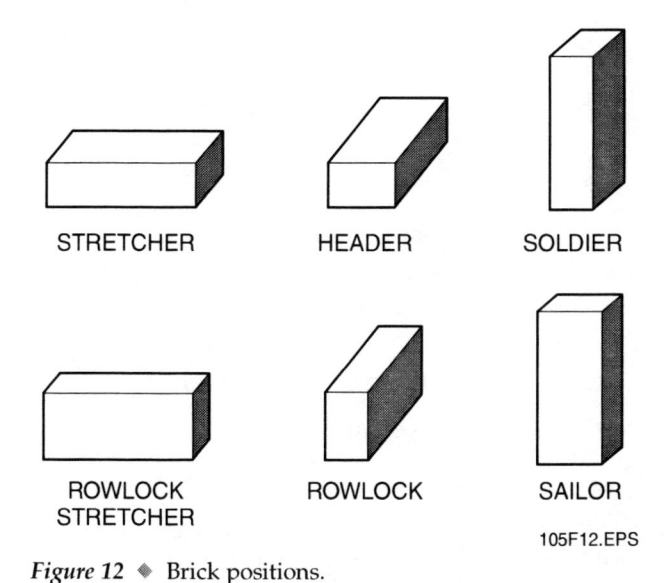

Hard-surfaced bricks and CMUs usually need to be covered on the job site so they do not get wet. Soft-surfaced bricks are usually very absorbent and may sometimes need to be wetted down before they are used.

The mason needs to determine whether the brick is too dry for a good bond with the mortar. The following test can be used to measure the absorption rate of brick:

Step 1   Draw a circle about the size of a quarter on the surface of the brick with a crayon or wax marker.

Step 2   With a medicine dropper, place 20 drops of water inside the circle.

Step 3   Using a watch with a second hand, note the time required for the water to be absorbed.

If the time for absorption exceeds 1½ minutes, the brick does not need to be wetted. If the brick absorbs the water in less than 1½ minutes, the brick should be wetted.

Wet brick with a hose played on the brick pile until water runs from all sides. Let the surface of the bricks dry before laying them in the wall.

3.2.0 Stone

As noted in the first module, rubble stone is irregular in size and shape. Ashlar stone has been cut into a rectangular unit. Stone is expensive to assemble and time-intensive to lay. It is rarely used today except for trim and detail.

Figure 12 ◆ Brick positions.

MASONRY UNITS AND INSTALLATION TECHNIQUES

Building Blocks

Looks Like Stone

The average person would not be able to tell that the stone in this decorative facing is really made of lightweight concrete. The material is available in a variety of sizes and shapes and is easy to cut and shape.

105SA06.EPS

Natural stone is used mostly for veneer walls, floors, and trim. Rubble and ashlar are used for dry stone walls, mortared stone walls, retaining walls, facing walls, slope protection, paving, fireplaces, patios, and walkways. Limestone ashlars still make the finest sill blocks. As they are one piece, there is no concern about water coming through into the wall underneath. They are also still used for **lintels** and as coping stones on top of brick walls.

Concrete masonry units are made in shapes and colorings to mimic every kind of ashlar. These units are called cast stone and are more regular in shape and finish than natural stone (*Figure 13*). They are lighter in weight and do not need as large a footing as natural stone. Cast stone has replaced natural stone in most commercial projects because it is less expensive. ASTM specifications cover cast and natural stone.

Masons may think stone work is a separate craft, but that is not entirely so. Masons still must know how to lay stone copings or sills or lintels. Masons must pattern, shape, and lay stone veneer for fireplaces or home walls.

105F13.EPS

Figure 13 ♦ Imitation stone made of concrete.

3.3.0 Other Masonry Materials

Masonry walls usually need material in addition to mortar to make the wall stronger, to hold it in place, or to handle moisture. The masonry contractor or general contractor will buy these materials and supply them as part of the job. Plans and specifications typically detail the locations and types of these materials to be used.

3.3.1 Metal Ties

Metal ties are used to tie cavity walls together and allow them to be loadbearing. The ties keep the walls from separating when weight is placed on them by the other parts of the structure. Metal ties are also used for composite walls. The ties equalize the loadbearing and also tie the two wythes together. Ties are made from 3/16" zinc-coated steel and are placed 24" apart. *Figure 14* shows rectangular and Z-shaped ties.

3.3.2 Veneer Ties

Veneer ties (*Figure 15*) are used to tie a masonry veneer wall to a backing wall or wythe. Unlike metal ties, veneer ties do not equalize loadbearing. They do keep the veneer wall from moving away from its backing. They are made of corrugated galvanized steel and placed about 12" apart.

3.3.3 Reinforcement Bars

Steel reinforcement bars come in different thicknesses and lengths. They are inserted in block cores, and then the cores are filled with grout. They add strength and weight-bearing capacity to block walls. Sometimes they are placed in the middle of cavity walls where the cavity is to be grouted.

3.3.4 Joint Reinforcement Ties

Joint reinforcement ties are made of two 10' lengths of steel bars welded together by rectangular or triangular cross bracing. *Figure 16* shows the ladder (rectangular) and truss (triangular) versions of these ties. They are used in horizontal joints every second or third course as specified.

3.3.5 Flashing

Flashing keeps water from leaking from the top of a masonry wall into the unit below. It is placed under masonry lintels, sills, copings, and spandrels. The most common flashing is made of copper, stainless, or galvanized metal. Bituminous flashing is made of fabric saturated with asphalt. Newer types of flashing are made of plastics. They are cheaper and easier to work with. *Figure 17* shows flashing in position under a sill and a lintel.

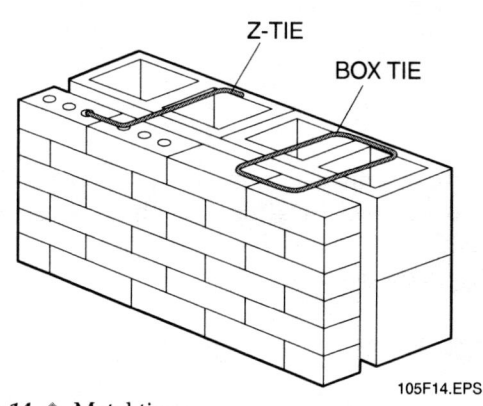

Figure 14 ◆ Metal ties.

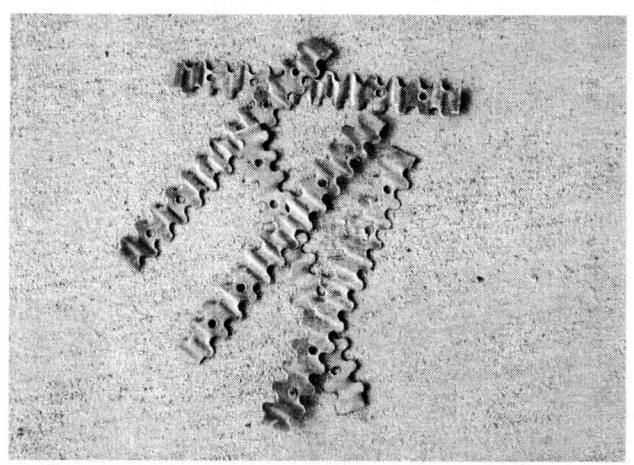

Figure 15 ◆ Veneer ties.

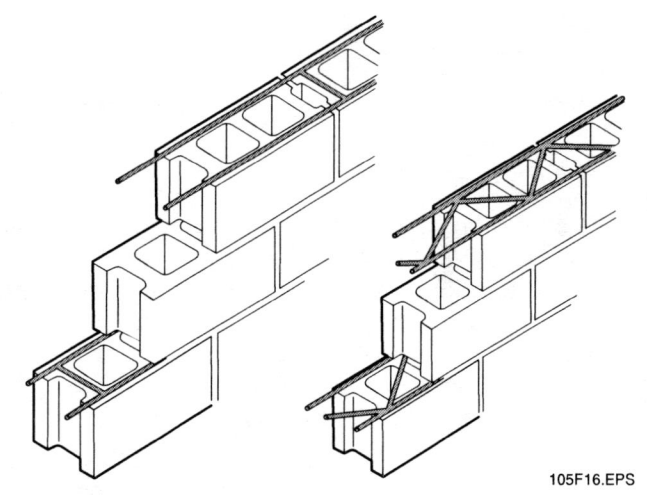

Figure 16 ◆ Joint reinforcement ties.

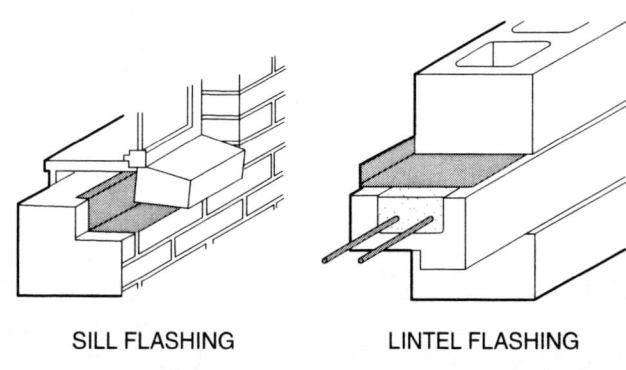

Figure 17 ◆ Flashing applications.

MASONRY UNITS AND INSTALLATION TECHNIQUES

BUILDING BLOCKS

Chimney Flashing

Proper flashing is extremely important around chimneys, especially in areas subject to snow buildup. On a new building, roofers will apply the flashing. If existing flashing is disturbed during a chimney repair, however, it is up to the mason to make sure the flashing is secure.

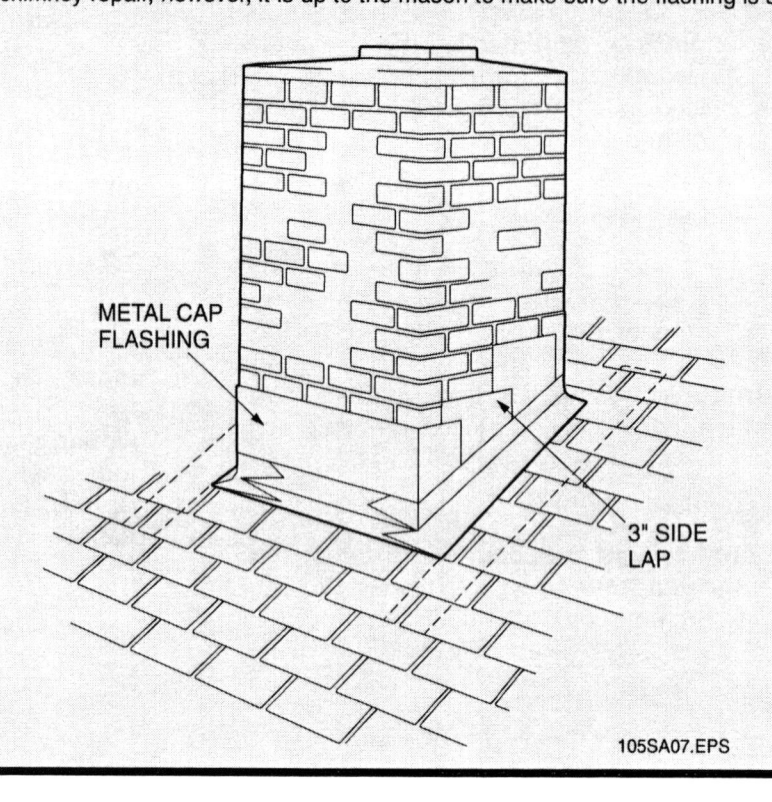

105SA07.EPS

3.3.6 Joint Fillers

Plastic or rubber joint fillers are used to replace mortar in expansion or contraction joints. They break the bond between adjacent masonry units and allow expansion and contraction of the wall. They fill the control or expansion joints in order to keep moisture out of the space. *Figure 18* shows molded joint fillers for CMUs.

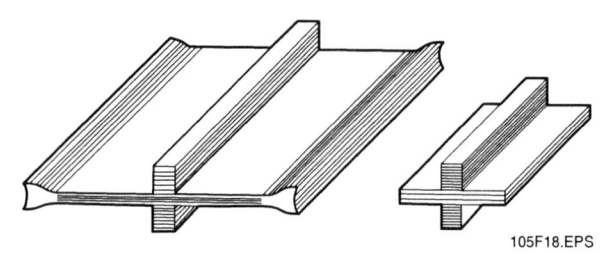

Figure 18 ◆ Joint fillers.

3.3.7 Anchors

Different kinds of metal bars, bolts, straps, and shaped ties are used to anchor a wall that meets another wall at a 90-degree angle. They are also used to tie different architectural elements to masonry walls. Anchors must be installed according to the specifications, as they affect the load-bearing of the wall. *Figure 19* shows several types of shaped anchors.

4.0.0 ◆ SETTING UP AND LAYING OUT

Setting the job up and laying the structure out are two distinct steps. Both must be complete before the mason can start to lay units. Setting up refers to materials and site preparation. Laying out refers to establishing the baseline for the masonry structure. The next sections give details for both of these tasks.

4.1.0 Setting Up

Masonry setup work starts when the contract for the job is signed. The first step for the masonry contractor is to read the contract, blueprints, and specifications. The next step is to review the schedule plus any standards and codes cited in the contract. After all of that, the masonry contractor is ready to estimate the workers, materials, and equipment needed for that job. Review the information in *Measurements, Drawings, and Specifications* to get a clearer idea of what this work entails.

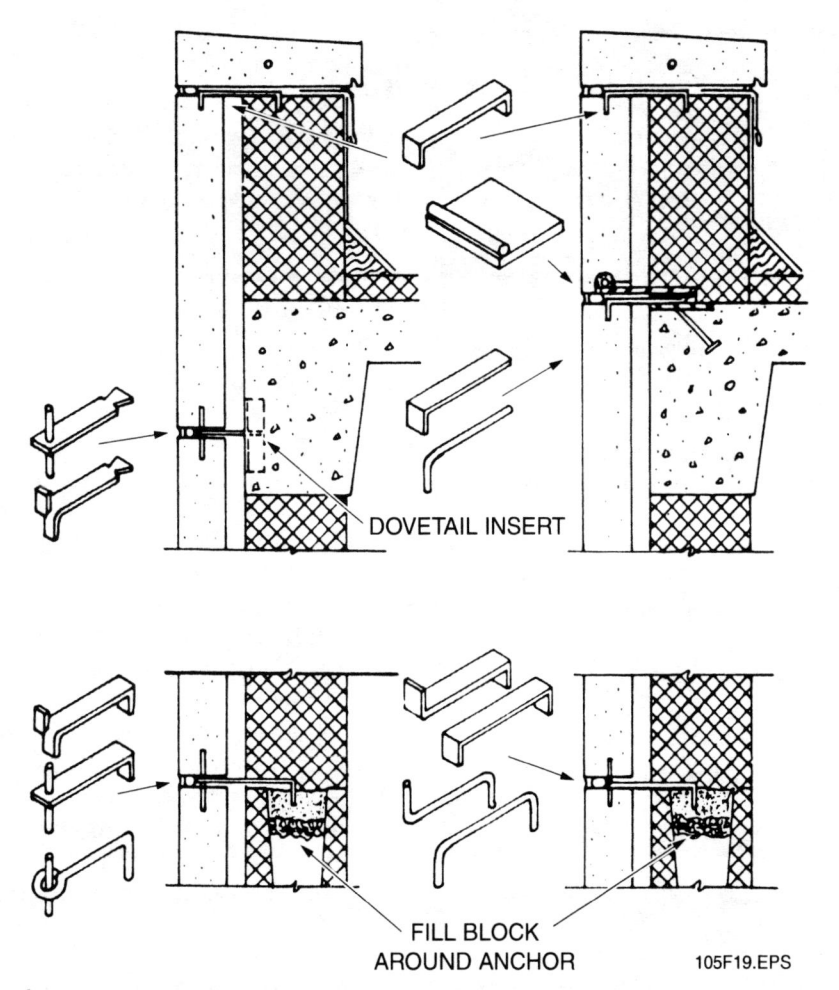

Figure 19 ◆ Shaped anchors.

The next step is to estimate again, check figures, and order the masonry equipment and materials. A visit to the job site and discussion with the engineer or construction foreman will give the masonry contractor an idea of where and how to store masonry materials. The masonry contractor must specify a delivery date and location on site. Materials must be stored close to where they will be used and protected from the weather. The crew must be hired and briefed. Then, the work is ready to begin, but there is still much to do before laying the first masonry unit. The following checklist shows some of the preliminary procedures:

- Check that all materials are stored close to work stations and protected from moisture. Masonry units must be laid dry in order to avoid shrinkage upon drying. Pile materials on pallets or planks off the ground, and cover with a tarpaulin or plastic sheet. Bagged materials can be stored in sheds or stacked on pallets and covered. Sand must be covered also, to protect it from moisture and dirt.
- Prepare mortar mixing areas within several feet of the work areas or as close as possible. Place water barrels next to them for water supply and for storage of hoes and shovels not in use. Be sure mixing equipment does not interfere with movement paths.
- Place mortar pans and boards by workstations. If you are using scaffolding, place the pans at intervals on the scaffolding near the point of final use.
- Stockpile units on each side of the mortar pans and at intervals along the wall line. If you are using scaffolding, place units along the top of the scaffold near the point of final use. Stockpiles should allow the mason to move block as little as possible once laying starts.
- Stack block in stockpiles with the bottom side down, just as they will be laid in the wall. The top of the block has a larger shell and web. Stack faced units with the faced sides in the direction they will go, just as they will be laid in the wall. Stack all units so the mason will move or turn them as little as possible.
- Check all scaffolding for proper assembly and position. Ensure that braces are attached and planks are secured at each end. Scaffolding should be level and no closer than 3" from the wall.

MASONRY UNITS AND INSTALLATION TECHNIQUES 5.15

> **BUILDING BLOCKS**
>
> *Cold Weather Considerations*
>
> Mortar temperature should be between 40°F and 120°F in order for proper hydration to occur. In temperatures below 40°F, it may be necessary to heat the water and/or sand in order to keep the mortar at a high enough temperature. One method often used to heat sand is to pile it over a large-diameter pipe, such as a culvert pipe, with a fire inside the pipe.

- Check all mechanical equipment, power tools, and hand tools. Make sure they are clean, in good condition, and the right size for the job.

The contractor may assign a helper to keep mortar pans full by supplying mortar from the mixer. A helper may also be assigned to keep the stockpiles refilled. The objective of all setup work is to make everything efficient and convenient for the masons once they begin laying.

4.2.0 Job Layout

Laying out the wall or other structural unit calls for a review of the plans and specifications. The first steps are to plan out the work, establish where it will go, and then to lay a **dry bond**.

4.2.1 Planning

Planning out the work means you need to check the plans for wall lengths, heights, door dimensions, and window openings. What pattern or bond is specified? What is the nominal size of the masonry unit? How are openings to be treated? Are the dimensions and the masonry units on the modular scale of 4" increments?

After answering these questions, the mason can draw a rough layout of the wall and lay out the bond pattern. If the job is sized on the modular grid, graph paper might be handy for the spacing drawing. This drawing can show where the bond pattern will start and how it will fit around the specified openings. From this drawing, the mason can count and calculate how many masonry units to cut.

The question of whether the designer did or did not use the modular grid becomes important. *Figure 20* shows door and window openings located in a running bond. Notice how the openings are set off the modular grid in the diagram on the left. The amount of cutting is enormous compared to the example on the right. Using many small units reduces wall strength as well. Sometimes the mason can persuade the designer or engineer to shift the openings slightly to avoid so much cutting. In other cases, the dimensions are critical and cannot be changed.

Drawing the bond pattern on the wall area may seem like a time-consuming exercise, but it can save a lot of time, especially with non-modular work. The starting point of the pattern determines how many masonry units will need to be cut. By adjusting the starting unit of a non-modular

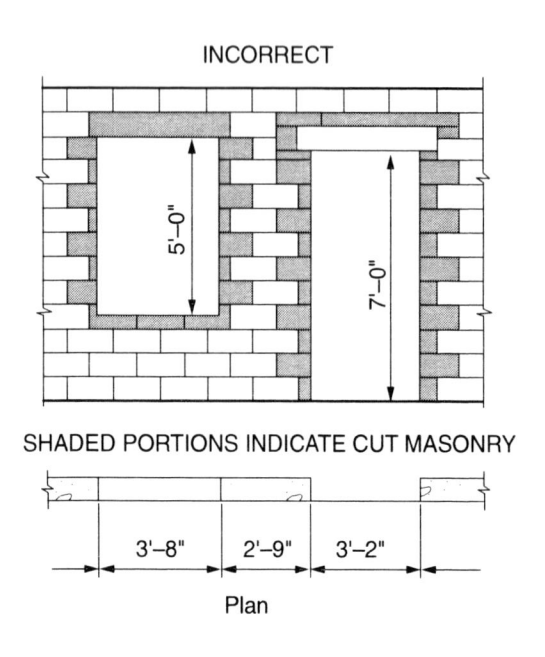

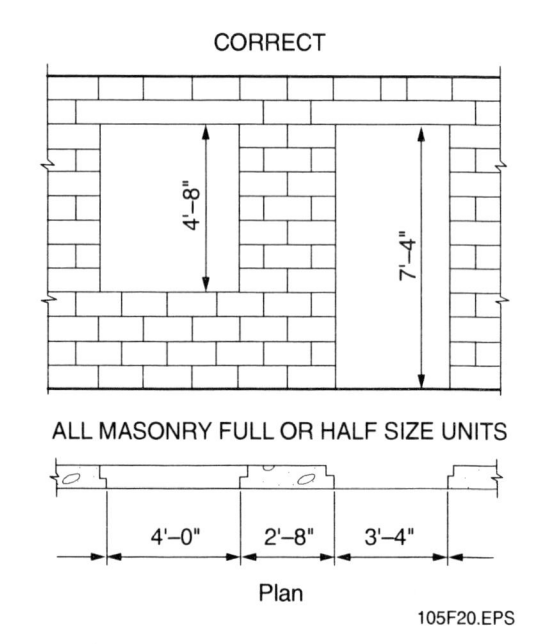

Figure 20 ◆ Door and window openings.

bond pattern, the mason can come up with a layout that calls for cutting the smallest amount of units. The mason will check these calculations by laying a dry bond before cutting any units.

4.2.2 Locating

The mason will check the location first. Masonry walls take a footing or support, usually made of concrete. The surveyor or foreman will mark the corners of the structure or slab. On some jobs, the foreman will drive nails into the wall footing to mark the building line. *Figure 21* shows a foundation layout with a footing plan for block foundation walls. At the job site, the first thing to do is to locate the footing. Next, brush it off. Remove any dried concrete particles or large aggregates to ensure a good bond between the footing and the first course.

Check that the footing is level. If the footing is not within an inch of level, it must be fixed. Do not apply a thick mortar joint to level the first course. This can result in a joint too thick to carry the load of the wall. If the footing is out of level, notify your supervisor.

The next step is to locate the walls. Take measurements from the foundation or floor plan and transfer them to the foundation, footing, or floor slab. All measurements on the plans must be followed accurately. Be sure the door openings are placed exactly and the corners are on the footings exactly as given on the detailed drawings. Check to see that you are not confusing the measurements for the interior and exterior walls. If it appears that the wall cannot be laid out exactly because of errors in the footing, notify your supervisor.

The next task is to establish two points, corner-to-corner or corner-to-door. Then, run a chalkline between the two points and snap it on the footing or foundation. Because a chalkline is easily erased, mark key points along the chalkline with a marking pencil, nail, or screwdriver. This will allow resnapping the chalkline without refinding the points.

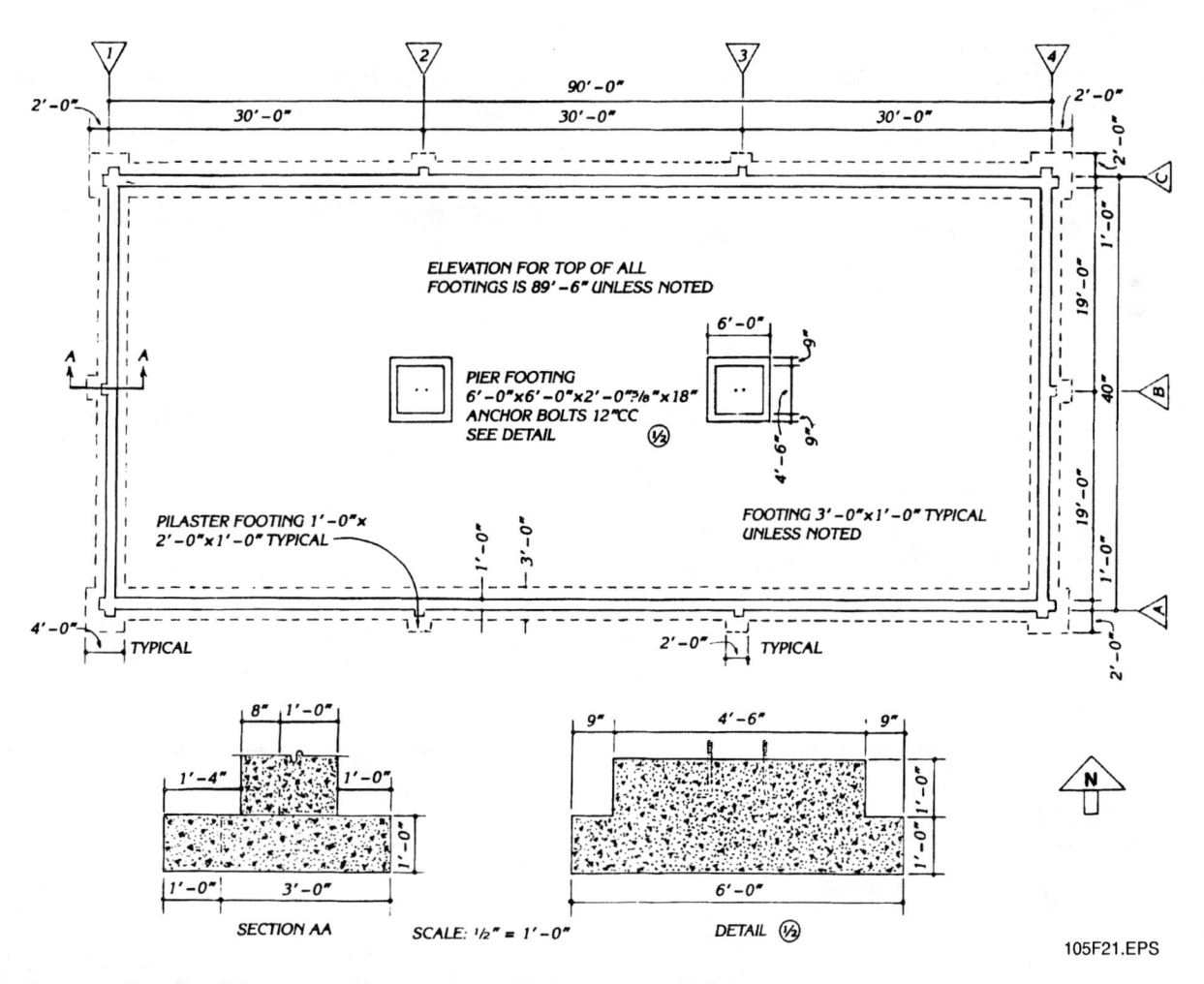

Figure 21 ◆ Foundation plan.

Mark the entire foundation for walls, openings, and control joints. After snapping the chalkline, mark over the chalk with a marker or nail. Once you have completed all markings, check the measurements of the markings against the foundation plan. Again, be sure you are reading the correct measurements. If there is to be a veneer wall, check that you are dimensioning the veneer, not the backing wythe. If everything does not fit precisely and exactly, it must be done over. It is easier to redo measurements than to redo a masonry wall.

4.2.3 Dry Bonding

Dry bonding is an alternative to measuring to establish the positioning of the masonry units. Starting with the corners, the mason can lay the first course with no mortar, or dry bond. This is a visual check of how the units will fit. It also checks the pattern bond drawing and the calculations for cut units. For CMUs, it provides a chance to check unit size and specifications.

From the corners, the mason lays units along the wall markings for the entire foundation, as in *Figure 22*. Since all mortar joints will be standard sizes, use a ⅜" or ½" piece of plywood or other material as a spacing jig. Check the specifications for the size of brick joints. Remember that all block is laid with a ⅜" joint. If you run into spacing problems, use a spacing jig, and mark any adjustments on the foundation and on the jig.

Figure 22 ♦ Example of a dry bond.

Lay the units through door openings to see how bond will be maintained above the doors. Then check spacing for openings above the first course, such as windows. Do this by taking away units from the first course and checking the spacing for the units at the higher level. These checks will show whether the joint width will work out for each course up to the top of the wall. Use the pattern bond diagram to help you. If spacing has to be adjusted slightly, mark it on the diagram and on the foundation.

After the units have been laid out correctly, mark the end of every other unit. Do this with a marking pencil directly on the foundation. This will guide you in laying mortar when the dry units are removed.

Once all of this has been done, the mason can use the steel square to mark the exact location and angle of the corners. The next step is checking the corner layout on the drawings.

The layout of the corner itself is important, especially when you are working with block and modular spacing. The architect will detail the corner layout on the working drawings. Different block layouts, as shown in *Figure 23*, are possible. Each layout takes up a slightly different amount of space. This will affect the modular spacing and determine whether any block will have to be cut. Building the corners as specified is the key to maintaining modular dimensions.

5.0.0 ♦ BLOCK HEAD JOINTS

Block is larger and heavier than brick and is not easy to lift one handed. Use two hands when lifting block to avoid strain. Block is more demanding than brick in that both blocks must be buttered to get a good head joint. Block is also more demanding than brick in that it calls for three different types of bed joints.

5.1.0 Buttering Block

Start by spreading and furrowing a bed joint. Position the first block in the mortar. Then stand two or three blocks on end next to their bed. Since block is wider at the top than at the bottom, stand

BUILDING BLOCKS

Dry Bonding

Some masons place their index finger between the bricks to account for the ⅜" mortar joint.

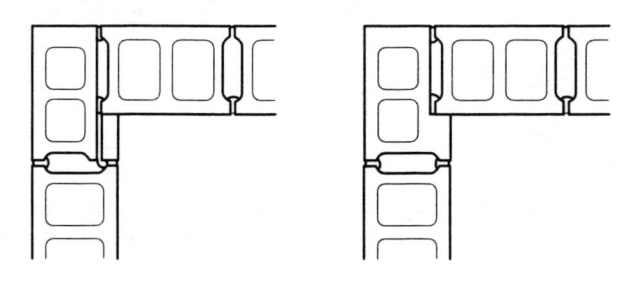

Figure 23 ◆ Block corner layouts.

the block so that the top sides will be on top when the block is placed. This makes it quick to butter several blocks at once.

You do not need to fill the trowel with mortar because one block does not take as much mortar as one brick. Butter the ear ends of the standing blocks. Wrap the mortar around the inside of each ear to help hold it in place. Then butter the ear end of the laid block. Lift the standing block by grasping the webs, or ends, with both hands, as shown in *Figure 24*. Do not jerk the block, or the mortar will fall off.

Place the block against the buttered, laid block. Tilt the block slightly toward you as you lay it into place, so that you can see the alignment of the cores and edges. Visually check that the edge of the block aligns with the block directly below.

Figure 24 ◆ Placing block.

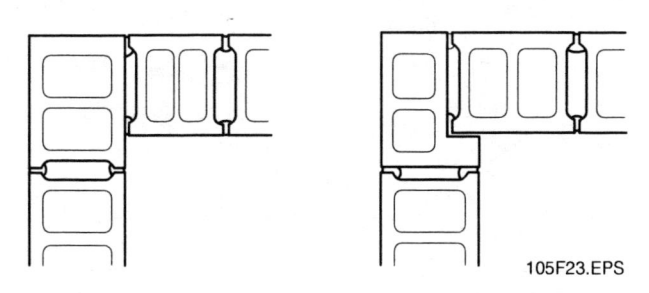

To seat the block, gently press down and forward, so that the mortar squeezes out at the joints. Do not drop the block, but ease it into place. Continue laying the pre-buttered blocks, being sure to butter the ear end of the laid block each time.

An alternative method of buttering block is to butter one end, lift it by the webs with one hand, and butter the other end. This method is not recommended for beginners.

After you place the block, cut off the excess mortar with the edge of your trowel. Check for level and plumb with your mason's level. Use the handle of your trowel to gently tap the block into place. After you place the block, the mortar joint spacing should be the standard ⅜" for both the bed and head joints.

Do not move the block after it is pushed against its neighbor. If you must move the block, take it off and remortar the bed joint and the head joint. Unlike brick with its solid and complete mortaring, block mortaring is fragile. Because its webs are so small in area compared to its size, block mortar joints are easily disturbed by movement. Do not take the chance of a weakened mortar joint developing a leak in the wall.

5.2.0 Block Bed Joints

While bricks have one type of bed joint, the full furrowed joint, blocks can use one of three types of bed joint depending on their purpose. Check the specifications before laying a block wall to confirm which type of bed joint to use. Consider the following:

- If the block is laid as the first course on a footing, it takes a full furrowed bed joint, as does brick.
- If the block is not to be in a reinforced wall, the bed joint has mortar on the face shells only. *Figure 25* shows this type of mortaring.
- If the block is part of a reinforced wall that will have reinforcing grout in some cores, the block needs a full block bed joint. This has mortar on the face shells and on the webs, as shown in the detail in *Figure 25*. Mortaring the webs will keep the grout from oozing out of the cores.

MASONRY UNITS AND INSTALLATION TECHNIQUES 5.19

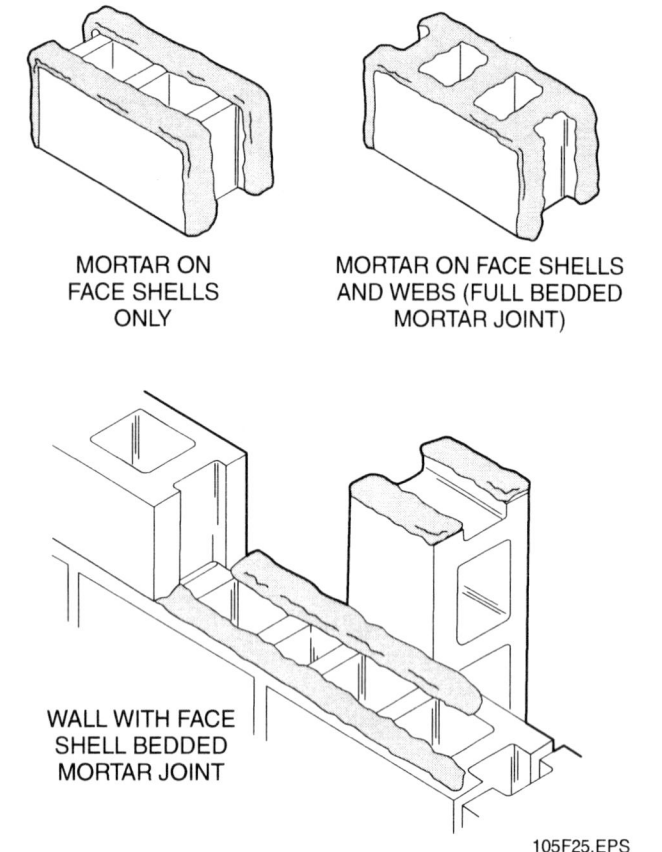

Figure 25 ◆ Buttering block.

Sometimes, the specifications will call for an unreinforced wall to be laid with a full block bed joint. Mortaring the webs as well as the shells increases the loadbearing strength of the wall. The architect or engineer may have calculated that a full block bed joint will do the job instead of reinforcement. If you use only a shell bed joint, the wall will not have the calculated strength. This is another reason why it is important to read the specifications.

5.3.0 General Rules

These guidelines were covered in an earlier module, but they bear repeating. The way you work the mortar determines the quality of the joints between the masonry units. The mortar and the joints form a vital part of the structural strength and water resistance of the wall. Learning these general rules and applying them as you spread mortar will help you build good walls:

- Use mortar with the consistency of mud, so it will cling to the masonry unit, creating a good bond.
- Butter the head joints thoroughly for brick and block; butter both sides of the head joints for block.
- When laying a unit on the bed joint, press down slightly and sideways, so the unit goes against the one next to it.
- If mortar falls off a moving unit, replace the mortar before placing the unit.
- Put down more mortar than the size of the final joint; remember that placing the unit will compress the mortar.
- Do not string a spread more than six bricks or three blocks long; longer spreads will get too stiff to bond properly as water evaporates from them.
- Do not move a unit once it is placed, leveled, plumbed, and aligned.
- If a unit must be moved after it is placed, remove all the mortar on it and rebutter it.
- After placing the unit, cut away excess mortar with your trowel, and put it back in the pan, or use it to butter the next joint.
- Throw away mortar after 2 to 2½ hours, as it is beginning to set and will not give a good bond.

6.0.0 ◆ BONDING MASONRY UNITS

Masons deal with four types of bonds:

- A simple mechanical bond is made by the joining of mortar and a masonry unit. The strength of this bond depends on the mortar. This is also called a mortar bond.
- A pattern bond is a pattern formed by masonry units and mortar joints on the face of a surface. Unless it is the result of a structural bond, a pattern bond is purely decorative.
- A structural bond is made by interlocking or tying masonry units together so they act as a single structural unit.
- A structural pattern bond is the result of a structural bond that forms a pattern as well as a bond. Most traditional pattern bonds are structural pattern bonds.

The next sections discuss these different types of bonds. Note that the distinction between a structural bond and a pattern bond is hard to make. The act of overlapping or interlocking masonry to create a structural bond also creates a pattern. Defining a particular pattern as a structural bond or a structural pattern bond depends on local custom.

6.1.0 Mechanical or Mortar Bond

On the basic level, a mechanical bond is formed between the masonry unit and the mortar. This bond ties the masonry in a wythe into a single unit.

BUILDING BLOCKS

Brick on Block
Adding a brick veneer to a reinforced concrete block wall is a common construction method.

For the majority of masonry construction, the most important property of mortar is bond strength. Mortar bond strength depends on the properties of the mortar and the bonding surface:

- The mortar must have the right proportions of ingredients for its use. It must stay wet enough to lay and level the masonry.
- The masonry surface should be irregular to provide mechanical bonding. It should be absorptive enough to draw the mortar into its irregularities.
- The masonry surface should not be so dry that it dries out the mortar. Slow, moist curing improves mortar bond and compressive strength.

The second most important property of mortar is bond integrity. The work of the mason defines the bond between masonry units. Bond integrity depends on the mason who does the following:

- Keeps tools and masonry units clean
- Butters every joint fully without air bubbles
- Does not move the masonry unit after it is leveled
- Levels units shortly after they are laid
- Uses mortar wet enough to dampen the masonry unit
- Keeps the mortar tempered
- Mixes fresh mortar after two hours

6.2.0 Pattern Bond

Pattern bonds add design but not strength to masonry walls. The stack bond (*Figure 26*) is only a pattern bond. It provides no structural bond as there is no overlapping of units. This pattern is more commonly used with block than brick.

If the stack bond pattern is used in a loadbearing wall, the wythe must be bonded to its backing with rigid steel ties. In loadbearing construction, this patterned wall should be reinforced with steel joint reinforcement ties.

Pattern and structural pattern bonding calls for placing brick in different positions in the wythes. *Figure 27* shows different ways of placing brick in order to make different kinds of patterns.

The stretcher is the everyday workhorse. Headers are used primarily for tying wythes together,

Figure 26 ◆ Stack bond.

MASONRY UNITS AND INSTALLATION TECHNIQUES

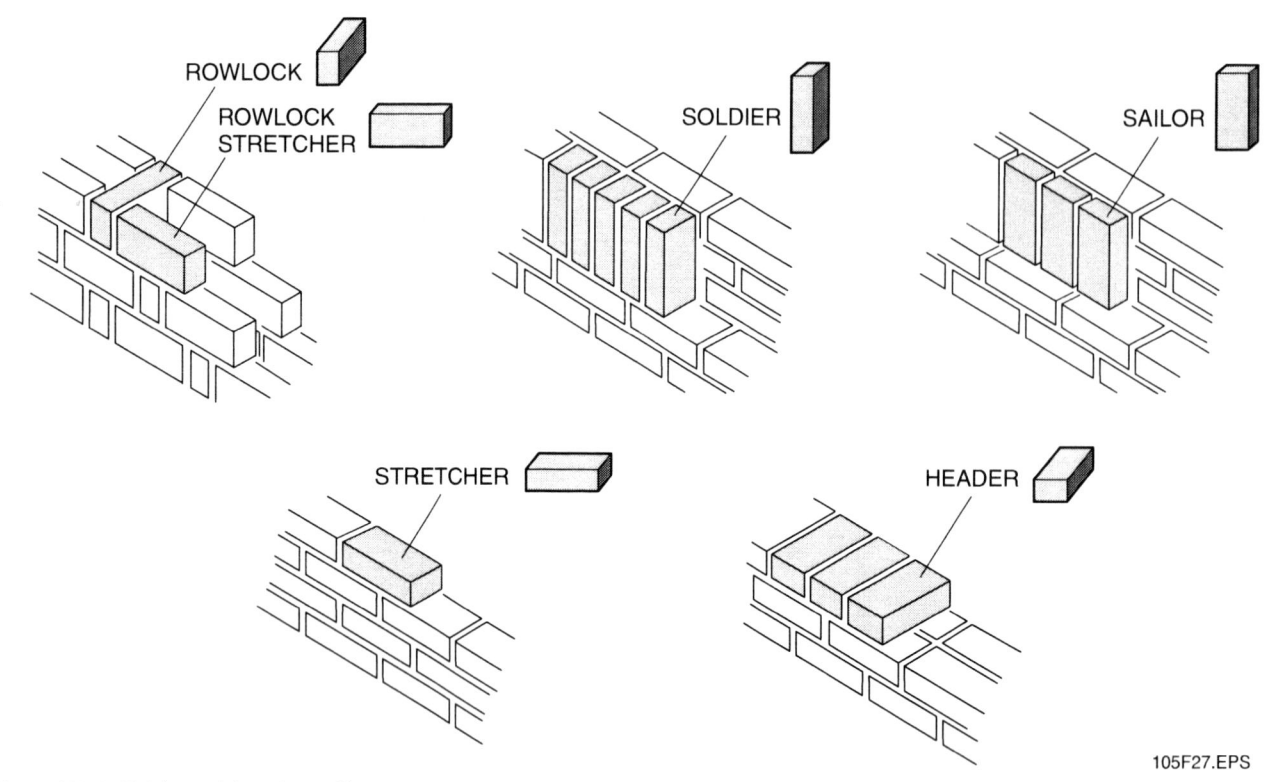

Figure 27 ◆ Brick positions in walls.

capping walls, flat windowsills, and pattern bonds. Soldiers are used over doors, windows, or other openings, and in pattern bonds. Shiners, or rowlock stretchers, are used in pattern bonds, in brick walks, and for leveling when a 4" lift is needed. Rowlocks are found in capping walls, windowsills, ornamental cornices, and pattern bonds. Sailors are rarely seen, except in pattern bonds and brick walks.

6.3.0 Structural Bond and Structural Pattern Bond

Wythes can be structurally bonded by using metal ties, joint reinforcements, anchors, grout, and steel rods. These engineering methods are used to increase strength and loadbearing by firmly tying masonry units and wythes together.

Another, older way to structurally bond a wythe is to lap masonry units. Lapping one unit halfway over the one under it provides the best distribution of weight and stress.

In a single-wythe wall, a structural bond is made by staggering the placement of the bricks. This results in the brick in one course overlapping the brick underneath. The structural pattern bond resulting from this simple overlap is the running bond, as shown in *Figure 28*. Common overlaps are the half lap and the one-third lap. Changing the proportion of the overlap changes the look of the pattern.

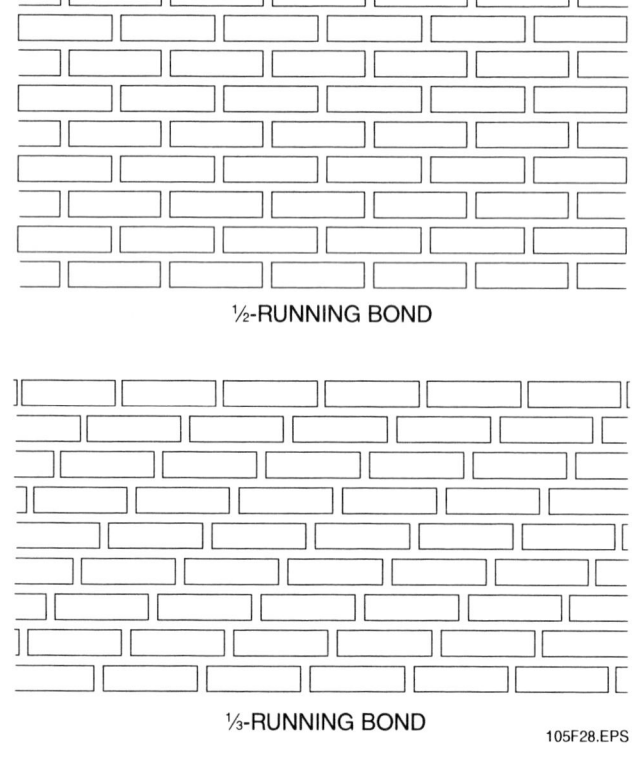

Figure 28 ◆ Running bonds.

In two-wythe walls, a structural bond is made between the wythes. This can be made by rigid steel ties that equalize loadbearing. It can also be made by overlapping a brick from the face wythe to the backup wythe. The overlap brick is turned into the header or the rowlock position. This results in a complex structural bond that is also a structural pattern bond, with different sizes of brick facing out. The results are the traditional English bond and Flemish bond shown in *Figure 29*.

The Flemish bond consists of alternating headers and stretchers in every course. The English bond consists of alternating courses of headers and stretchers. If the headers are not needed for structural bonding, cut bricks are used. Brick can be laid to show different faces and cut in different ways.

The combination of the Flemish and English bonds with the running bond results in the common or American bond. As shown in *Figure 30*, the common bond is a running bond with headers every sixth course. The headers are in the Flemish or English pattern, according to the specifications.

The English cross, or Dutch bond, uses a structural pattern bond that repeats every four courses. The pattern courses are all stretcher, all header, and a course of three stretchers and one header. The last pattern course is all header again. *Figure 31* shows the Dutch bond.

The Dutch bond may seem complicated until you look at a traditional garden wall bond. *Figure 32* shows two variations on the garden wall struc-

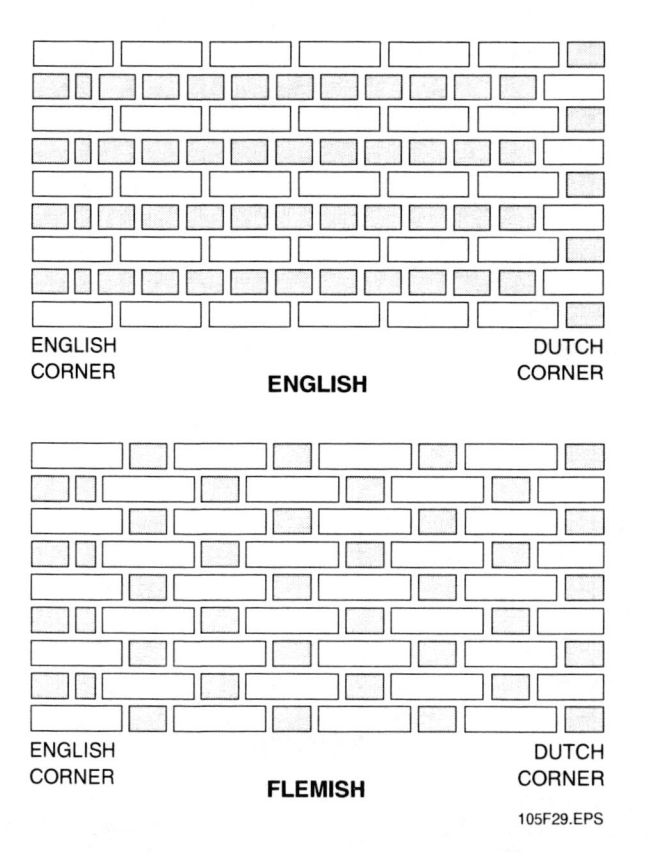

Figure 29 ◆ English and Flemish bonds.

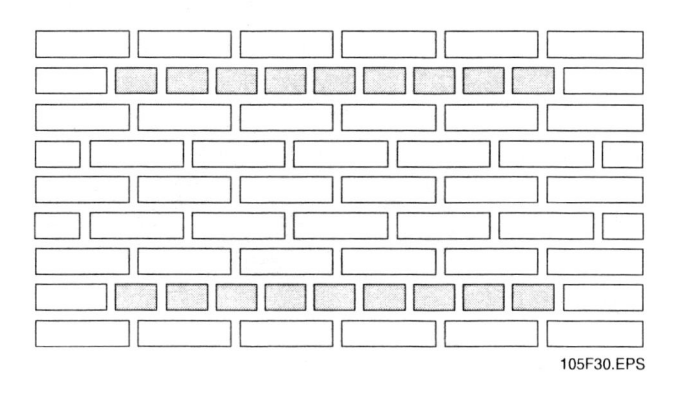

Figure 30 ◆ American, or common, bond.

Figure 31 ◆ English cross, or Dutch bond.

tural pattern bond. The double stretcher garden wall pattern shown in *Figure 32A* repeats every five courses. The dovetail garden wall pattern shown in *Figure 32B* repeats every 14 courses. More variations are possible. The only limit on patterning is the skill and ingenuity of the mason.

6.4.0 Block Bond Patterns

Block has its own set of commonly used bond patterns. *Figure 33* shows common block bonds, some of which, such as the herringbone, are also seen in brickwork. As with brick, the stack bonds do not provide any structural strength. With block, however, it is simple to reinforce stack bond with grout and steel or grout in the cores.

The other block bond patterns add structural strength. To get a solid face, block can only be laid in the stretcher mode. Pattern variations, such as the coursed ashlar, can be made by using different sizes of block. Modern block walls can also add visual interest through texture and surface designs.

Many designers rely on textures and surface designs, alone or in combination with bond patterns, to enhance block walls.

MASONRY UNITS AND INSTALLATION TECHNIQUES   5.23

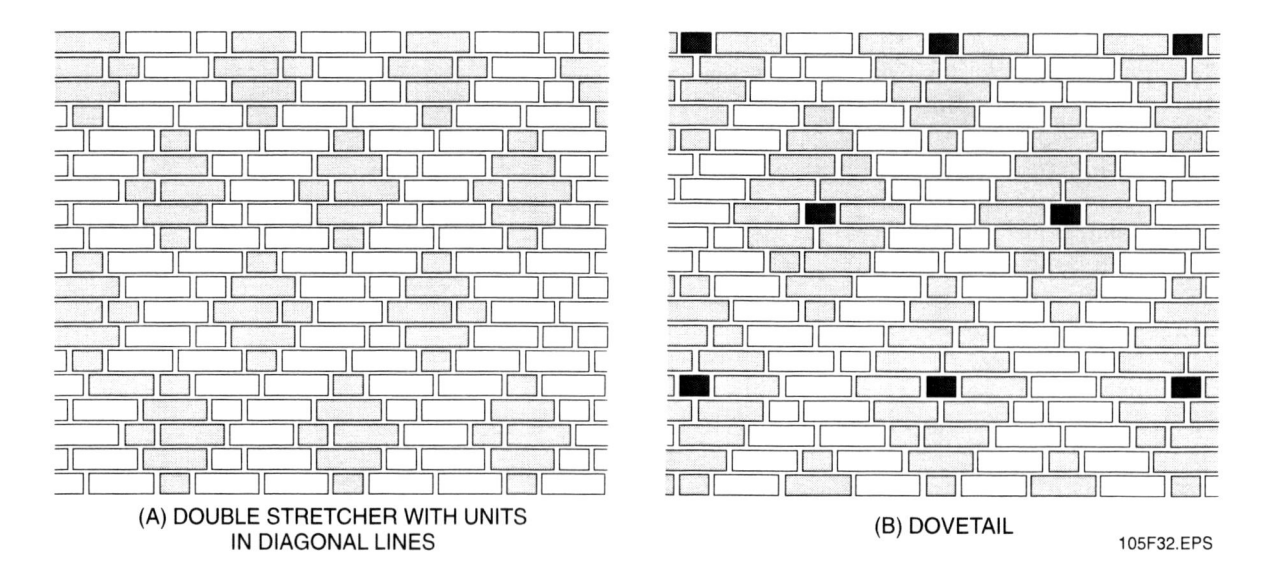

(A) DOUBLE STRETCHER WITH UNITS IN DIAGONAL LINES

(B) DOVETAIL

Figure 32 ◆ Garden wall bonds.

RUNNING BOND

HORIZONTAL STACK BOND

VERTICAL STACK BOND

HERRINGBONE

DIAGONAL

SINGLE BASKET WEAVE

DOUBLE BASKET WEAVE

COURSED ASHLAR

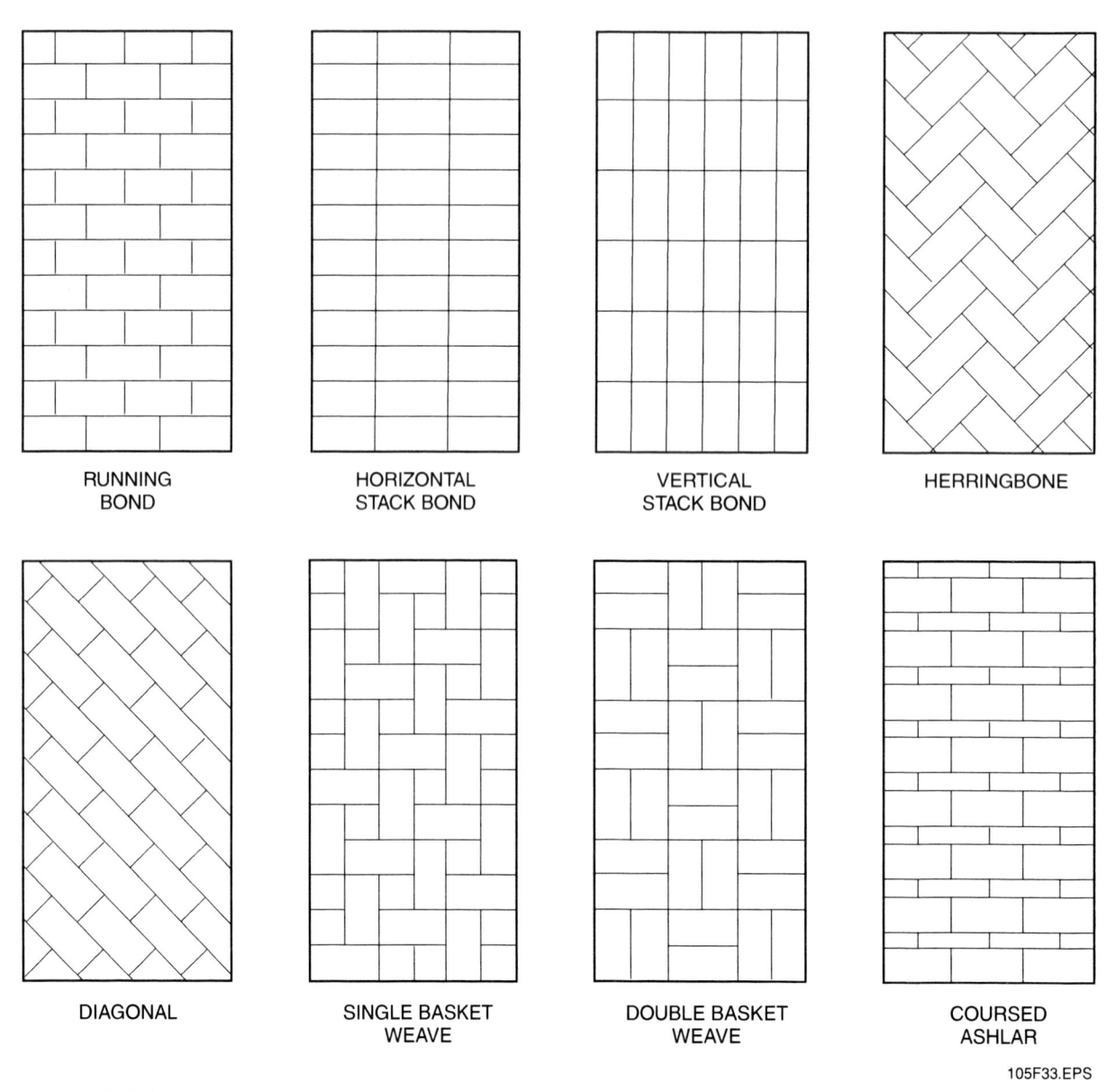

Figure 33 ◆ Block bonds.

7.0.0 ♦ CUTTING MASONRY UNITS

Masonry units often need to be cut to fit a specific space. Even when building on a modular grid, structural bond patterns, door and window openings, and corners usually call for some cut masonry units. English and Dutch corners specifically call for cut masonry units as part of the patterning.

On a large job, the masonry contractor or foreman will figure the pattern layouts and calculate the number of masonry units to be cut. Someone will be assigned to cut the units with a masonry saw or a splitter before they are needed. Sometimes, masons need to cut a few more units or cut to a slightly different size. This is when you need to know how to cut masonry with hand tools.

7.1.0 Brick Cuts

Brick can easily be cut by hand tool, masonry saw, or splitter. Sometimes you will need cut bricks for finishing corner patterns or for pattern bonds. *Figure 34* shows the common cut brick shapes and names. The king and queen closures are used for cornering.

7.2.0 Block Cuts

Block is usually cut in several standard ways. It can be cut across the stretcher face, both horizontally and vertically, as shown in *Figure 35*. You may easily make these cuts by hand. Blocks cut across the face horizontally are called splits or rips. If the block is cut exactly in half, it is called a half-high rip. Rip blocks are often used under windows. They act as a filler to reach a height of 8", so normal coursing can continue.

Blocks also get their webs cut out. Taking one end off a block makes an opening easily slipped over a pipe. Fitting the block to its location may take more cuts. You might make these cuts using a masonry saw. The cuts have their own names to save time and confusion (*Figure 36*).

The following three types of cuts are shown in *Figure 36* on a three-cell block:

- The quarterback cut has the end of one cell cut out, leaving two cells.

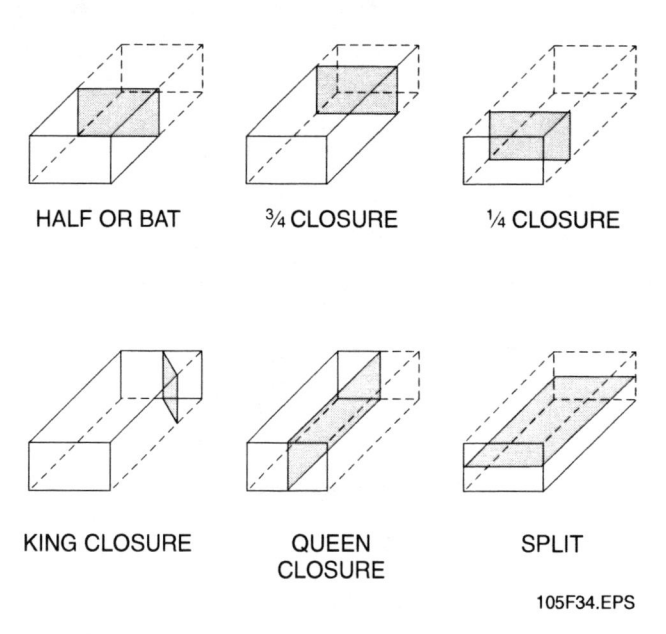

Figure 34 ♦ Common brick cuts.

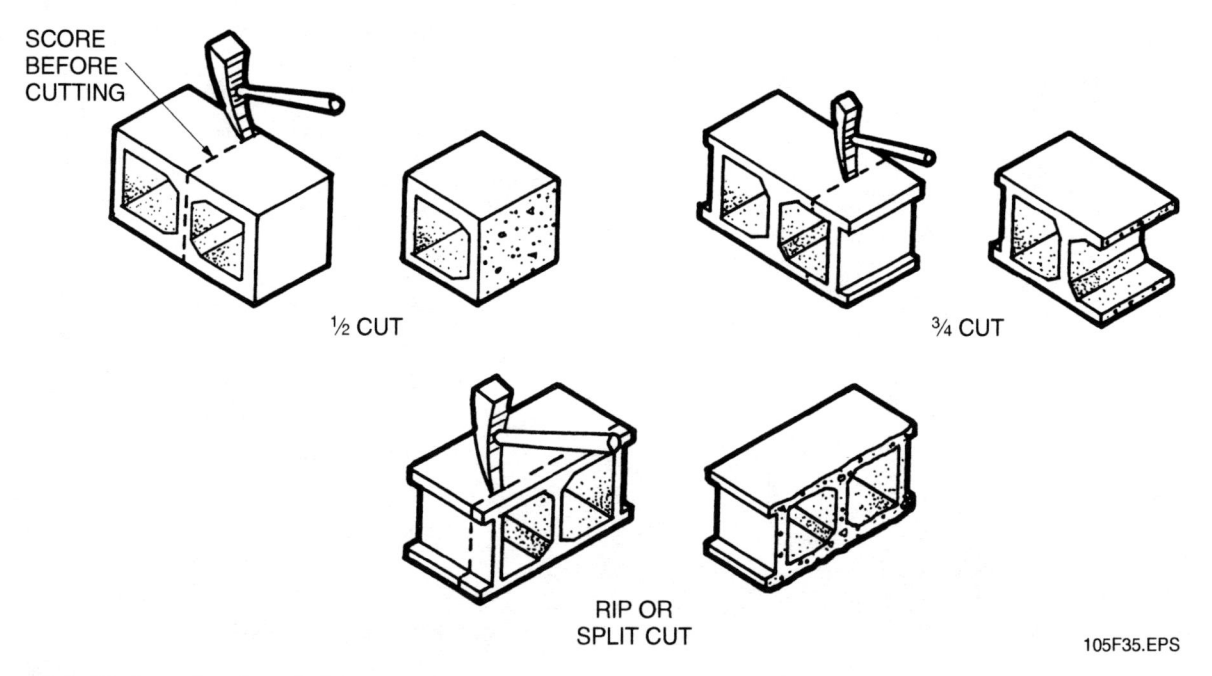

Figure 35 ♦ Horizontal and vertical face cuts.

MASONRY UNITS AND INSTALLATION TECHNIQUES

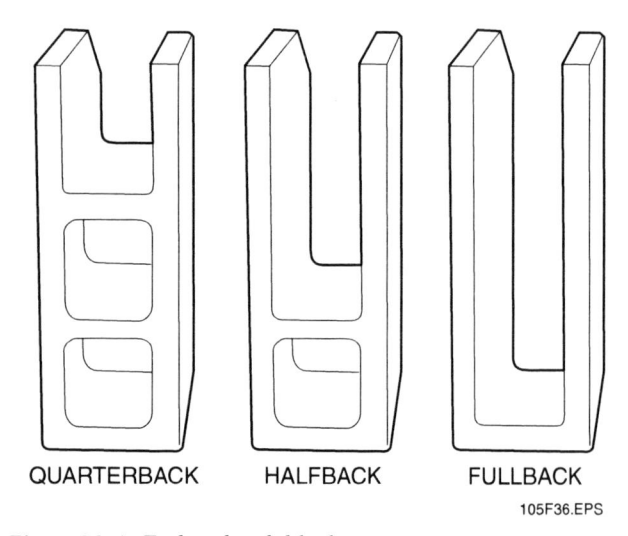

Figure 36 ♦ End and web block cuts.

- The halfback cut has the end of one cell and the web of the next cell cut. This leaves one cell.
- The fullback cut has one end and both internal webs cut. This leaves only one end to hold the block together.

If a block has only two cells, the cuts are halfback and fullback; there is no quarterback cut for a two-cell block.

The bond beam block has the ends and inside webs of the block cut down about three fourths of the way. This cut (*Figure 37*) can be used for a lintel over an opening. The cuts give room for the reinforcement on top of the opening.

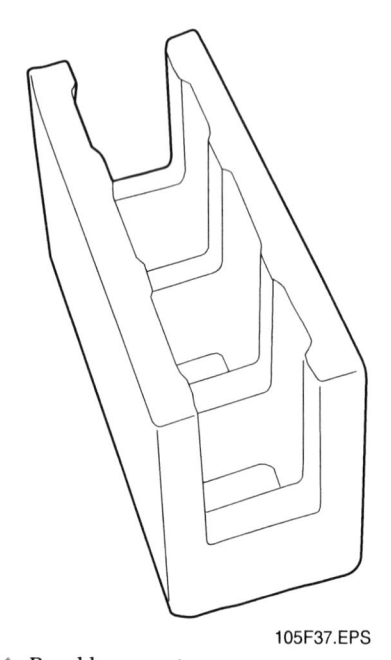

Figure 37 ♦ Bond beam cut.

7.3.0 Cutting with Hand Tools

Brick and block can be cut with chisels or a brick hammer. Brick can also be cut with the edge of a trowel. The procedures are detailed in the following sections.

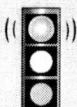

> **WARNING!**
> Remember to wear a hard hat and eye protection when cutting with hand tools. Never cut masonry over the mortar pan or near other workers. Chips may fly off, causing injury.

7.3.1 Cutting with Chisels and Hammers

Using the chisel and hammer can result in a smooth cut for block and brick. This procedure works well for both types of units:

Step 1 Check the tools you will use. Cutting edges should be sharp, and the hammer handle should be firmly attached.

Step 2 Put on your hard hat and safety goggles or other eye protection.

Step 3 Put the brick or block on a bag of sand, a board, or the ground to make a safe cutting surface. Make sure it is resting flat and plumb on a surface with some give to it.

Step 4 Use a steel square and a pencil to mark the cut all the way around the masonry unit.

Step 5 Hold the blocking chisel (for blocks) or the brick set (for bricks) vertically on the marked line. The flat side of the chisel should face the finished cut, or the part you want to keep.

Step 6 Give the chisel end several light taps with the striking end of the hammer to score the masonry unit. Move the hammer and chisel all around the unit, scoring all along the cut mark. Be sure to keep your fingers above the cutting edge of the chisel. *Figure 38* shows this step.

Step 7 Turn the unit so the finished cut is toward you and the waste part is away from you.

Step 8 Place the chisel on the scored line, with the flat side facing the finished cut. Deliver a hard blow to the chisel head with the hammer. Sometimes two blows are needed.

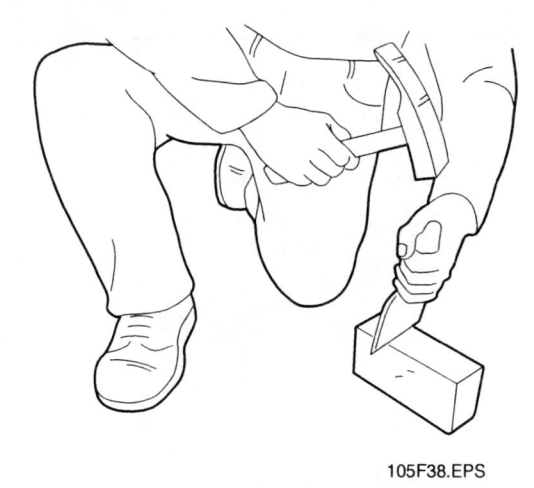

Figure 38 ◆ Cutting with a hammer and chisel.

Cutting in this way gives an accurate and clean cut. You can also make an accurate cut by using a hammer, instead of a chisel, for the final step. If you are cutting with the mason's hammer, follow Steps 1 through 7 just listed, then continue with these steps:

Step 1 Place the scored block on top of another block so that the waste part hangs free. You can hold the wanted part secure with your foot.

Step 2 Strike the waste end of the block with the striking end of the hammer. This knocks off the waste end, leaving a clean, finished cut.

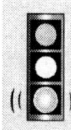

NOTE
Hammer and chisel cutting may not be permitted on some commercial jobs. Check the project specifications.

7.3.2 Cutting with Masonry Hammers

Cutting with the chisel end of the masonry hammer gives a rougher cut. The steps for cutting brick in this way are as follows:

Step 1 Check the tool you will use. Cutting edges should be sharp, and the hammer handle should be firmly attached.

Step 2 Put on your hard hat and safety goggles or other eye protection.

Step 3 Use a steel square and a pencil to mark the cut all the way around the masonry unit.

Step 4 Hold the brick in one hand and your hammer in the other. Hold the part of the brick you want to keep with the waste part down.

Step 5 Strike the brick lightly with the chisel end of the hammer to score it along the marks on all sides. As you turn the brick, be sure to keep your fingers and thumb off the side of the brick being scored. *Figure 39* shows this step.

Step 6 Strike the face of the brick sharply with the chisel end of the hammer. Let the waste part fall to the ground.

Step 7 If necessary, use either end of the hammer to dress out any small, rough edges left by the cut.

The same procedure can be used for block, except that block is not usually held in your hand. Set the block on sand, the ground, or a board for a safe cutting surface. Follow Steps 1, 2, 3, and 5. Then tilt the block face away and prop it with another block. Hold it with your foot and apply a sharp blow with the hammer. Blocks may need to be struck on both faces. Finish by dressing out any rough edges.

7.3.3 Cutting with Trowels

Cutting with a trowel is a last resort when your hammer is not available. It is not recommended for block or very hard brick. Cutting with a trowel in cold weather can break the blade. After you have mastered cutting with the hammer and brick set, this method will be easier to learn.

Figure 39 ◆ Cutting with a brick hammer.

MASONRY UNITS AND INSTALLATION TECHNIQUES

Mark the brick for cutting. Hold the brick in one hand by the part you want to keep. Keep your fingers well under the brick to avoid cutting them. Strike the brick using the upper edge of the trowel, close to the heel. Strike hard with a quick sharp blow and a sharp snap of your wrist. If the brick does not break, use a brick hammer and follow the steps previously outlined.

7.4.0 Cutting with Saws and Splitters

Cutting masonry units with power saws or splitters takes two kinds of awareness: you must be aware of how to operate the machinery, and also be aware of the masonry units and cuts.

7.4.1 Saws and Splitters

Masonry saws are available in freestanding and portable models (*Figure 40*). They use either diamond or carborundum blades. Diamond blades are irrigated to prevent fire. The water wets the masonry unit, which must dry out before it can be laid. Carborundum blades are not irrigated, but they make clouds of dust. The dust must be blown or vented away from the saw and nearby workers. Smaller handheld saws use dry blades and also make clouds of dust, which must be blown or vented away.

Splitters (*Figure 41*) do not use water or generate dust. They do, however, exert tremendous force through gearing and hydraulic power.

Figure 41 ◆ Brick splitter.

As with any potentially dangerous equipment, follow these general safety rules:

- Do not operate any saw or splitter until you have had specific instructions in handling that equipment.
- Check the condition of the equipment before using it.
- Follow all safety rules for using power equipment or otherwise dangerous equipment.
- Wear a hard hat, respiratory protection, goggles, gloves, and other appropriate personal protective equipment as needed.
- Never force the equipment.
- For bedded saws, use conveyor carts, pushers, or blocks to move the unit under the blade.
- For handheld saws, secure and brace the unit before cutting it.
- Do not operate equipment when you are feeling ill or are taking any medication that may slow your reaction time.

Review the safety rules as well as the operating instructions before operating any equipment.

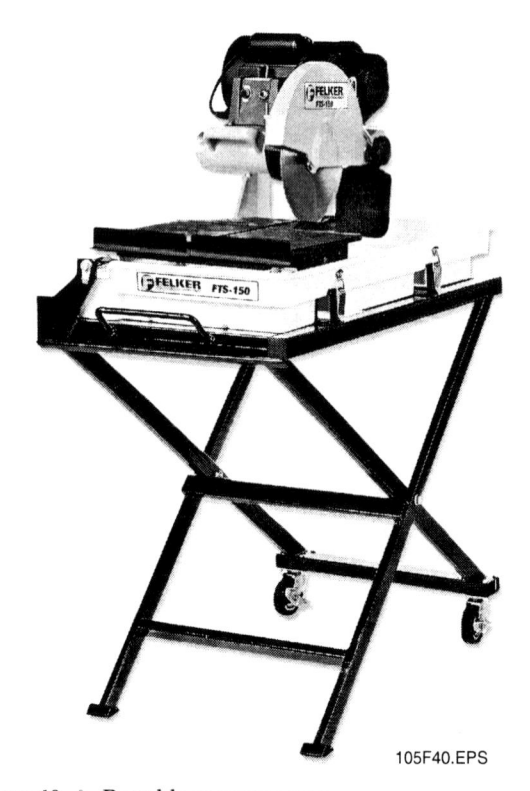

Figure 40 ◆ Portable masonry saw.

> **WARNING!**
> Silicosis is a serious lung disease that is caused by inhaling sand dust. Silica is a major component of sand and is therefore present in concrete products and mortar. Silica dust is released when cutting brick and cement, especially when dry-cutting with a power saw. Any time you are involved in the cutting or demolition of concrete or masonry materials, be sure to wear approved respiratory equipment.

7.4.2 Units and Cuts

After you have checked out the equipment, the safety procedures, and the operating procedures, check out the masonry units.

- Know what the finished item should look like.
- Mark all cutting lines before the blade starts running.
- Mark cutting lines in grease pencil for wet-cut saws.
- Do not cut a cracked masonry unit.

If you are not clear about the cuts to be made, ask your supervisor for more direction.

8.0.0 ♦ LAYING MASONRY UNITS

Laying masonry units is a multi-stage process. As discussed in previous sections, the first step in any masonry job is reading the specifications. This is followed by planning the layout of the job. The next tasks are to locate and lay out the wall, then do the dry bonding. Dry bonding assures that the layout will be correct and that the minimum number of cut blocks will be needed. Then, calculate the number of units to cut, and cut them. For the purpose of this module, assume all work is done in running bond on the modular grid system. Now you are ready to mix the mortar and start the actual laying.

The next tasks are to spread mortar, lay masonry units in place, and check their positioning.

An earlier module gave detailed procedures for spreading and furrowing bed joints and buttering head joints. The following sections give procedures for positioning individual masonry units, laying to the line, and building corners and leads.

8.1.0 Laying Brick in Place

Laying brick will be less of a strain if you use as few motions as possible. One way to make things easier is to have your materials close by. If you are working on a veneer wall, use both hands to pick up bricks and stack them on the completed back section of the wall. Place them along the length you will be laying before you start spreading mortar. This will eliminate the need to bend to pick bricks up off the ground as you go.

Using both hands is another efficient practice. Keep your trowel in one hand, and use the other for picking, holding, and placing bricks. This will make the work easier and faster.

Use your fingers efficiently as well. When you pick up a brick, hold it plumb. Pick it up so that your thumb is on the face of the brick. Let your fingers and thumb curl down over the top edges of the brick, slightly away from the face. In this position, your fingers will not interfere with the line as you place the brick on the wall. *Figure 42* shows the proper position of fingers and thumb for holding and laying the brick in place.

Keep your mason's level close by. After laying every six bricks, check them for position. This means checking for height, level, plumb, and straightness as you go. If you cannot adjust a unit to meet the four measures of height, level, plumb, and straightness, you must take the brick out and start over.

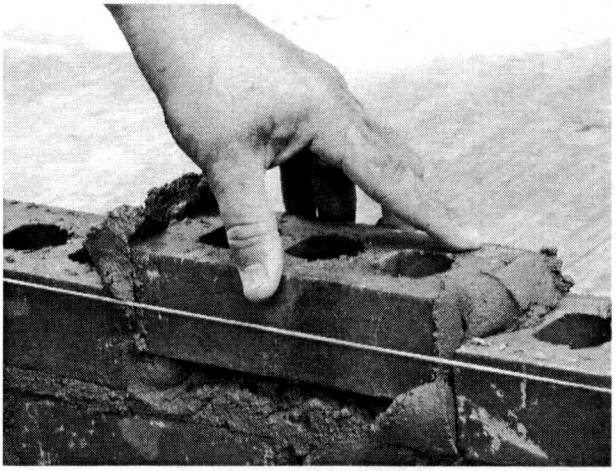

Figure 42 ♦ Proper hand position.

8.1.1 Placing Brick

The most important placing rule is to place gently. Do not drop the brick or block onto the bed joint; lower it down gently. Press the brick forward at the same time so that it will butt against the unit next to it. Mortar should ooze out slightly on both head and bed joints to show that there has been full contact.

Align the latest brick with the brick next to it as you place it. Line it up with the mason's line if you are using one. By standing slightly to one side, you will be able to sight down the wall. This will help maintain plumb head joints by sighting the brick below the newly laid unit.

You may need to slightly adjust the unit in its bed. First try pressing downward on the brick with the heel of your hand. Keep part of the heel of your hand on the brick next to the one you are adjusting. If this is not sufficient, you may need to tap the brick with the handle of your trowel. After adjusting the brick, cut off the extruded mortar with your trowel, and lay the next unit. Cutting mortar as you go will help you to keep the masonry clean.

8.1.2 Checking the Height

The first check is always course height. If this is off, there is no use checking anything else. Use your modular or standard course spacing rule to check the height of the brick. Follow these steps:

Step 1   After the bricks are laid on the wall, unfold the rule (*Figure 43*), and place it on the base or footing used for the mortar and brick.

Step 2   Hold the rule vertically. Check that the end of the rule is flat on the base, so the reading is accurate. If you are using standard brick, the first course should be even with number 6 on the modular rule. If you are using a different size of brick, check the appropriate scale on the modular or course rules.

Step 3   If the height of the course does not hit the right place on the rule, take the bricks out, and clear off the bed joint. Lay the bed joint again, and replace the buttered bricks. Recheck the height of the course. Then replumb and relevel.

The height, or vertical course spacing, depends on the thickness of the mortar joints. Practicing laying full bed and head joints is the fastest way to learn to make standard size joints.

If more than one course is laid, always set the modular rule on the top of the first course to measure. The base may have been irregular, and a large joint may have been used to level the first course.

8.1.3 Checking Level

After checking the height for your string of six bricks, check with your mason's level for levelness using the following steps:

Step 1   Remove any excess mortar on top of the bricks.

Step 2   Place your mason's level lengthwise on the center width of the six bricks to be checked.

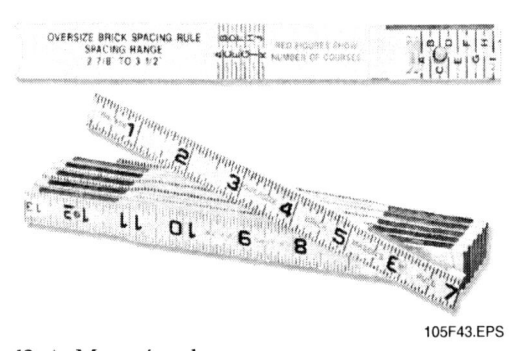

Figure 43 ◆ Mason's rules.

Step 3   Use your trowel handle to gently tap down any bricks that are high with relation to the mason's level (*Figure 44*). Do not tap them so hard they sink too low.

Step 4   If bricks are low, pick them up. Clean and mortar the bed and the head joint again, and reposition the brick. Reposition the mason's level again, and get it level.

8.1.4 Checking Plumb

The next step after leveling is to check for plumb, or vertical straightness. Follow these steps:

Step 1   Hold the level in a vertical position against the end of the last brick laid (*Figure 45*).

Step 2   Tap the brick with the trowel handle to adjust the brick face either in or out.

Step 3   Move the level to the end of the first brick laid, and repeat the process.

Figure 46 gives profiles and names for bricks that are plumb and out of plumb. The large black dot represents the mason's line. By looking and touching, you can train your hand and eye to know bricks that are plumb and bricks that are not.

Figure 44 ◆ Leveling the course.

Figure 45 ◆ Checking plumb.

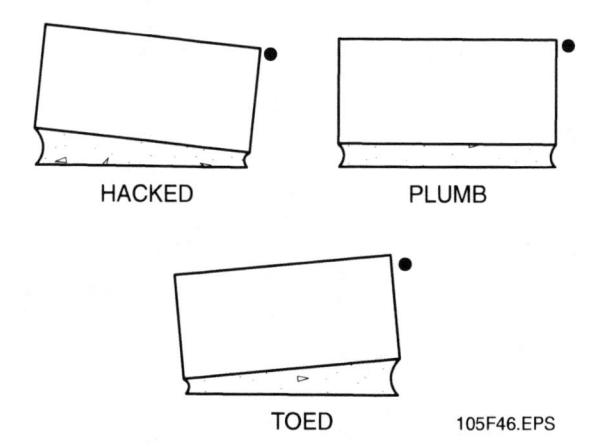

Figure 47 ◆ Checking for straightness.

HACKED   PLUMB

TOED

Figure 46 ◆ Plumb and out-of-plumb bricks.

8.1.5 Checking Straightness

After the first and last bricks in the string have been plumbed, check the rest for straightness:

Step 1 Hold the mason's level in a horizontal position against the top of the face of the six bricks, as shown in *Figure 47*.

Step 2 Tap the bricks either forward or back until they are all aligned against the mason's level. Be careful not to move the end plumb points while you are aligning the middle four bricks.

By sighting down from above, you can train your eye to know bricks that are straight and bricks that are not.

8.1.6 Laying the Closure Unit

The last unit in a course is called the **closure unit**. Masons lay corners of a wall first then work from each corner toward the middle. The last unit, or closure unit, must fit in the gap between the masonry units that have already been laid (*Figure 48*). The closure unit should fall toward the middle of a wall. The space left for it should be large enough for the unit and its two head joints.

The process for laying the closure unit is the same for block and brick:

Step 1 Butter the closure unit on both head joints.

Step 2 Butter the adjacent units on their open head joints.

Step 3 Gently ease the unit into the space.

Step 4 If any mortar falls out of a closure unit joint, remove the unit, and reset it in fresh mortar.

If the head joints have been properly spaced, the closure unit will slide in with the specified

Figure 48 ◆ Placing a closure unit.

MASONRY UNITS AND INSTALLATION TECHNIQUES

> ### Making Adjustments
>
> If bricks are not in line, some adjustment can be made as long as the mortar has not begun to set. Moving a brick after hydration has started will weaken the bond. If it is necessary to move a brick after hydration has begun, the brick and mortar must be removed and new mortar laid. Do not lift a brick or pull it from the head joint because this will destroy the bond.
>
> *Source: Brick Institute of America*

joint spacing. Otherwise, the closure unit will have head joints that are too large or too small. If this is the case, remove the last three or four units that were laid on either side of the closure unit. Remortar them, and relay them to correct for the closure head joint size. The objective is to avoid a sudden jump in the size of a head joint. A big change in joint size will catch the eye and can also skew the pattern bond. If you must move bricks, be sure to check them again for height, level, plumb, and straightness.

8.2.0 Placing Block

Placing block is similar to placing brick except that it is not a one-handed job (*Figure 49*). Butter the units on both sides of head joints. Use both hands to place the block. Do not drop it, but move it slowly down and forward so it butts against the adjacent unit. Slightly delaying release allows the block to absorb moisture from the mortar, which makes a good mechanical bond. If the mortar does not ooze from the joints, you are not using enough mortar. There will be voids in the joints, and the wall will eventually leak.

Figure 49 ◆ Placing a block.

> ### Masonry as Art
>
> Look around, and you will see many examples of the creative side of masonry, like the one in this photo. More than any other construction craft, masonry provides the opportunity to design eye-catching structures.

Tilt the block toward you as you position it. Look down over the edge and into the cores to check alignment with the block underneath. If a block is set unevenly, check to see if a pebble or other material is wedged between the mortar and the block. If so, take the block up and clean it. Rebutter fresh mortar for both bed and head joints, and reset it.

If the block requires adjustment, lightly tap it into place. You may want to use your mason's hammer to tap the block as it may require stronger taps.

Check each block for position as described for brick. If the block cannot be adjusted, take it up, remortar it, and reset it.

8.3.0 Laying to the Line

To keep masonry courses level over a long wall, masons lay the units to a line. Working to a line allows several masons to work on the same wall without the wall moving in several directions. The line is set up between corner poles or corner lead units. The poles or leads must be carefully checked for location, plumb, level, and height. The line is placed on the outside of the course, so that will be the most precisely laid side. The mason usually works on the same side as the line but can also work from the other side depending on job conditions and experience.

Each masonry unit is placed with its outside top edge level with the line and 1/16" away from it. The distance is the same for all types of masonry units. Your eye will get trained to measure that distance automatically after some practice laying to the line.

A mason's line needs to be tied to something that will not move. It must be tied taut at a height that can be measured precisely. The mason's line is attached to corner poles or corner leads by means of line stretchers, line blocks, or line pins.

8.3.1 Setting Up the Line Using Corner Poles

Corner poles allow masons to lay to the line without laying the corners first. Attach the corner pole securely. It must not move as you pull a mason's line from it. For brick veneer walls, the corner pole can be braced against the frame or backing wall. You must check the placement of the corner poles before you string the line. If the pole has course markings, check that they are the correct distance from the footing. If the pole has no markings, transfer markings from your course rule. Make sure you start the measures from the footing.

Step 1   Attach the line to the left pole to start. If the pole has no clamps or fasteners, attach the line with a hitch or half hitch knot. Stretch it to the right pole and gradually tighten it until it is stretched. Use a hitch or half hitch to secure it, tightening it as you tie. Check that it is at the proper height before you start laying.

Step 2   After laying each course, move the line up to the next course level. Stretch and measure it again. It is critical to make sure the line is at the proper height for each course.

Step 3   Use your modular rule or course spacing rule to check the line height at each end for every course.

8.3.2 Setting Up the Line Using Line Blocks and Stretchers

A mason's line can be set between corner leads or corners laid to mark the ends of the wall. The line can be attached by line blocks, line stretchers, or line pins.

Line blocks (*Figure 50*) have a slot cut in the center to allow the line to pass through. It takes two sets of hands to set up line blocks. The procedure is as follows:

Step 1   Pass the line through the slot of the block. Tie a knot, or tie a nail on the end of the line, to keep it from passing through the slot.

Step 2   Have one person hold the line block aligned with the top of the course to be laid. Traditionally, mason's lines start on the left side as you face the wall.

Step 3   Place the line block so that it hooks over the edge of the masonry unit, and hold it snug.

Step 4   The second person will walk the line to the right end of the wall.

Figure 50 ♦ Line block.

MASONRY UNITS AND INSTALLATION TECHNIQUES

5.33

Step 5 The second person then pulls the line as tight as possible and wraps the line three or four turns around the middle of the line block.

Step 6 The second person hooks the tensioned line block over the edge of the corner.

Step 7 Both parties check that the line is at the correct height.

Line stretchers are put in place following the same steps. The line stretcher slips over the top of the blocks, not the edges (*Figure 51*). Line stretchers are useful when the corner lead is not higher than the course to be laid.

8.3.3 Setting Up the Line Using Line Pins

Steel line pins hold a mason's line in place. The line pin is less likely to pull out of the wall because of its shape. The peg end of the line, or the starting end, is traditionally started at the left as the mason faces the wall.

Step 1 Drive the line pin securely into the lead joint (*Figure 52*). Make sure that the top of the pin is level with the top of the course to be laid. Place the pin at a 45-degree downward angle, several units away from the corner. This will prevent the pin from coming loose as the line is pulled.

Step 2 Tie the line securely to the pin using the notches on the pin. Give the line a few very sharp, strong tugs. This tests whether the pin will come out as the line is tightened and helps to prevent injuries caused by flying line pins.

Step 3 Walk the line to the other lead. Drive the second line pin securely into the lead joint even with the top of the course. Check and measure that the pin is secure and in the correct position before applying the line.

Step 4 Wrap the line around the pin, and start tensioning. Pull the line with your left hand, and wrap it around the line pin with your right hand. Use a clove hitch or half hitch knot to secure the taut line to the pin. Be careful not to pull the line so tight that it breaks.

Step 5 When you move the line up for another course, immediately fill the pin holes with fresh mortar. If you wait until later to fill the line pin holes, you will need to mix another batch of mortar. Taking care of the holes as you move the pins saves many steps at the end of the project.

8.3.4 Setting Up the Line Using Line Trigs

To keep a long line from sagging, masons set trigs to support the line midstring.

Step 1 Set the trig support unit in mortar in position on the wall. Be sure that this unit is set with the bond pattern of the wall, close to the middle of the wall.

Step 2 Check that the unit is level and plumb with the face of the wall. Check that the unit is at the proper height with a course rule or course pole.

Step 3 Sight down the wall to be sure that the trig unit is aligned with the wall and is set the proper distance from the line. The trig support unit is a permanent part of the wall, so place it carefully.

Step 4 After the trig support unit is in the proper place, slip a trig or clip over the taut line. Check that the line is still in position. Lay the trig on the top of the support unit with the line holder on the bottom side. Place another masonry unit on top of the trig to hold it in place. *Figure 53* shows the use of a trig.

Step 5 Check the line for accuracy once more. The line should just be level with and slightly off the corner of the trig support unit and the standard $\frac{1}{16}$" away from it.

8.3.5 Laying Brick to the Line

After you string the line and put on a trig, you can begin to lay masonry units to the line. The advantage of laying to the line is that it cuts down on the need for the mason's level.

Figure 51 ♦ Using a line stretcher.

Figure 52 ◆ Using a line pin.

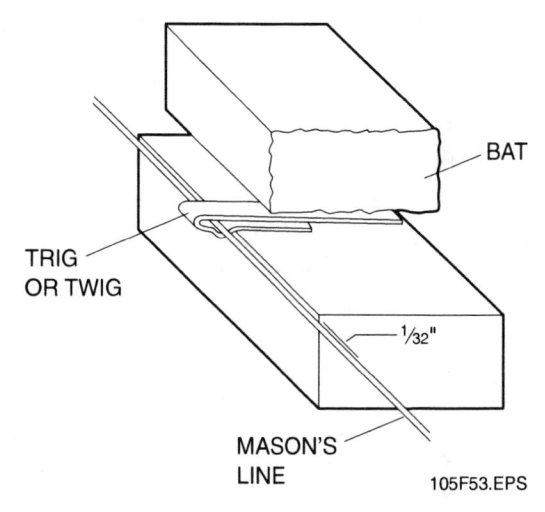

Figure 53 ◆ Using a line trig.

You must lay to the line without disturbing the line. If the line is hit, other masons on the line will have to wait for the line to stop moving. Hitting the line is called **crowding the line**. Even experienced masons will crowd the line occasionally. To avoid crowding the line, hold your brick from the top, as shown in *Figure 54A*. As you release the brick, roll your fingers or thumb away from the line, then press the brick into place, as shown in *Figure 54B*.

The brick must come to sit 1/16" inside the line. The top of the brick must be even with the top of the line. Looking at the brick from above, you should be able to see a sliver of daylight between the line and the brick. Brick set too close is crowding the line. Brick set too far is **slack to the line**. When laid correctly, the bottom edge of the brick should be in line with the top of the course under it, and the top edge of the brick should be 1/16" back and even with the top edge of the line.

Adjust the brick to the line by pressing down with the heel of your hand. Check that the brick is not hacked or toed. While you are learning to lay to the line, it is a good idea to check your placement with the mason's level. After you have gained some skill in working to the line, you will find you will not need to use the mason's level so often.

With your trowel, cut off the mortar just squeezed out of the joints. Apply the mortar to the head of the brick just laid. By buttering the head joint like this, you will not have to return to the mortar pan for each individual head joint. When buttering the head, hold the trowel blade at an angle so as not to move or cut the line.

MASONRY UNITS AND INSTALLATION TECHNIQUES

5.35

(A) (B)

Figure 54 ♦ Laying brick to the line.

Most of the mason's time is spent laying brick to the line. Practice will improve your ability to lay precisely without disturbing the line constantly. As you learn to do this, there are some additional habits you should pick up to save yourself time and energy:

- Always pick up a brick with the face out so that it is in the same position in which you will lay it. Limit turning brick in your hand as this slows and tires you.
- Pick up frogged brick with the frog down because this is the way it will be laid.
- Fill head and bed joints plump and full. This cuts down the time you will need to strike joints later. This also ensures stronger, waterproof walls.
- Stock your brick within arm's length, or approximately two feet away. When working on a veneer or cavity wall, stock your brick on the backing wythe.

8.3.6 Laying Block to the Line

The difference in laying block and brick to the line is the difference in handling the units.

Blocks should be kept dry at all times, as moisture will cause them to expand. If they are used wet, they will shrink when they dry and cause cracks in the wall joints. To cut down on handling, stack them close to the work sites with the bottom (smaller) shells and webs down.

Practicing laying block will let you discover the easiest methods for yourself. Find a way to hold the buttered block that is comfortable for you. Lift the block firmly by grabbing the web at each end of it, and lay it on the mortar joint. Keep the trowel in your hand when laying block to save time.

As you place the block, tip it toward you a little. You can look down the face to align the block with the top of the block in the course below. Then, roll the block back slightly so that the top is in correct alignment to the line. At the same time, press the block toward the last block laid. Moving the block slowly is key to this process. Do not release the block quickly, or you will have to remortar and reposition it.

You can adjust the block by tapping. Be sure to tap in the middle of the block, away from the edges. Block face shells may chip if you tap on them. Using your trowel handle on the block is not recommended because the block roughens up the end of the handle. Use your mason's hammer.

8.4.0 Building Corners and Leads

Corners (*Figure 55*) are called leads because they lead the laying of the wall. They set the position, alignment, and elevation of the wall by serving as guides for the courses that fill the space between them. Building corners requires care as well as accurate leveling and plumbing to ensure that the corner is true.

As you learn to build corners, practice technique and good workmanship. Speed will follow. Be certain that each course is properly positioned before going on to the next. Once a corner is out of alignment, it is difficult to straighten it.

In addition to corners being leads, masons also have leads that are not corners. The next sections discuss both types of leads.

Figure 55 ◆ Building corner leads.

8.4.1 Rackback Leads

Sometimes it is necessary to build a lead or guide between corners on a long wall. This is a lead without corner angles, or **returns**. It is merely a number of brick courses laid to a given point. A **rackback** lead is **racked**, or stepped back a half brick on each end. This means that the lead is laid in a half-lap running bond with one less brick in each course.

The first course is usually six bricks long, the length of the mason's level. Each course is one brick less until the sixth course has only a single brick, pyramid fashion. *Figure 56* shows a completed racked lead.

Building a lead starts with marking the exact place in line with the corners and properly located for the bond pattern. Use a chalkline between the corners to locate the place. Lay brick in the rackback lead by following standard techniques. As each course is laid, check the course spacing on each course with the modular or spacing rule. Then check the course level, plumb, and straightness with your mason's level.

When the lead is complete, it needs to be checked for diagonal alignment. Do this by holding the mason's level at an angle along the side edges of the end bricks. As shown in *Figure 57*, hold the rule in line with the corner of each brick. This lets you check that no bricks are protruding.

Next, **tail** the diagonal by laying your mason's level on the points of the end bricks, as shown in *Figure 58*. If the rule touches the edges of all the bricks, the head joint spacing is correct. If the head joint spacing is not correct, take out one or two units in that course and the courses above it. Clean and reset them with fresh mortar.

The rackback lead can now be used to anchor a line as detailed previously. Learning to build rackback leads will teach you three-quarters of what you need to know about building corners.

8.4.2 Brick Rackback Corners

A rackback corner is a rackback lead with a return or bend in it. The return must be a 90-degree angle unless the specifications say otherwise. Placement and alignment of the corner are crucial

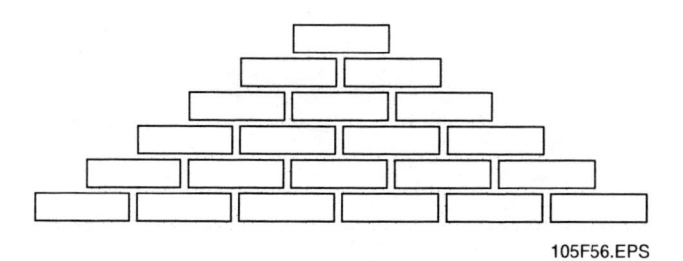

Figure 56 ◆ A rackback lead.

Figure 57 ◆ Checking alignment.

MASONRY UNITS AND INSTALLATION TECHNIQUES

5.37

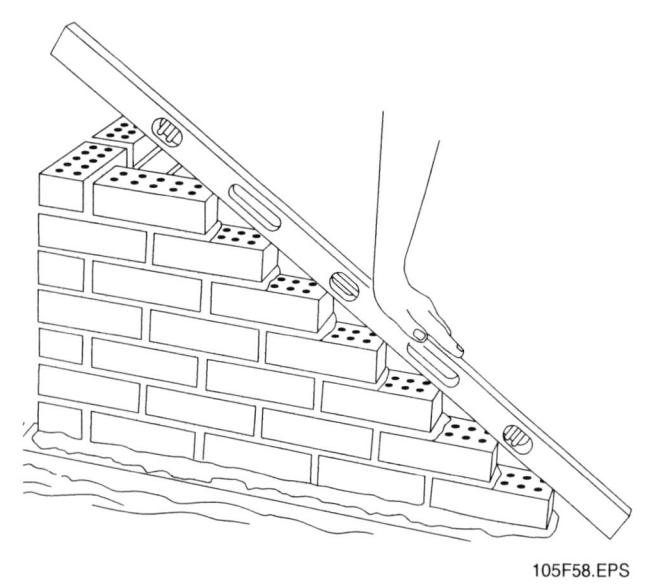

Figure 58 ◆ Tailing the diagonal.

because the corner will set the location of the remainder of the wall. Laying a corner can be intricate, demanding work if there is a pattern bond to follow. The following steps are for building an unreinforced outside rackback corner in a half-lap running bond pattern.

Step 1 Check the specifications for the location of the corner. Determine how high the corner should be. The corner should reach halfway to the top of a wall built with no scaffold, or halfway to the bottom of the scaffold. Subsequent corners are built as the work progresses.

Step 2 Determine the number of courses the corner will need. Use your course rule to calculate the courses in a given height; then calculate the number of bricks in each course. The sum of the stretchers in the first course must equal the number of courses high the corner will reach. If, for instance, you need a corner 11 courses high, the first course will have 6 bricks on one leg, 5 bricks on the other.

Step 3 Locate the building line and the position of the corner on the footing or foundation. Clean off the footing, and check that it is level. Lay out the corner with a steel square. Mark the location directly on the footing. Check the plan or specifications to determine which face of the corner gets the full stretcher and which face gets the header. Some plans have detailed drawings of corners.

Step 4 Dry bond the units along the first course in each leg of the corner. Mark the spacing along the footing.

Step 5 Lay the first course in mortar, and check the height, level, plumb, and straightness. Be sure to level the corner brick and the end brick before the bricks in the center. Check height, level, plumb, and straightness for each leg of the corner.

Step 6 Check level on the diagonal, as in *Figure 59*. Lay the mason's level across each diagonal pair of bricks. This will let you make sure that the corner continues level across the angle.

Step 7 Remove excess mortar along the bed and head joints and from the leg ends. Also remove excess mortar from the inside of the corner.

Step 8 **Range** the bricks. Ranging is sighting along a string to check horizontal alignment. Ranging is done after the first course is laid. Fasten one end of the line to the edge of one corner leg, and wrap the line around the outside of the corner. Fasten the other end of the line to the outside edge of the other corner leg. Adjust any bricks in the line that are not in perfect horizontal alignment with the line. Then take the line off.

Step 9 Lay the second course, reversing the placement of the bricks at the corner. The leg that had the full stretcher before now gets the header. Use one less brick in the second course to rack the ends. Since each course is racked, stop spreading mortar half a brick from the end of each course.

Figure 59 ◆ Leveling the diagonal.

Step 10 After placing the bricks, check height, level, plumb, and straightness. Check across the diagonal as well. To train your eye, sight down the outermost point of the corner bricks from above to check plumb. Remove excess mortar from the outside and inside of the corner and from each exposed edge brick.

Step 11 Continue until you have reached the required number of courses. Use one less brick in each course. Lay and check each course. Remove excess mortar from each course, inside and outside. Be sure that aligning the bricks does not disturb the mortar bond. If the mortar bond is disturbed, take up the brick, clean it, rebutter it, and replace it.

Step 12 If the corner does not measure up at each course, take the course up, and do it again. This is easier than taking the wall up and doing that again.

Step 13 When the corner is at the required number of courses, check each leg of the corner for diagonal alignment just as you did for the rackback lead. Hold the mason's level at an angle along the sides of the end bricks as shown in *Figure 57*. This lets you check that no bricks are protruding. Also set the mason's level on top of the racked edges as shown in *Figure 58*. This lets you check that each one touches the level and that the head joint spacing is correct.

Step 14 Check the mortar, and strike the joints. Brush the loose mortar carefully from the brick. Check the height of the corner with the modular spacing rule. Recheck the corner to make sure it is plumb, level, and straight. If it does not measure up, take the corner down, and start over.

Because the corner is so important to the wall, speed is not half as important as accuracy. Learn to be accurate, and the speed will follow.

8.4.3 Block Rackback Corners

Speed is the main advantage in building with block. It takes an experienced mason about 40 minutes to lay a block corner compared to 180 minutes to lay a brick corner. Because they are larger and heavier, blocks require some special handling. Blocks chip easily when moved or tapped down in place; therefore, they must be eased slowly into position. Their size makes them harder to keep level and plumb. Each block needs to be checked for position in all dimensions. But even with these disadvantages, they do save time and money.

The procedure for laying a block rackback outside corner follows the same steps as for brick. The main difference is that block requires different

BUILDING BLOCKS

Make Sure the Corner Is Square

You can use a framing square to make sure the corner is square. Each brick in the lead should touch the framing square.

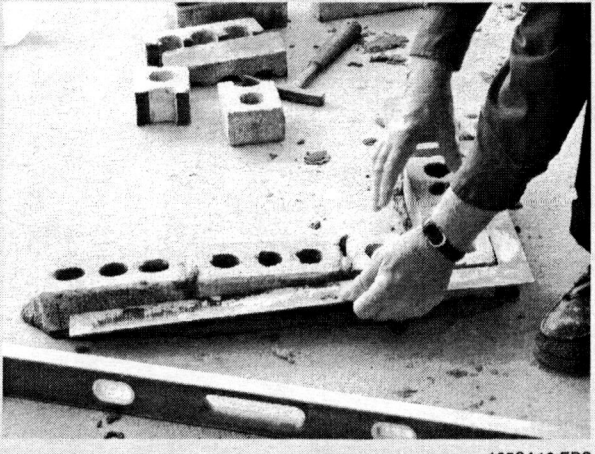

105SA10.EPS

MASONRY UNITS AND INSTALLATION TECHNIQUES

mortaring and more checking with the mason's level. The following steps do not repeat location material previously covered:

Step 1 Clean the foundation, and dampen it. Locate the point of the corner. Snap a chalkline from this point across the wall location to the opposite corner. Repeat the procedure on the other side of the corner. This aligns the corner with the other corners. Check the accuracy of the chalkline with a steel square before laying any block.

Step 2 Check that the footing is level. Lay the first course as a dry bond. Because actual sizes of block may vary, space out the dry bond with bits of wood for the joints. Check that the dry bond is plumb to find any irregularities in the footing. If the footing is too high in places, make an adjustment by cutting off some of the bottom of the block. If the footing is too low, add some pieces of block to bring the first course to the correct level.

Step 3 Lay the corner block first. Use a full bedded mortar joint without a furrow. Check the corner block for height, level, plumb, and straightness. Check for plumb on both sides of the block.

Step 4 Continue with the leg of the corner. Line up each block with the chalkline, checking the placement of each block for height, level, plumb, and straightness. Check for alignment as well. When both legs are finished, check them again, and check the diagonals.

Step 5 Do not remove excess mortar immediately, because this could cause the block to settle unevenly. Remove excess mortar from the first course after the second course is laid.

Step 6 For subsequent courses, apply mortar in a face shell bedding on top of the previously laid course. Check each block for height, level, plumb, and straightness.

Step 7 Check for diagonal alignment; then tail the rack ends. If the edges of all blocks do not touch the level, the head joints are not properly sized. Adjust the block if the mortar is still plastic enough, or rebuild the courses, as required.

When measuring blocks, each one must touch and be completely flush with the mason's level. They must also be completely in line with the chalk marks on the footing. Repeat all measurements often to prevent bulges or depressions in the wall and to keep the courses in line.

9.0.0 ◆ MORTAR JOINTS

Mortar joints between masonry units serve the following functions:

- Bonding units together
- Compensating for differences in the size of the units
- Bonding metal reinforcements, grids, and anchor bolts
- Making the structure weathertight
- Creating a neat, uniform appearance

Mortar joints are made by buttering masonry units with mortar and laying the units. The mason controls the amount of mortar buttered so that it fills a standard space between the units. Excess mortar oozes out between the units, and the mason trims it off. But this is not the last stage in making a mortar joint. After it dries partially, the mortar left between the masonry units must be tooled to be a proper mortar joint.

It is this last step, the tooling, that gives the mortar joints their uniform appearance and weathertight quality.

9.1.0 Joint Finishes

Mortar joints can be finished in a number of ways. *Figure 60* shows some standard joint finishes. Usually, the joint finish will be part of the detailed specifications on a project. The process of tooling the joint compresses the mortar and thereby

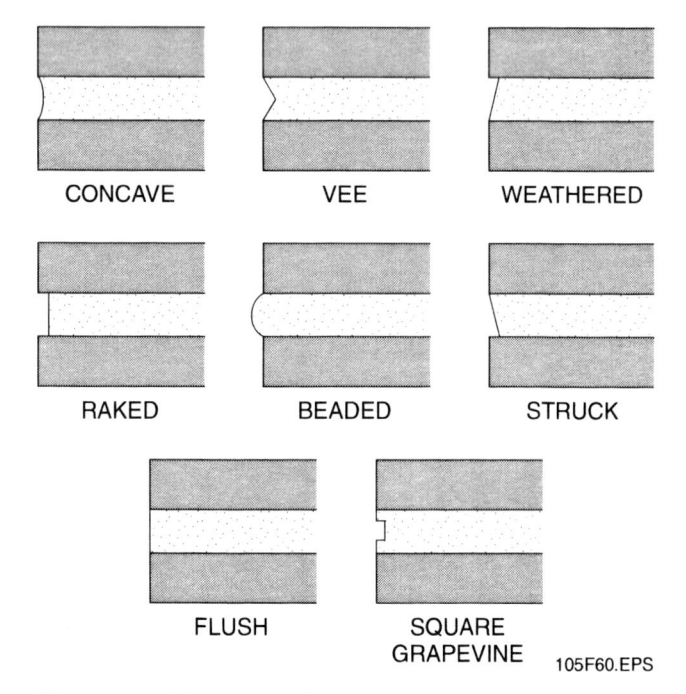

CONCAVE   VEE   WEATHERED

RAKED   BEADED   STRUCK

FLUSH   SQUARE GRAPEVINE

Figure 60 ◆ Mortar joint finishes.

increases its water resistance. Tooling also closes any hairline cracks that open as the mortar dries. Joints are tooled by shaped jointers. Raked joints are made by rakers. Struck, weathered, and flush joints are tooled by a trowel.

The raked joint is scooped out, not compressed. It does not get the extra water resistance, so it is not recommended for exterior walls in wet climates. The struck joint collects dirt and water on the ledge, so it is not recommended for exterior walls. Flush joints are not compressed, only struck off. They are recommended for walls that will be plastered or parged.

One additional joint type is the extruded or weeping joint (*Figure 61*). This joint is made when the masonry unit is laid. When the unit is placed, the excess mortar is not trimmed off. The mortar is left to harden to become the extruded joint. Since the mortar is not compressed in any way, this joint is not recommended for exterior walls.

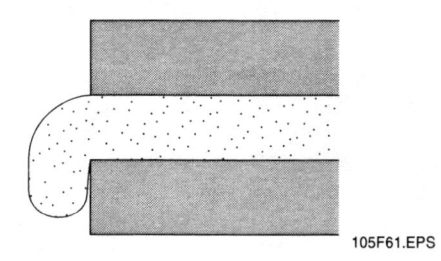

Figure 61 ◆ Extruded or weeping joint.

9.2.0 Striking the Joints

Working the joints with the jointer, raker, or trowel is called striking the joints. Whichever tool you use to strike the joint, the procedure is the same. The first step is to test the mortar.

9.2.1 Testing the Mortar

After you have laid the masonry units, the mortar must dry out before it can be tooled. The test equipment for checking proper dryness is the mason's thumb. Press your thumb firmly into the mortar joint:

- If your thumb makes an impression, but mortar does not stick to it, the joint is ready.
- If the mortar sticks to your thumb, it is too soft. The mortar is still runny, and the joint will not hold the imprint of the jointing tool.
- If your thumb does not make an impression, the mortar is too stiff. Working the steel tool will burn black marks on the joints.

The best time for striking the joints will vary because weather affects mortar drying time. Test the mortar repeatedly to find the right window of time to do the finishing work.

9.2.2 Striking

When the mortar is ready, the next step is the striking. The tool should be slightly larger than the mortar joint to get the proper impression.

- Hold the tool with your thumb on the handle, so it does not scrape on the masonry unit.
- Apply enough pressure so that the runner fits snugly against the edges of the masonry units. Keep the runner pressed against the unit edges all the way through the strike.
- Strike the head joints first (*Figure 62*). Strike head joints upward for a cleaner finish.
- Strike the bed joints last. The convex sled runner striking tool (*Figure 63*) is most commonly used. To keep the joints smooth, walk the jointer along the wall as you strike. The joints should be straight and unbroken from one end of the wall to the other. If the head joints are struck last, they will leave ridges on the bed joints.

If you are making a raked joint, follow the same order of work. Some joint rakers, or skate rakers (*Figure 64*), have adjustable set screws that set the depth of the rake-out. Do not rake out more than ½", or you will weaken the joint and possibly expose ties or reinforcements in the joint. Be sure you leave no mortar on the ledge of the raked unit.

If you are making a troweled joint, follow the same order of work. Ensure that the angle of the struck or weathered joint faces the same way on all the head joints. If you are striking flush joints with your trowel, strike up rather than down.

Figure 62 ◆ Using a jointer.

MASONRY UNITS AND INSTALLATION TECHNIQUES

Figure 63 ♦ Using the convex sled.

Figure 64 ♦ Skate raker.

9.2.3 Cleaning Up Excess Mortar

After you strike the joints, you must clean up the excess mortar. Dried mortar sticks to masonry and is difficult to clean. Cleaning is much easier when you do it immediately after striking. Follow these steps to clean up excess mortar:

Step 1 Trim off mortar burrs by using a trowel. Hold the trowel fairly flat to the wall, as shown in *Figure 65A*. As you trim the burrs, flick them away so they do not stick to the units.

Step 2 Dress the wall after trimming off burrs. Dressing can be done with a soft brush, as shown in *Figure 65B*. In can also be done with coarse fabric, such as burlap or carpet, wrapped around a wood block. Flush joints need the fabric dressing, as a brush will not smooth them.

Figure 65 ♦ Trimming and dressing mortar burrs.

Step 3 Brush the head joints vertically first. Then brush the bed joints horizontally. If necessary, restrike the joints after brushing to get a sharp, neat joint.

Step 4 After you finish cleaning the wall, clean the floor at the foot of the wall.

10.0.0 ♦ PATCHING MORTAR

Two common methods of patching mortar are pointing and tuckpointing. Pointing is the act of putting additional mortar into a soft mortar joint. This type of patch does not require much preliminary work. Tuckpointing is the act of replacing hardened mortar with fresh mortar. This type of patch requires some preparation.

10.1.0 Pointing

Despite the best workmanship, mortar can fall out of a head joint or crack when a unit settles. A unit may get a chipped edge or a lost corner or get moved by some accident. Line pins and nails leave holes in mortar joints. Because pointing is easier than tuckpointing, it is a good idea to continuously check the surface of mortar joints. Perform the following tasks as you work:

- Fill line pin and nail holes as you move the line.
- Use mortar of the same consistency as was used for laying the units.
- Force the mortar into the holes with the tip of a pointing trowel or a slicker.
- Push all the mortar with a forward motion, in one direction for each hole.
- If the hole is deep, fill it with several thin layers of mortar, each no more than ¼" deep. This will avoid air pockets in the pointed joint.
- Clean excess mortar off masonry units with your trowel.

Inspect the condition of mortar joints after you finish a section of wall. Inspect them again before you strike them with the jointer. If there is a void, force additional mortar into the joint. If the back of the unit can be reached, use a backstop, such as the handle of a hammer, to brace the unit. This will prevent the unit from moving as you point the joint.

10.2.0 Tuckpointing

Patching or tuckpointing after the mortar has hardened is more complex. Follow these steps for proper tuckpointing:

Step 1   Mix some mortar, and let it dry out for about an hour to get partly stiff. This will reduce shrinkage after it is put into the joint. While it is stiffening, clean out the damaged joints.

Step 2   With a joint chisel or a tuckpointer's grinder, dig out the bad mortar to a depth of about ½". The damaged area may be deeper due to cracks or shrinkage. Be sure you have cleaned down to solid mortar. As a rule, the depth of mortar removed should be at least as deep as the joint is wide. *Figure 66* shows a properly excavated joint along with examples of improperly excavated joints.

Step 3   Remove all loose mortar with a stiff brush or with a jet of water from a hose.

Step 4   Thoroughly wet the surrounding masonry with water, but do not saturate it. Wetting will slow setting time and produce a better bond. However, excess moisture will bead up and prevent a good bond in the joint.

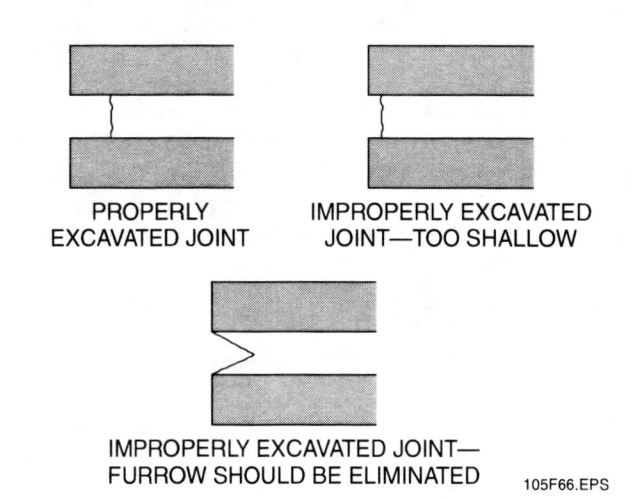

Figure 66 ◆ Preparation of joints for tuckpointing.

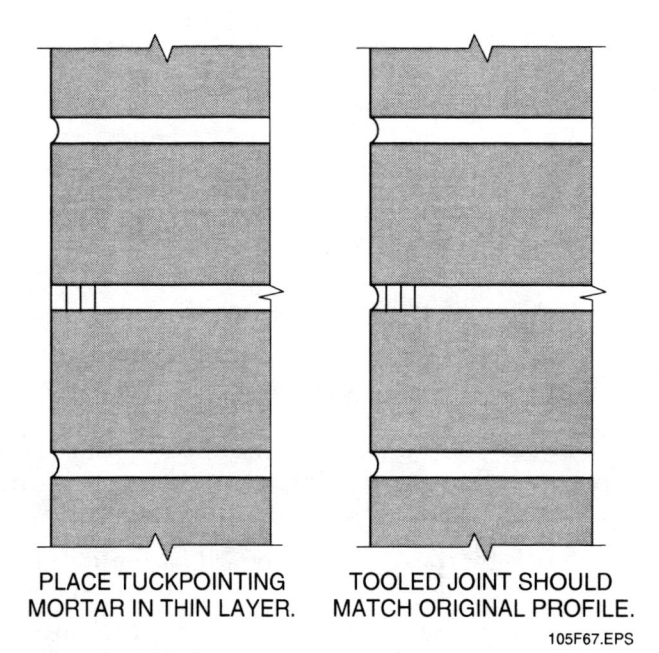

Figure 67 ◆ Tuckpointing.

Step 5   Force fresh mortar into the damp joint. Use a trowel or slicker with a point narrower than the joint, and press the mortar hard. If the damaged area is deep, fill it with several thin layers of mortar, each no more than ¼" deep (*Figure 67*). This will prevent air pockets from forming in the pointed joint.

Step 6   Clean excess mortar off of the units with your trowel.

Step 7   Retool the joint after the mortar has set long enough.

Step 8   Clean the pointed areas after you retool the joints. Use the cleaning procedure described previously under in the section on striking the joints.

11.0.0 ♦ CLEANING MASONRY UNITS

Cleaning masonry units marks the end of a particular project. The finishing touch on all masonry work is the removal of any dirt and stains. This can be a wet, difficult task. The best way to minimize this effort is by cleaning as the project goes along.

11.1.0 Clean Masonry Checklist

The hardest soil to clean off masonry units is dried, smeared mortar that has worked its way into the surface of the masonry unit. This seriously affects the appearance of the finished structure. Your best approach is to avoid smearing and dropping mortar during construction. These guidelines will help you clean as you work:

- When mortar drops, do not rub it in. Trying to remove wet mortar causes smears. Let it dry to a mostly hardened state.
- Remove it with a trowel, putty knife, or chisel. Try to work the point under the mortar drop, and flick it off the masonry.
- The remaining spots can usually be removed by rubbing them with a piece of broken brick or block, then with a stiff brush.

You should spend some time cleaning every day, removing stray mortar from the wall sections as you complete them.

In addition to cleaning dropped mortar, there are other things to do. To keep masonry clean during construction, practice these good work habits:

- Stock mortar pans and boards a minimum of two feet away from the wall to avoid splashes.
- Temper the mortar with small amounts of water, so it will not drip or smear on the units.
- After laying units, cut off excess mortar carefully with the trowel.
- Wait until mortar hardens for striking, to avoid smearing wet mortar on masonry units.
- After tooling joints, scrape off mortar burrs with your trowel before brushing.
- Avoid any motion that rubs or presses wet mortar into the face of the masonry unit.
- Keep materials clean, covered, and stored out of the way of concrete, tar, and other staining agents. Do not store materials under the scaffolding.
- Turn scaffolding boards on edge with the clean side to the wall at the end of the day. This will prevent rain from splashing dirt and mortar on the wall.
- Always cover walls at the end of the day to keep them dry and clean.

Following these practices should reduce the amount of time you spend cleaning the masonry units after construction is complete. The clean masonry checklist is especially important when working with CMUs. Because of their rougher surface texture, mortar spilled on them is harder to clean.

11.2.0 Bucket Cleaning for Brick

The best method of cleaning any new brick masonry is the least severe method. If the daily cleaning practices listed previously are not enough, the next step is bucket and brush hand cleaning. This may include using a proprietary cleaning compound or an acid wash. Acid affects brick over time, so it should be your last resort. *Table 1* lists cleaning methods for different types of brick as developed by the Brick Institute of America.

Any chemical compound you use should first be tested on a 4' × 5' inconspicuous section of wall. Sometimes, minerals in the brick may react with some chemicals and cause stains. Read the brick manufacturer's material safety data sheet (MSDS) for recommended cleaning solutions. Read the MSDS, and follow the manufacturer's directions for mixing, using, and storing any chemical solution.

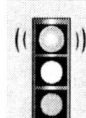

> **WARNING!**
> Wear appropriate personal protective equipment when using chemical solutions.

When cleaning, you will need a hose, bucket, wooden scraper, chisel, and stiff brush. Follow these guidelines:

- Do not start cleaning until at least one week after the wall is finished. This gives the mortar time to cure and set. Do not wait longer than six months because the mortar will be almost impossible to remove.
- Dry scrub the wall with a wooden paddle. Go over large particles with a chisel, wood scraper, or piece of brick or block. This should remove most of the mortar.
- Before wetting, protect any metal, glass, wood, limestone, and cast stone surfaces. Mask or cover windows, doors, and fancy trim work.
- Prepare the chemical cleaning solution. Follow manufacturer's directions. Remember to pour chemicals into water, not water into chemicals.
- Presoak the wall with the hose to remove loose particles or dirt.

Table 1 Cleaning Guide for New Masonry

BRICK CATEGORY	CLEANING METHOD	REMARKS
Red and red flashed	Bucket and brush hand cleaning High-pressure water Sandblasting	Hydrochloric acid solutions, proprietary compounds, and emulsifying agents may be used. *Smooth texture:* Mortar stains and smears are generally easier to remove; less surface area is exposed; easier to pre-soak and rinse; unbroken surface, thus more likely to display poor rinsing, acid staining, and poor removal of mortar smears. *Rough texture:* Mortar and dirt tend to penetrate deep into textures; additional area for water and acid absorption; essential to use pressurized water during rinsing.
Red, heavy sand finish	Bucket and brush hand cleaning High-pressure water	Clean with plain water and scrub brush, or lightly applied high-pressure and plain water. Excessive mortar stains may require use of cleaning solutions. Sandblasting is not recommended.
Light colored units, white, tan, buff, gray, specks, pink, brown, and black	Bucket and brush hand cleaning High-pressure water Sandblasting	*Do not use muriatic acid!* Clean with plain water, detergents, emulsifying agents, or suitable proprietary compounds. Manganese colored brick units tend to react to muriatic acid solutions and stain. Light colored units are more susceptible to acid burn and stains, compared to darker units.
Same as light colored units, plus sand finish	Bucket and brush hand cleaning High-pressure water	Lightly apply either method (see notes for light colored units). Sandblasting is not recommended.
Glazed brick	Bucket and brush hand cleaning	Wipe glazed surface with soft cloth within a few minutes of laying units. Use a soft sponge or brush plus ample water supply for final washing. Use detergents where necessary and acid solutions only for very difficult mortar stain. Do not use acid on salt-glazed or metallic-glazed brick. Do not use abrasive powders.
Colored mortars	Method is generally controlled by the brick unit	Many manufacturers of colored mortars do not recommend chemical cleaning solutions. Most acids tend to bleach colored mortars. Mild detergent solutions are generally recommemded.

- Start working from the top. Keep the area immediately below the space you are scrubbing wet also to prevent the chemicals from drying into the wall.
- Scrub a small area with the chemical applied on a stiff brush. Keep the scrub area small enough so that the solution does not dry on the wall as you are working.
- To remove stubborn spots, rub a piece of brick over them. Then scrub the spot again with more chemical solution. Repeat this as needed.
- As you complete scrubbing in each area, rinse the wall thoroughly. Rinse the surrounding wall area above and below, all the way to the bottom of the wall, to keep chemicals from staining the wall.
- Flush the entire wall for ten minutes after you finish scrubbing. This will dilute any remaining chemical and prevent burns.

High-pressure water washing, steam cleaning, and sandblasting are also used to clean new and old masonry. Because these techniques can damage masonry surfaces, they require trained operators. If you have not been trained, do not use this equipment.

11.3.0 Cleaning Block

Cleaning block is difficult because the surface on standard block is very porous. It is important to follow the clean masonry checklist procedures previously described. If block is stained with mortar at the end of a job, rub it with a piece of block.

For further cleaning, it is important to check the manufacturer's MSDSs for recommended chemical cleaners and cleaning procedures. Acid is very destructive to block and cannot be used without protective countermeasures. Read the block manufacturer's MSDSs for recommended cleaning solutions. Read the MSDSs, and follow manufacturer's directions for mixing, using, and storing any chemical solution. Detergents and surfactants are often recommended for use on block. If no cleaning procedures are given, follow the bucket and brush procedures listed earlier.

Any chemical compound used should first be tested on a 4' × 5' inconspicuous section of wall. Sometimes, minerals in the block may react with some chemicals and cause stains.

Summary

Masons are skilled craft professionals whose work often stands, not only for many years, but for many decades after completion. It is up to each individual mason to learn the skills necessary to create strong, properly built structures. These skills take practice and patience to master.

In order to have a long, productive career as a mason, you must work hard toward mastering your craft. You must learn all of the relevant specifications and standards and keep yourself informed of changes in the regulations governing the masonry industry. As well, you need to be familiar with the specific challenges and special project types common to your particular region.

Learning the basic skills in this module is essential to becoming a skilled mason. However, being a successful mason is about much more than understanding how to choose masonry units, lay a strong course, and properly mix mortar. It is about having pride in your workmanship and about taking the time to do each job right. By mastering the craft, having real pride in your work, and always continuing to learn, you can ensure yourself a long, productive career in masonry.

Review Questions

1. Random cracking in CMU walls can be controlled by _____.
   a. contraction joints
   b. changes in wall thickness
   c. pilasters
   d. chases

2. Grout is used with concrete masonry units to _____.
   a. strengthen joints
   b. reinforce walls
   c. remove stains
   d. add color

3. Which of the following is a correct statement about calcium silicate brick?
   a. It is banned in the United States.
   b. It is not fired in a kiln.
   c. It cannot be used for fireplaces and chimneys.
   d. It is made without sand.

For Questions 4 through 8, refer to *Figure 1*, and match the brick placement to its name.

4. _____ Stretcher

5. _____ Sailor

6. _____ Soldier

7. _____ Rowlock

8. _____ Header

9. Solid masonry units have 75% or less of their surface as a void.
   a. True
   b. False

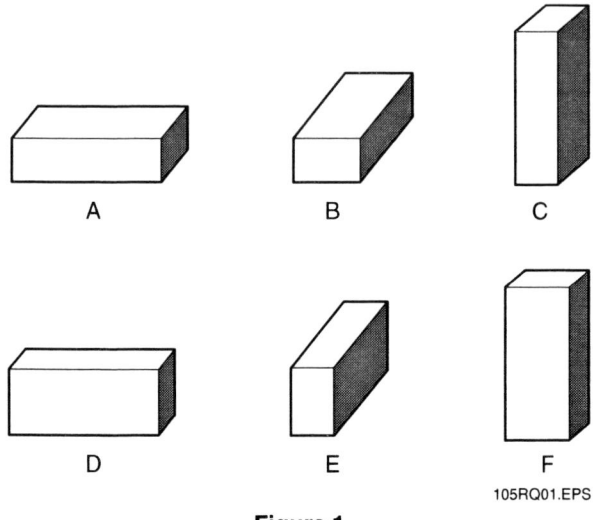

Figure 1

10. Cavity walls can be loadbearing when they _____.
    a. have flashing and expansion joints in the cavity
    b. are placed on a modular grid to avoid cuts
    c. are tied with metal ties to equalize bearing forces
    d. have both wythes of the same material

11. Laying the first course with no mortar is called _____.
    a. leading
    b. rackbacking
    c. hot ranging
    d. dry bonding

12. Moving a CMU after it is in place will _____.
    a. chip its edges
    b. weaken the bond
    c. set mortar firmly
    d. retemper the mortar

13. All of the following are true of pattern bonds *except* pattern bonds _____.
    a. call for cutting brick to fit
    b. add a visual design to masonry walls
    c. add structural strength to masonry walls
    d. call for placing brick in different positions

14. Flemish, common, and running are _____.
    a. stack bonds
    b. defined brick cuts
    c. corner patterns
    d. structural pattern bonds

15. Blocks cut horizontally across the face are known as _____.
    a. rowlocks
    b. rips
    c. half-highs
    d. ¾ cuts

16. Masonry is plumbed and leveled after the joints are struck.
    a. True
    b. False

17. Laying to the line calls for positioning the _____ edge of the masonry unit level with the line and _____ away from it.
    a. outside bottom; 3⁄16"
    b. outside top; 1⁄16"
    c. inside top; 3⁄32"
    d. inside bottom; 1⁄16"

18. Poles, blocks, pins, and trigs are used to _____.
    a. level masonry units
    b. strike mortar joints
    c. secure the mason's line
    d. shield masonry from mortar smears

19. Which of the following allows the mason to lay to the line without building corner leads?
    a. Line trig
    b. Corner block
    c. Line pin
    d. Corner pole

20. A rackback lead is _____.
    a. stepped back a half brick on each end
    b. used to check plumb
    c. slack to the line
    d. crowding the line

21. A rackback lead with a return in it is a _____.
    a. furrow
    b. rackback corner
    c. string
    d. ranged line

22. Joints should be struck when the mortar _____.
    a. brushes off the masonry units
    b. has set firm and hard
    c. will hold a thumbprint
    d. still oozes

23. When striking flush joints with a trowel, you should strike _____.
    a. upward
    b. downward

24. Laying the mason's level on the points of the end bricks laid in a rackback lead is called _____.
    a. laying the line
    b. tailing the diagonal
    c. ranging the angle
    d. leveling the corner

25. Which of the following is *not* a standard joint finish?
    a. struck
    b. convex
    c. vee
    d. concave

PROFILE IN SUCCESS

Zachary Reinert, Apprentice

Nester Brothers Masonry
Pennfield, PA

While still a high school student, Zack Reinert took first place in the 2004 masonry competition at the Associated Builders and Contractors Craft Olympics held in Hawaii.

How did you get started in the Masonry trade?
I enrolled in the three-year masonry program at my high school. This is a co-op program, so I work part time at Nester Brothers getting real job experience. I started as a laborer, but was given the opportunity to lay block. Fortunately, I did well at it, so I was able to continue doing it.

Describe your job.
I'm still in high school, so I attend school most of the time. In my job at Nester Brothers, I lay brick, block, and some stone.

What do you like most about your job?
I like working outdoors, and I like working with my hands. I also like that I can see the results of my work every day and take pride in what I've accomplished. I like the company I work for. They have helped and encouraged me along the way and sponsored me to compete in the Craft Olympics.

What would you say to someone just entering the trade?
Get into a training program. It is difficult to learn correct technique on the job. You will probably start out as a laborer, but when the opportunity comes to work on a wall, you have to be prepared. If you don't do well, it could be a long time before you get another chance. Taking some training, especially in high school, will help you decide if you really like the work. I like doing masonry work. I learned pretty quickly that if you enjoy what you do, it keeps you motivated and makes the time go quickly.

GLOSSARY

Trade Terms Introduced in This Module

Autoclave: A pressurized, steam-heated tank used for sterilizing and cooking.

Closure unit: The last brick or block to fill a course.

Crowding the line: A person touching the mason's line, or a masonry unit too close to the line.

Dry bond: Laying out masonry units without mortar to establish spacing.

Lintel: The support beam over an opening such as a window or door. Also called a header.

Rackback: A lead or other structure built with each course of masonry shorter than the course below it.

Racking: Shortening each course of masonry by one unit so it is shorter than the course below it, resulting in a pyramid shape.

Ranging: Aligning a corner by using a line. Corners can be ranged around themselves or from one corner to another.

Return: A corner in a structure or lead.

Slack to the line: Masonry units set too far away from the mason's line.

Tail: To check the spacing of head joints by checking the diagonal edges of the courses on a lead or corner.

REFERENCES

Additional Resources

This module is intended to be a thorough resource for task training. The following reference works are suggested for further study. These are optional materials for continued education rather than for task training.

Bricklaying: Brick and Block Masonry. Reston, VA: Brick Institute of America.

Building Block Walls—A Basic Guide, 1988. Herndon, VA: National Concrete Masonry Association.

Concrete Masonry Handbook. Skokie, IL: Portland Cement Association.

The ABCs of Concrete Masonry Construction, Videotape, 13:34 minutes. Skokie, IL: Portland Cement Association.

ACKNOWLEDGMENTS

Figure Credits

National Concrete Masonry Association	105F01, 105F03, 105F33
Topaz Publications, Inc.	105F02, 105SA03, 105F10, 105SA04–105SA06, 105F13, 105F15, 105F22, 105F39, 105F42, 105F44, 105F45, 105F47, 105F48, 105SA09, 105F54, 105F55, 105F59, 105SA10, 105F62–105F64
Used with permission of the Brick Industry Association, Reston, Virginia, www.gobrick.com	105F04, 105F05, 105F12, 105F16, 105F26, 105F27, 105F29–105F32, 105F34, 105F38, 105F67, 105T01
Portland Cement Association	105SA01, 105F09, 105F11, 105F20, 105F24, 105SA08, 105F49–105F53, 105F61, 105F65, 105F66
ChemGrout, Inc.	105SA02
ICD Corporation www.selecticd.com	105F06
Air Vol Block, Inc.	105F08
Indiana Limestone Institute of America, Inc.	105F19
Granite City Tool Company	105F40, 105F41
Bon Tool Company	105F43

CONTREN® LEARNING SERIES — USER UPDATES

The NCCER makes every effort to keep these textbooks up-to-date and free of technical errors. We appreciate your help in this process. If you have an idea for improving this textbook, or if you find an error, a typographical mistake, or an inaccuracy in NCCER's Contren® textbooks, please write us, using this form or a photocopy. Be sure to include the exact module number, page number, a detailed description, and the correction, if applicable. Your input will be brought to the attention of the Technical Review Committee. Thank you for your assistance.

Instructors – If you found that additional materials were necessary in order to teach this module effectively, please let us know so that we may include them in the Equipment/Materials list in the Instructor's Guide.

Write: Product Development and Revision
National Center for Construction Education and Research
P.O. Box 141104, Gainesville, FL 32614-1104

Fax: 352-334-0932

E-mail: curriculum@nccer.org

Craft _____ Module Name _____

Copyright Date _____ Module Number _____ Page Number(s) _____

Description
_____
_____
_____
_____
_____

(Optional) Correction
_____
_____
_____

(Optional) Your Name and Address
_____
_____
_____

Masonry Level One

Index

Index

Accidents, 1.28–1.29, 1.45, 1.48. *See also* Safety
Acetylene, 1.52
ACI. *See* American Concrete Institute
Acid
   avoidance on concrete block, 5.46
   hydrochloric, 5.45
   muriatic, 4.24, 5.45
Acoustic barriers, 1.7
Adaptability, 1.21
Adhesion, 4.11, 4.13, 4.21, 4.22
Admixtures, 1.57, 4.2, 4.5, 4.7–4.9, 4.22–4.23
Adobe, 1.1, 1.57
Aesthetics and design, 1.3, 4.8, 4.14, 5.32
Aggregate, 1.9, 1.12, 1.14, 1.57, 4.2, 5.3
Air-entraining admixtures, 4.3, 4.5, 4.7, 4.8, 4.22, 4.27
AISC. *See* American Institute of Steel Construction
Alkaline, 1.31, 1.57
American Concrete Institute (ACI), 3.32
American Institute of Steel Construction (AISC), 3.32
American Society for Testing and Materials (ASTM), standards
  aggregates, 4.5
  brick, 1.4, 1.5
  concrete masonry units, 1.9, 5.2, 5.5, 5.8
  hydrated lime, 4.3
  masonry cement, 4.5
  masonry construction, 3.31–3.32, 3.41–3.42
  mortar, 1.14, 4.2, 4.5
  stone, 1.13
American Society of Civil Engineers (ASCE), 3.32
Anchoring devices, 1.42–1.43, 5.14, 5.15
Anchors, wire, 1.7
Antifreeze, 4.7, 4.23
Apprenticeship, 1.4, 1.18–1.19, 5.48
Arches, 1.3, 1.4
Architect, 3.33, 5.18
Architectural abbreviations, 3.21, 3.25
Architectural symbols, 3.21, 3.22, 3.27
Area
   calculation of, 3.14, 3.15, 3.16, 3.18
   units for, 3.5, 3.13
Arresting force, 1.39, 1.40, 1.41
ASCE. *See* American Society of Civil Engineers
Ashlar, 1.1, 1.13, 5.12, 5.24
ASTM. *See* American Society for Testing and Materials

Autoclave, 5.49
Autoclave-cured calcium silicate units, 5.9
Autoclave expansion test, 4.14

Bag, grout, 2.10, 2.11
Bagged materials, storage, 1.33, 5.15
Barrow, brick and tile, 2.20
Beam, bond, 1.12, 5.9, 5.26
Bed joint
   definition and use of term, 1.15, 2.39
   jointing of, 2.8, 5.40
Belt, body, 1.41
Blacksmith, 2.6
Blade, carborundum, 1.35, 5.28
Block
   concrete. *See* Concrete block
   line or corner, 2.15, 2.16, 5.33–5.34
   and tackle, 2.27
Blueprints, 3.27, 5.14
Body, measurement using the, 3.4
Bolt cutter, 2.10, 2.11
Bolts, 1.42, 2.32, 5.15
Bonds
   dry. *See* Dry bonding of a structural unit
   four types of, 5.20–5.24
   mechanical or mortar, 5.20–5.21
   pattern, 1.5, 5.7, 5.20. *See also* Patterns
   running, 5.22, 5.24
   stack, 5.21, 5.24
   strength of, 4.4, 4.7, 4.8, 4.10, 4.13–4.14
   structural and structural pattern, 1.5, 5.7, 5.20, 5.22–5.24. *See also* Patterns
Boom, 1.47, 1.48, 2.28, 2.29
Booster, 1.34, 1.35
Box
   chalk, 2.16–2.17
   cubic-foot measuring, 2.19, 4.16, 4.17, 4.20
   mortar, 1.22, 2.18–2.19, 4.16, 4.20
Brick
   as an architectural style, 1.3
   calculation of amount needed, 3.15, 3.16
   classifications and positions, 1.7, 1.8, 5.11, 5.22
   cleaning, 4.24, 5.45
   clinker, 1.7
   concrete. *See* Concrete brick

Brick (continued)
  cored, 1.7
  cutting, 5.25, 5.26
  facing, 1.5, 1.7
  fire, 1.5, 1.7
  glazed, 1.2, 1.3, 1.7, 5.45
  handmade clay, 1.2
  how to stack, 2.2, 5.1
  manufacture, 1.4
  overview, 1.5–1.6, 1.53
  palletized, 1.33–1.34, 1.57, 2.26
  pressed, 1.7
  sand-lime. *See* Calcium silicate units
  shapes, 1.6
  sizes, 1.4, 1.5, 3.9
  spacing of, 2.12, 2.13
  standard types, 1.5
  symbol for, 3.22, 3.27
  veneer of, 5.21
Brick cleaners, 1.32
Brick Institute of America, 5.44
Brick laying
  adjustments, 5.29, 5.32, 5.35
  design elements in, 4.8, 5.32
  expected number per day, 1.1
  general procedure, 1.22–1.28, 5.29–5.32
  to the mason's line, 5.33–5.36
  rackback corners, 5.37–5.39
  rackback leads, 5.37
  world's record for, 1.1
Brick spacing rule or system, 3.7–3.8
Bridge (putlog), 2.33–2.34
Brownstone, 1.13
Brushes, 2.9–2.10, 5.42, 5.45
Buffers, 1.35
Buggy, motorized, 2.30
Building codes, 3.29, 3.32, 3.33, 5.14
Bull header. *See* Header, bull
Bull stretcher. *See* Stretcher, bull
Burlap, 5.42
Buttering the joints, 1.26–1.27, 1.57, 5.18–5.19, 5.20, 5.31

Cables, in cranes and derricks, 2.28
Calcium chloride, warning against, 4.23
Calcium silicate units, 5.8, 5.9
Canadian Standards Association (CSA), 3.31, 3.32
Cap, concrete, 1.12, 5.9
Capacity chart, 1.45, 1.48
Capacity rating, for hoist or scaffold, 2.27
Capital, 1.7, 1.57
Carabiners, 1.42
Carbon black, 4.8
Carbon monoxide, 1.35
Cart, brick, 2.20
Cast stone, 1.13, 5.12
Catch basins, drainage, 1.11, 1.12, 5.3, 5.8, 5.10
Caulking, 3.31
Caulking gun, 2.10, 2.11
Cell or core, of a hollow concrete block, 1.11, 5.4
Celsius, conversion to Fahrenheit, 3.13
Cement. *See also* Portland cement
  dust, 1.31, 1.32, 4.5, 4.20
  hydration crystals in, 4.13
  masonry, 1.14, 4.2, 4.4–4.5, 4.16, 4.27
  portland. *See* Portland cement
  safety, 1.23, 1.31–1.32
  storage of bags, 4.15, 5.15

Center of gravity, 1.49, 1.50
Certifications, 1.19, 1.35, 1.45
Chalkline and chalk box, 2.16–2.17, 5.17–5.18, 5.37
Channel, concrete block, 5.7
Chimney stacks, 5.9, 5.14
Chisel, 2.6–2.8, 5.2, 5.26–5.27, 5.43, 5.44
Chock, wheel, 1.51
Circle, 3.15–3.16
Circumference, 3.15
Clamp, pre-mounted conduit, 1.34, 2.24
Clay masonry units, 5.3, 5.10–5.11. *See also* Brick
Cleaning equipment, pressurized, 2.25–2.26, 5.45
Cleaning procedures for masonry, 5.44–5.46
Clean up of work site, 1.30, 1.32, 1.38, 1.52, 4.21
Closure unit, 5.31–5.32, 5.49
Clothing, 1.29, 1.30, 1.32, 1.36
CMU. *See* Concrete masonry units
Codes. *See* Building codes
Cohesion, 4.11, 4.22
Cold weather, safety considerations, 1.36–1.37, 1.41
Colorants, in mortar, 4.2, 4.8–4.9, 4.23, 5.45
Color-coding system for powder load charges, 1.34, 2.23
Columns, 1.25
Combustible liquids, 1.52
Compressive strength, 4.10, 4.13, 5.2
Compressor, air, 2.23, 2.26
Concrete
  safety, 1.31–1.32
  symbols for, 3.22
  vs. mortar, 4.2, 4.11, 4.12
Concrete block
  aerated. *See* Calcium silicate units
  cleaning, 5.46
  cutting, 5.25–5.26
  general procedure for installation, 5.18–5.20, 5.32–5.36
  hollow. *See* Hollow masonry units/tiles
  interlocking (hardscape), 5.7
  laying rackback corners, 5.39
  laying to the mason's line, 5.33–5.36
  overview, 1.11, 1.53, 5.3–5.5, 5.7–5.8
  parts of, 1.11, 5.4
  shapes and dimensions, 1.10, 5.5–5.7
  stockpiles, 1.33
  structural *vs.* nonstructural, 1.11, 1.57, 5.7–5.8
Concrete brick
  overview, 1.11–1.12, 5.8
  sand-lime. *See* Calcium silicate units
  shapes and dimensions, 1.10, 5.6, 5.8
  slump, 1.11, 1.12, 5.8, 5.9
Concrete masonry units (CMU)
  block. *See* Concrete block
  brick. *See* Concrete brick
  cutting, 5.25–5.26
  definition and use of term, 1.1, 1.9, 1.53, 1.57
  glazed, 1.12, 5.9
  hollow. *See* Hollow masonry units/tiles
  manufacture technique, 1.9
  overview, 1.8–1.12, 1.53
  pre-faced or pre-coated, 1.12–1.13, 5.9, 5.10
  reinforcement for, 1.44, 5.3, 5.5, 5.13
  standards, 1.9, 5.2, 5.5
  in technical specifications, 3.31
Construction drawings, 3.27
Contract, 3.28, 3.29, 5.14
Contractor, 1.20, 5.14, 5.15, 5.16, 5.25
Conversion, mathematical, 3.2
Conveyors, 2.30–2.31

Corner pole, 2.16, 2.17, 2.39, 5.33
Corners
    block rackback, 5.39–5.40
    brick rackback, 5.37–5.39
    concrete block for, 1.10, 1.12, 5.6, 5.7, 5.9
    Dutch, 5.23, 5.25
    English, 5.23, 5.25
    how to build, 5.36–5.40
    layout and planning, 5.18, 5.19
    in order of work on a course, 5.31
Corner two-piece, 1.6
Cornice, 1.7, 1.57, 5.22
Costs, 1.29, 4.4
Coupling, eye, 1.34, 2.24
Course, 1.7, 1.8, 1.57, 5.23, 5.30, 5.37
Course system of measurement, 3.7–3.8
Cove base, 1.12, 5.9
Cracks, 4.4, 4.13, 4.14, 5.2, 5.41
Cradle, on a forklift, 1.48
Cranes
    hand signals for operations, 1.31
    mobile, 2.29, 2.30
    safety, 1.31, 2.28–2.29
    tower, 2.29
Crowding the line, 5.35, 5.49
Crushing injury, 1.31, 1.33, 1.47, 1.48, 1.49, 1.53
CSA. *See* Canadian Standards Association
Cube, of bricks or blocks, 1.4, 1.57, 2.26, 5.8
Cubic-foot box, 2.19, 4.16, 4.17, 4.20
Curing, 4.14
Cutting implements and techniques
    bolts, 2.10, 2.11
    brick, 5.25, 5.26
    concrete block, 5.25–5.26
    with hand tools, 5.26–5.28
    safety, 2.22, 5.2, 5.26
    with saws or splitters, 2.21, 5.28–5.29
Cylinder, 3.17

Deadman, 2.16, 2.17
Deceleration devices, 1.39, 1.41
Deck, on scaffolding, 1.45, 2.32, 2.33, 2.35
Demolition projects, 5.28
Denominate numbers, 3.1, 3.2–3.6, 3.37
Dependability, 1.21
Dermatitis, 1.31, 4.5
Derricks, 2.28
Design elements, 4.8, 4.14, 5.32
Designer, 5.16
Diesel fuel, 1.51–1.52
Dimensions
    in drawings, 3.24, 3.26
    nominal, 1.4, 3.8, 3.37
Distance, free-fall, 1.39, 1.41
Docks, 1.51
Documentation
    forklift operator's daily checklist, 1.46
    technical specifications form, 3.30
Dogleg, 1.6
Domes, 1.2–1.3
Doors, 5.16, 5.17, 5.18
Drainage for walls, 1.15, 1.57
Drawings
    architectural abbreviations, 3.21, 3.25
    architectural symbols, 3.21, 3.22, 3.27
    dimensions used in, 3.24, 3.26
    electrical symbols, 3.21, 3.23
    elevation, 3.27–3.28
    geometry symbols used in, 3.17, 3.18, 3.21
    job layout from, 5.16–5.18
    lines as symbols, 3.19–3.20, 3.21
    metric, 3.28, 3.29
    plumbing symbols, 3.21, 3.24
    residential, 3.26–3.28
    scales used in, 3.21
    section, 3.28
    types, 3.1–3.2
Dressing the mortar burrs, 5.42
Drill, power, 2.23
Drill bits, 2.23
D-ring, 1.40
Drivers, powder-actuated, 1.34, 2.23, 2.24
Dry bonding of a structural unit, 5.16, 5.18, 5.29, 5.38, 5.40, 5.49
Durability of mortar, 4.13, 4.22–4.23
Dust
    cement, 1.31, 1.32, 4.5, 4.20
    control of, 1.35, 2.21, 5.28

Ear protection. *See* Hearing protection
Education, 1.18–1.20
Efflorescence, 4.23–4.24, 5.9
Electrical plans, 3.28
Electrical symbols, 3.21, 3.23
Electrical utilities, 1.47
Elevation drawings, 3.27–3.28
Emergency escape route, 1.52
Emergency medical services, 1.37
Emergency response procedures, 1.32, 1.44
Emergency stop signal, 1.31
Ends or ears, of a hollow concrete block, 1.11, 5.4
Engineer, 5.16
Equipment
    connectors for, 1.42–1.43
    fall protection, 1.30, 1.37–1.38
    hand-powered mortar, 2.18–2.20
    lifting, 2.26–2.31
    personal fall-arrest, 1.33, 1.39–1.43, 1.44, 2.32
    power, 2.23–2.26
    rigging, 1.48, 1.50
Escape route, 1.52
Ethane, 1.52
Ethical principles, 1.22
Evacuation, 1.52
Evaporation, 4.23
Exhaust system, 1.35
Explosion, 1.32, 1.35, 1.51–1.52
Extender, scaffolding, 2.33
Eye injuries, 1.29, 2.22, 4.19, 4.20, 5.26. *See also* Dust, cement; Goggles
Eye wash station, 1.32

Face mask, 1.30, 4.20
Face shell, of a hollow concrete block, 1.11, 5.4
Facing
    brick, 1.7
    concrete block which mimics stone, 1.12, 5.9, 5.12
    definition and use of term, 1.5, 1.57
    stone, 1.13, 5.12
    structural clay tile, 1.7
Fahrenheit, conversion to Celsius, 3.13
Falling objects, 1.30–1.31, 1.48, 1.50, 1.53
Fall protection equipment, 1.30, 1.37–1.38. *See also* Guardrails; Nets, safety; Personal fall-arrest equipment

Falls, 1.38, 1.43, 1.44
Farmers Home Administration (FmHA), 3.29, 3.30
Fastening tool, powder-actuated, 1.34, 2.23, 2.24
Federal standards, 3.11
Fire, 1.32, 1.51–1.52
Fire extinguisher, 1.32, 1.34, 1.52
Fire fighting, 1.52
Fireproofing partitions, 1.6
Firewalls, 1.11, 5.3
First aid, 1.37
Flagstone, 1.13
Flammable liquids, 1.32, 1.51–1.52
Flashing, 1.15, 1.17, 5.13–5.14
Flexibility, 4.4
Floor blocks, concrete, 1.10, 5.6
Floor openings, 1.31
Floor plan, 3.27, 5.17
Flues, 5.9
Fly ash, 5.3
FmHA. See Farmers Home Administration
Footing, 1.17, 1.57, 5.17, 5.38, 5.40
Foreman, 2.38, 5.15, 5.17, 5.25
Forklift
    fire and explosion hazards, 1.51–1.52
    handling loads with a, 1.48–1.50, 1.51
    safety, 1.44–1.53, 2.29–2.30, 4.16
    traveling in a, 1.47–1.48
    working on ramps and docks, 1.51
    before you operate a, 1.45–1.47
Forklift operator, 1.45
Foundation plan, 3.27, 5.17, 5.18
Fractions, 3.6–3.7
Fractures, 1.33
Frames, in scaffolding, 2.31–2.32
Frog, in a concrete brick, 5.6, 5.8, 5.36
Frostbite, 1.36–1.37
Furring partitions, 1.6
Furrowing, 1.25, 1.26, 1.57, 5.18

Gallon, 3.6
Gases. See Vapors and gases
Gasoline, 1.32, 1.34, 1.51–1.52
General conditions, in a specification, 3.28–3.29
General Services Administration, 3.11
Geometry, 3.14–3.18
GFCI. See Ground fault circuit interrupter
Glazes, 1.5, 1.7. See also Brick, glazed; Concrete masonry
    units, glazed
Gloves, 1.29, 2.2, 2.10, 2.22
Goggles, safety
    during mixing processes, 1.32, 4.20
    with powder-actuated tools, 1.35
    when cutting masonry, 2.1, 2.22, 5.2, 5.26
Government projects, 3.11
Grab, rope, 1.41
Granite, 1.13
Grinders, tuckpoint, 2.22–2.23, 5.43
Grip, protective, 2.7
Ground fault circuit interrupter (GFCI), 1.30
Grounding of power tools, 2.21
Grout
    to anchor veneer, 1.7
    definition and use of term, 1.14, 1.57
    elimination of air holes, 2.25
    pumped into concrete block walls, 1.15, 5.3, 5.5, 5.13
    safety, 2.2
    slump test and water content, 4.12
Guardrails, 1.31, 1.38–1.39, 2.32, 2.34

Hair, 1.29
Hammer, 2.4–2.6, 5.26–5.27
Hand signals for crane operations, 1.31, 2.29
Hand tools, 2.2–2.11
Hard hat, 1.29, 1.30, 2.22, 5.26, 5.28
Hardscape, 5.7
Harness, body, 1.40, 1.42–1.43
Hawk, 2.18
Header
    brick, 1.7, 1.8, 5.11, 5.22
    bull, 1.7, 1.8
    concrete, 1.12, 5.6, 5.9
    definition, 1.7, 1.8
    ogee step tread, 1.6
Head joint
    buttering the, 1.26–1.27, 5.18–5.19
    in concrete block, 1.15, 5.18–5.19
    definition and use of term, 1.17, 1.57
Healing, autogenous, 4.4
Hearing protection, 1.30, 1.35, 2.22
Heat cramps, 1.37
Heat exhaustion, 1.37
Heat stroke, 1.37
Height, of the course, 5.30
High-pressure water cleaning, 2.26, 5.45
Historical background
    building codes, 3.32
    denominate measures, 3.5, 3.6
    hollow concrete blocks, 1.8–1.9
    masonry, 1.1, 1.2–1.4, 1.53
    metric system, 3.11, 3.12
    mortar, 1.2, 1.3, 1.13–1.14, 4.1
Hod, 2.18, 2.19
Hoe, mortar, 1.22, 1.23, 2.19, 4.20–4.21
Hoists
    brick, 2.28
    mounted, 2.27
    personnel lift, 2.28, 2.35
    portable materials, 2.28
Hollow masonry units/tiles
    clay, 1.6, 1.7, 5.10–5.11
    parts of, 1.11, 5.4
    percent which is hollow, 5.3, 5.10
    standards for concrete, 5.5
    symbols, 3.22
Hopper, 1.33, 2.27, 4.18
Horn, 1.47
Horseplay, 1.48
Hot weather, safety considerations, 1.37
Housekeeping, 1.30, 1.32, 1.52
Houses, residential drawings, 3.26–3.28
Hydration, 4.2, 4.23, 4.27
Hydrogen gas, 1.52
Hygroscopic, 1.31, 1.57
Hypotenuse, 3.18

IBC. See International Building Code
Illness, 1.36–1.37, 5.28
Inch, 3.5
Inspection
    of construction projects, 3.32–3.34
    personal fall-arrest equipment, 1.43
    pre-shift, for forklifts, 1.45, 1.46
    of scaffolding, 1.39, 5.15
    tools, 2.3, 2.4, 2.6, 2.15, 2.20, 5.16, 5.28
Inspector, 3.32–3.33
Insurance considerations, 1.28

International Building Code (IBC), 3.32
International Bureau of Weights and Measures, 3.12
International System (SI), 3.1, 3.2, 3.11–3.13, 3.37

Jack
  pallet, 2.30
  screw, 2.33
Jamb, 1.10, 5.6, 5.7
Jewelry, removal of, 1.29
Jig, 2.22, 5.18
Job site. *See* Work site
Jointer, 2.8–2.9, 5.41
Joint fillers, 5.3, 5.14
Joint reinforcement ties, 5.13
Joints. *See also* Bed joint; Head joint
  buttering the, 1.26–1.27, 5.18–5.19, 5.20, 5.31
  color, and hardening of mortar, 4.12
  control (contraction and expansion), 5.2–5.3, 5.4, 5.14
  definition, 1.57
  pointing, 2.3, 2.39, 5.42–5.43
  striking the, 5.41–5.42
  terms, 1.7, 1.8
  tool to shape. *See* Jointer
  types of finishes for, 2.8, 5.40–5.41
  width or thickness, 1.25, 3.9, 5.18, 5.31
Journeyman, 1.4, 1.19–1.20

King closure, 5.25

Ladders, 2.35
Lanyard, 1.38, 1.40–1.41
Laying to the line, 5.33–5.36
Layout of structural unit, 5.16–5.18
Lead
  definition and use of term, 2.14, 2.39
  how to build a, 5.36–5.40
  rackback, 5.37
Length units, 3.5, 3.11, 3.12, 3.13
Level (geometric)
  how to check for, 1.27, 2.12, 5.29, 5.30, 5.38
  laying to the line to maintain, 5.33–5.36
Level (tool), 2.11–2.12., 5.30, 5.37, 5.38, 5.39
Level line, 2.11
Lifeline, 1.38, 1.40, 1.41–1.42, 1.43
Lifting equipment, 2.26–2.31
Lime
  hydrated, 1.14, 4.2, 4.3–4.4, 4.6, 4.9, 4.15
  slaked, 4.6, 4.22
Lime putty, 4.6, 4.22. *See also* Quicklime
Limestone, 1.13, 4.6, 5.12
Line
  chalk, 2.16–2.17
  crowding the, 5.35, 5.49
  laying to the, 5.33–5.36
  mason's, 2.14–2.16, 5.33–5.36, 5.38
  slack to the, 5.35, 5.49
Lines
  symbols for, 3.19–3.20, 3.21
Lintels
  definition, 5.49
  flashing for, 5.13
  masonry units for, 1.10, 1.12, 5.6, 5.9
Liquid measurements, 3.11, 3.12, 3.13
Liquid propane gas (LP), 1.51–1.52
Load limit, 1.33, 1.48, 2.27
Lowest common denominator, 3.6
LP. *See* Liquid propane gas

Maintenance
  hand tools, 2.4, 2.6, 2.12, 2.14, 2.17
  mortar equipment, 2.19, 2.25
  safety nets, 1.44
  tools, 2.2
Mallet, rubber, 2.6
Manholes, 1.11, 1.12, 1.13, 5.3, 5.11
Man lift, 2.28, 2.35
Marble, 1.13
Mashes, 2.5
Masonry
  advantages of, 1.14
  building codes, 3.29, 3.32, 3.33, 5.14
  historical background, 1.1, 1.2–1.4, 1.53
  modern techniques, 1.17–1.18
  quality of workmanship, 1.22, 3.28, 3.31, 4.23
  setting up and laying out, 5.14–5.18
  standards for quality of work. *See* Standards
Masonry cement. *See* Cement, masonry
Masonry unit, 1.1, 1.8, 1.53, 1.57
Masons
  attitude and work, 1.21–1.22
  career stages for, 1.4, 1.18–1.20, 1.53
  definition, 1.57
  historical background, 1.4
  skills needed by, 1.20–1.21, 3.1–3.2, 3.34, 5.46
Mason's line and fasteners, 2.14–2.16, 5.33–5.36
Materials
  estimate of needs, 5.16, 5.25
  handling of, 1.33–1.34, 2.2, 2.27, 5.1, 5.15
  storage of, 1.33, 4.15–4.16, 4.17, 5.2, 5.15
Material safety data sheet (MSDS), 5.44, 5.46
Math
  addition, 3.2–3.3, 3.6–3.7
  conversion between units, 3.2–3.6, 3.12–3.13
  denominate numbers, 3.1, 3.2–3.6, 3.37
  division, 3.7
  fractions, 3.6–3.7
  geometry
    plane figures and area measures, 3.14–3.17
    solid figures and volumes, 3.17–3.18
  mason's denominate measures, 3.7–3.10
  metric measurements, 3.1, 3.2, 3.11–3.13
  multiplication, 3.7
  necessary skills, 3.1
  subtraction, 3.4–3.5, 3.7
Mauls, 2.6
Medication, 5.28
Metal ties, 1.15, 5.13
Metric drawings, 3.28, 3.29
Metric system of measurement, 3.1, 3.2, 3.11–3.13
Mineral oxides, 4.8
Mixer, mortar, 1.32, 2.24–2.25, 4.15, 4.17–4.21, 4.22
Mixing accessories for mortar, 2.19
Modular system of measurement, 3.8–3.10
Moisture barrier, 4.23
Momentum, 1.50
Mortar. *See also* Joints; Mixer, mortar
  cleaning off of excess, 2.9–2.10, 5.42, 5.44
  colored, 4.2, 4.8–4.9, 4.23, 5.45
  cutting off or edging, 1.25–1.26, 5.29, 5.35
  definition and use of term, 1.1, 1.57
  furrowing, 1.25, 1.26, 1.57, 5.18
  general rules, 1.27–1.28
  historical background, 1.2, 1.3, 1.13–1.14, 4.1
  masonry, 1.23
  materials in, 1.14, 4.2–4.9
  measuring of materials for, 4.16–4.17, 4.18

Mortar (continued)
    mixing
        by hand, 4.20–4.21
        by machine, 4.19. *See also* Mixer, mortar
        problems, 4.21–4.24
        ratios and proportions of ingredients, 4.16, 4.22
    overview, 1.13–1.14, 4.1–4.2
    patching. *See* Pointing joints; Tuckpointing
    percent of face of masonry structure, 4.2, 4.14
    performance specifications, 1.14
    picking up, 1.23–1.25
    pre-mixed, 4.5
    preparation, 1.22–1.23
    properties of, 4.10–4.14
    pumping and elimination of air holes, 2.25
    retempering of, 2.19, 2.39, 4.12, 4.23
    safety, 1.31–1.32, 2.2
    salts in. *See* Efflorescence
    setting time, 1.28, 4.7, 4.11, 4.15, 5.20
    spreading, 1.23, 1.25
    types of, 1.14, 4.9–4.10, 4.13, 4.16
    vs. concrete, 4.2, 4.11, 4.12
    waterproof, 1.3, 1.14, 2.8
Mortar board, 1.24, 2.18, 4.21
Mortar bonds. *See* Bonds
Mortar box, 1.22, 2.18–2.19, 4.16, 4.20
Mortar equipment, hand-powered, 2.18–2.20
Mortar pan, 2.18, 4.21
MSDS. *See* Material safety data sheet
Mud, 1.13, 4.1

National Building Code, 3.32
National Center for Construction Education and Research (NCCER), 1.20
National Concrete Masonry Association (NCMA), 3.32
National Fire Protection Association (NFPA), 3.32
National Institute for Occupational Safety and Health (NIOSH), 1.53
NCCER. *See* National Center for Construction Education and Research
NCMA. *See* National Concrete Masonry Association
Nets, safety, 1.43–1.44, 1.45
NFPA. *See* National Fire Protection Association
911 emergency services, 1.37
NIOSH. *See* National Institute for Occupational Safety and Health
Nipple, 2.32
Noise, 1.30
Nose mask, 4.20

Occupational Safety and Health Administration (OSHA), 1.29
    forklift training and certification, 1.45
    gasoline-powered tool safety, 1.34
    materials stockpiling and storage, 1.33
    powder-actuated tool safety, 1.35, 2.23
    scaffolding, 2.35
Oil drum, 1.50
OSHA. *See* Occupational Safety and Health Administration
Outrigger, 2.35
Oxygen, 1.52

Pan, mortar, 1.24–1.25, 2.18
Paneling or screening, 1.31
Parallelogram, 3.14
Parapet, 1.14, 1.57
Parge, 2.4, 2.39
Partition blocks, concrete, 1.10, 5.6, 5.7

Partitions, 1.6, 4.10
Patterns
    for brick, 1.2, 1.5, 5.21, 5.22, 5.23
    for cement block, 5.23–5.24
Pedestrians. *See* Vehicles, and pedestrian safety
Personal fall-arrest equipment, 1.33, 1.39–1.43, 1.44, 2.32
Personal-positioning systems, 1.40
Personal protection equipment (PPE)
    with grinders, 2.22
    with masonry saws, 2.22
    with mortar mixer, 2.24
    overview, 1.28, 1.29–1.30
    while cleaning, 2.26, 5.44
pi, 3.15–3.16
Pier, 1.10, 1.25, 5.6
Pigments, 4.8, 4.9
Pilaster, 1.25, 1.57, 5.3
Pinch bar, 2.10, 2.11
Pins
    drive, 1.34, 2.23, 2.24
    line, 2.15, 5.34, 5.35, 5.43
    steel, in scaffolding, 2.32
Pitching tool, 2.7
Plane figures, 3.14–3.17
Plasticity, 4.27
Plasticizers, 4.7, 4.8
Platform, on scaffolding, 1.45, 2.32, 2.33, 2.35
Plot plan, 3.27
Plumb, how to check for, 5.30–5.31
Plumb bob, 2.17–2.18
Plumbing symbols, 3.21, 3.24
Plumb line, 2.11
Pneumatic tools, 1.35
Pointing joints, 2.3, 2.39, 5.42–5.43
Polygons, 3.16
Portland cement
    bags, 4.16, 4.17
    five types of, 4.3
    in mortar, 1.22, 4.2, 4.3, 4.9, 4.22
    overview, 1.14, 1.57
    storage, 4.17
Powder-actuated tools, 1.34–1.35, 2.23, 2.24
Power equipment, 2.23–2.26
Power tools, 2.20–2.23, 2.24
Pozzolan, 4.7, 4.8, 4.27
PPE. *See* Personal protection equipment
Pressure washing, 2.26, 5.45
Pride in your work, 1.21
Prism, 3.17
Profiles in success
    apprentice Zachary Reinert, 5.48
    field superintendent Arnold Shueck, 1.55
    foreman Garett Hood, 2.38
    president Sam McGee, 4.26
    vice president Kenneth Cook, 3.36
Project management, 5.14–5.15
Pug mill, 2.24–2.25
Pulley system, 2.27
Pumice, 5.3
Pump, mortar or grout, 2.25, 5.5
Putlog, 2.33–2.34
Pythagorean theorem, 3.18

Quality control, mortar, 4.16
Quality of materials, 4.22
Quality of workmanship, 1.22, 3.28, 3.31, 4.23
Queen closure, 5.25
Quicklime, 1.14, 4.2, 4.6, 4.23

Rackbacks, 5.37–5.40, 5.49
Racking, 1.53, 5.37, 5.49
Radial, internal, 1.6
Raker, 2.8– 2.9, 5.41, 5.42
Ramps, 1.51
Ranging the bricks, 5.38, 5.49
Rectangle, 3.14, 3.17
Reinforcing steel bars (rebar), 1.44, 5.3, 5.5, 5.13
Repair work, 1.18. *See also* Tuckpointing
Residential drawings, 3.26–3.28
Resin, epoxy, 1.32
Respiratory protection, 1.30, 1.35, 2.22, 5.28
Responsibility, 1.21
Restoration, 1.18
Retempering of mortar, 2.19, 2.39, 4.12, 4.23
Returns, 5.37, 5.49
Rigging equipment, 1.48, 1.50
Right triangle, 3.18
Rods
   backer, 5.4
   reinforced steel, 1.44, 5.3, 5.5, 5.13
Roofboards, 2.32
Rowlock
   brick, 1.7, 1.8, 5.11, 5.22
   definition, 1.7, 1.8
   ogee, 1.6
   water table, 1.6
Rubble stone, 1.13
Rules and rulers, measuring
   based on the human body, 3.1
   brick spacing, 3.8, 5.38
   how to read, 3.8, 3.9, 3.10
   overview, 2.12–2.14

Safety
   air quality. *See* Ventilation
   the cost of job accidents, 1.28–1.29
   falling objects, 1.30–1.31, 1.53, 5.8
   fall protection, 1.37–1.38, 1.53
   filling gasoline cans, 1.32, 1.52, 1.53
   flammable liquids, 1.32
   forklift, 1.44–1.53, 4.16
   health effects of cement dust, 1.31, 1.32, 4.5, 4.20, 5.28
   material handling, 1.33–1.34, 2.2, 2.27
   mortar and concrete, 1.23, 1.31–1.32, 2.2
   mortar mixers, 4.19, 4.20
   overview of hazards, 1.30, 5.1
   personal protective equipment, 1.28, 1.29–1.30
   powder-actuated tools, 1.34–1.35, 2.23, 2.24
   rotating mixers, 1.32
   scaffolding, 1.38, 1.41, 2.35
   tools, 1.34, 1.35, 2.1–2.2, 2.20, 5.1, 5.2, 5.28
   weather hazards, 1.36–1.37, 1.41
   while cutting masonry units, 2.22, 5.2, 5.26
Safety grate, 4.19, 4.20
Sailor, 5.22
Salts in mortar. *See* Efflorescence
Sand
   gradations, 4.6
   measurement of, 2.19
   in mortar, 1.22, 4.5–4.6, 4.9, 4.16, 4.17, 4.22
   in portland cement, 1.14
   safety, 1.31
   siltation test for, 4.6, 4.7, 4.14
   storage, 4.16
   symbol, 3.22
Sandblasting, 2.26, 5.45
Sandstone, 1.13

Saw blade, 1.35, 5.28
Saws
   cutting depth of, 2.21
   masonry, 2.21–2.22, 5.28
   safety with, 5.2, 5.28
Scaffolding
   hydraulic personnel lift, 2.35
   overview, 2.31
   safety, 1.38, 1.41, 2.27, 2.35
   steel tower, 2.34
   swing stage, 2.34–2.35
   tubular steel sectional, 2.31–2.34
Scale of a drawing, 3.21
Schedules, 3.28
Scope of work, 3.31
Scoria, 5.3
Scoring, on concrete block, 1.12, 5.9, 5.25
Scraper, 2.10, 5.44
Sealant on control joints, 5.3, 5.4
Seat belts, 1.47
Section drawings, 3.28
Set accelerators or retarders, 4.7, 4.8
Setting time, 1.28, 4.7, 4.11, 4.15, 5.20
Setup for a masonry project, 5.14–5.18
Shapes
   brick, 1.5, 1.6
   concrete block, 1.10, 5.5–5.7
   concrete brick, 1.10, 5.6, 5.8
Sheathing, 1.15
Shiner, 5.22
Shoes, 1.30, 1.32, 1.38, 2.22
Shovel, 2.19, 4.16, 4.17, 4.18, 4.20
Shrinkage, 4.4, 5.2
SI. *See* International System
Side bracket, mason's, 2.33
Sighting the bricks, 5.38
Signal person, 1.47, 1.49
Silicon, 5.3
Silicosis, 1.31, 4.5, 5.28
Sill return, ogee rowlock, 1.6
Sills
   flashing for, 5.13
   masonry units for, 1.12, 5.9
   mechanical bond for, 5.22
   mud, on scaffolding, 2.33, 2.35
Silos, 1.33
Silt, 4.6
Site plan, 3.27
Size
   of brick, 1.3, 1.5, 3.9
   of concrete block or brick, 1.10, 5.5, 5.6–5.7, 5.8
   use of term in drawings, 3.24
Skatewheel, 2.8– 2.9
Skewback, 2.14, 2.39
Skin protection, 1.23, 1.30, 1.31–1.32, 1.36–1.37, 2.2, 4.5
Slack to the line, 5.35, 5.49
Slate, 1.13
Sled, 5.41, 5.42
Slings, 1.48
Slipping hazards, 1.38, 1.48
Slump brick, 1.11, 1.12, 5.8, 5.9
Slump test, 4.11, 4.12
Smoking, 1.36
Snaphook, double locking, 1.42, 1.43
Soffit floor block, concrete, 1.10, 5.6
Soldier, 1.7, 1.8, 5.11, 5.22
Solid masonry unit. *See* Brick
Solvents, organic, 1.32

Southern Standard Building Code, 3.32
Specialty plans, 3.28
Specifications, 3.2, 3.28–3.31
Spills, 4.22
Splitters, masonry, 2.22, 2.23, 5.28
Spotter, 1.47
Sprains, 1.33
Spread, 1.25, 1.28, 1.57
Spreader bar, 1.48
Spreading. *See* Mortar, spreading
Square (geometric), 3.14
Square (tool), 2.14, 5.18
Standards, 3.11, 3.31–3.32, 3.41–3.42. *See also* ASTM
Static electricity, 1.32, 1.52, 1.53
Steam cleaning, 2.26, 5.45
Steel bars. *See* Reinforcing steel bars (rebar)
Stockpiles, 1.33–1.34, 4.15–4.16, 5.1, 5.2, 5.8, 5.15
Stone
 facing, 1.13, 5.12
 overview, 1.13, 5.11–5.12
 symbol for, 3.22
 tools for, 2.7
Storage
 of flammable liquids, 1.32, 1.53
 and guardrails, 1.31
 of materials, 1.33, 4.15–4.16, 4.17, 5.2, 5.15
 of tools, 2.2
Straightness, how to check for, 5.31
Stretcher (masonry)
 brick, 1.7, 1.8, 5.11, 5.22
 bull, 1.7, 1.8
 concrete, 1.10, 1.12, 5.4, 5.6, 5.7, 5.9
 definition, 1.7, 1.8
 single bullnose, 1.6
Stretcher, line, 2.15–2.16, 5.33–5.34
Striking the joints, 5.41–5.42
Stringing the mortar, 1.25, 1.57
Studs, threaded, 1.34, 2.23, 2.24
Sulfates, 4.13, 5.9
Superintendents, 1.20, 1.55
Supervisors, 1.20
Surveyor, 5.17

Tag lines, 1.49, 1.50
Tailing the diagonal, 5.37, 5.38, 5.49
Take-up drum or wheel, 2.27
Tanks, 1.33
Tape, measuring, 2.12–2.14
Tapping the masonry unit, 1.26, 1.27, 4.14, 5.19, 5.30, 5.31
Tarpaulins, 4.15, 5.2, 5.15
T-bevel, sliding, 2.14
Team approach, 1.18
Technical specifications, 3.29–3.31
Temperature measurements, 3.13
Tempering of mortar, 2.19, 2.39
Terra-cotta, 1.7
Tests
 autoclave expansion, 4.14
 brick absorption rate, 5.11
 cleaner, 5.44, 5.46
 consistency of mortar, 4.21
 construction projects, 3.32–3.34
 correct dryness of mortar, 5.41
 purity of mortar, 4.6, 4.7, 4.14
 safety nets, 1.44
 siltation, 4.6, 4.7, 4.14
 slump, 4.11, 4.12
Tie-off points, 1.42–1.43

Ties
 joint reinforcement, 5.13
 metal reinforcement, 1.15, 5.13
 veneer, 5.13
Tile. *See* Hollow masonry units/tiles
Tipping of a forklift, 1.50
Toeboard, 1.31, 2.32, 2.33, 2.34
Tongs, brick, 2.10
Tool bag, 2.10–2.11
Tool belt, 2.11
Tools
 gasoline-powered, 1.34
 hand, 2.2–2.11, 5.26–5.28, 5.41–5.42
 lifting or lowering, 1.31, 5.1
 maintenance, 2.2
 measuring, 2.11–2.18, 3.4
 overview, 1.17, 2.36
 pneumatic, 1.35
 powder-actuated, 1.34–1.35, 2.23, 2.24
 power, 2.20–2.23, 2.24, 5.28–5.29
 safety, 1.31, 2.1–2.2, 5.1
Torque, 2.23
Tower, steel, 2.34
Training, 1.18–1.20, 1.45, 2.35
Tray, carrier or conveyor, 2.21
Trestle, 2.32, 2.33, 2.39
Triangle, 3.14–3.15, 3.18
Trig (twig), line, 2.16, 5.34, 5.35
Tripping hazards, 1.38
Trolley, steel, 2.27
Trough, concrete block, 5.7
Trowel(s)
 cutting masonry units with, 5.27–5.28
 how to hold the, 1.23–1.24, 2.3, 5.35
 parts of a, 2.3
 striking the joints with a, 5.41
 types, 1.17, 2.3–2.4
Tuckpointing, 1.18, 1.57, 2.3–2.4, 5.43

Uniform Building Code, 3.32
U.S. Customary system, 3.1, 3.2, 3.12, 3.37
U.S. Department of Labor, 1.18
U.S. Department of Veterans Affairs (VA), 3.29, 3.30

VA. *See* U.S. Department of Veterans Affairs
Valves, 1.35, 2.26
Vapors and gases, 1.32, 1.35, 1.52
Vehicles
 chock the wheels of, 1.51
 hydraulic lift material truck, 2.28
 and pedestrian safety, 1.32, 1.45, 1.47, 1.52–1.53
Veneer
 adhesion ceramic, 1.7
 anchored ceramic, 1.7
 of brick, 5.21
 concrete brick used for, 5.8
 of stone, 1.13, 5.12
 tools for, 2.16, 2.17
 for walls, 1.15, 5.12, 5.18, 5.21
Veneer ties, 5.13
Ventilating partitions, 1.6
Ventilation, 1.30, 1.32, 1.34, 1.35, 1.52
Vibrator, masonry, 2.25
Visibility, 1.47, 1.49, 1.50, 1.51, 2.26
Volume
 calculation of, 3.17–3.18
 change in mortar, 4.14
 units for, 3.5, 3.6, 3.11, 3.12, 3.13

Walkways, 5.22
Wall cap, 1.6, 5.22
Walls
    basement, 4.23
    cavity, 1.9, 1.15, 5.36
    composite, 1.15
    control joints in, 5.2–5.3, 5.14
    curtain, 1.16
    dimension lines in drawings, 3.24, 3.26
    drainage, 1.15, 1.57
    firewall, 1.11, 5.3
    garden, 5.24
    hollow, 1.15
    planning, 5.16, 5.17
    reinforced, 1.15, 1.16
    types of mortar for, 4.10
    veneer, 1.15, 5.12, 5.18, 5.36
Water
    in brick, 5.11
    flow in a pressure washer, 2.26
    in mortar, 4.4, 4.6, 4.7, 4.11–4.12, 4.14, 4.22–4.23
    use with masonry saws, 2.21
Water barrel, 2.19, 4.15, 4.21
Water bucket, 2.19
Water content of mortar, 4.11–4.12
Water reducers, 4.7, 4.8
Water repellents, 4.7, 4.8
Water retention, 4.4, 4.11, 4.27
Weather considerations
    and control joints, 5.3
    health hazards from cold or heat, 1.36–1.37
    ice, 1.41
    resistance of masonry units to, 5.9
    resistance of mortar to, 4.4, 4.7, 4.13, 4.22–4.23, 5.41
    sand storage, 4.16
    wind, 2.17
Web, of a hollow concrete block, 1.11, 5.4, 5.25
Weepholes, 1.15, 1.57
Weight units, 3.5, 3.11, 3.13
Wheelbarrow, 1.22, 1.23, 2.20, 4.15, 4.19, 4.21
Winch, 2.34
Windows, 5.16, 5.18
Workability of mortar, 4.3, 4.11, 4.27. *See also* Retempering
    of mortar
Workmanship, 1.22, 3.28, 3.31
Work site
    air quality. *See* Ventilation
    clean up, 1.30, 1.32, 1.38, 1.52, 4.21, 5.44
    clear aisles or pathways in, 1.33, 2.27, 2.29, 4.19
    job layout, 5.16–5.18
    mortar area, 4.15, 4.19, 5.15
    setting up for the project, 5.14–5.16
World's record for bricklaying, 1.1
Wythe(s)
    definition, 1.7, 1.8, 1.57
    in different bond types, 1.15, 5.20, 5.21, 5.22